REVIEWS in
MINERALOGY
Volume 10

JOHN M. FERRY, Editor

CHARACTERIZATION of METAMORPHISM through MINERAL EQUILIBRIA

The Authors:

Donald M. Burt

Department of Geology
Arizona State University
Tempe, AZ 85287

Eric J. Essene

Department of Geological Sciences
University of Michigan
Ann Arbor, MI 48109

John M. Ferry

Department of Geology
Arizona State University
Tempe, AZ 85287

Jack M. Rice

Department of Geology
University of Oregon
Eugene, OR 97403

Douglas Rumble III

Geophysical Laboratory
2801 Upton Street, N.W.
Washington, D.C. 20008

Frank S. Spear

Department of Earth & Planetary Sciences
Massachusetts Institute of Technology
Cambridge, MA 02139

James B. Thompson, Jr.

Department of Geological Sciences
Harvard University
Cambridge, MA 02138

Robert J. Tracy

Department of Geology & Geophysics
Yale University
New Haven, CT 06520

Series Editor:
Paul H. Ribbe

Department of Geological Sciences
Virginia Polytechnic Institute &
State University
Blacksburg, Virginia 24061

MINERALOGICAL SOCIETY OF AMERICA

QE
475
.A2 C52

REVIEWS IN MINERALOGY

(Formerly: SHORT COURSE NOTES)
ISSN 0275-0279

VOLUME 10: CHARACTERIZATION OF META-
MORPHISM THROUGH MINERAL EQUILIBRIA

ISBN 0-939950-12-X

Additional copies of this volume as well as those
listed below may be obtained at moderate cost from

Mineralogical Society of America
2000 Florida Avenue, NW
Washington, D.C.   20009

CHARACTERIZATION of METAMORPHISM through MINERAL EQUILIBRIA

FOREWORD

Characterization of Metamorphism through Mineral Equilibria is the
second book in the *REVIEWS in MINERALOGY* series to be dedicated to a subject other than a distinct group of minerals. The first, *Kinetics of Geochemical processes,* edited by A.J. Lasaga and R.J. Kirkpatrick (1980),
was distinguished by its emphasis on thermodynamics, rate laws, and
kinetics and mechanisms of crystallization, diffusion, weathering and
diagenesis. This volume is oriented primarily toward application of
methods of determining the history of metamorphism through chemical
and mineralogical studies. To obtain a thorough overview of its
thrust, read John Ferry's preface on the following pages.

This is the eleventh volume of *REVIEWS in MINERALOGY,* and its
publication coincides with that of Volume 9B: *Amphiboles: Petrology
and Experimental Phase Relations,* which also contains a primary emphasis
on metamorphism: more than sixty percent of it is devoted to the phase
relations of metamorphic amphiboles. At one time it was suggested that
a series with a distinctly petrologic emphasis, entitled *REVIEWS in
PETROLOGY,* be initiated by the Mineralogical Society of America. But
it now appears that despite its name, the present series has expanded
its purview and embraced petrology as well as mineralogy and crystallog-
raphy. This is entirely appropriate inasmuch as the Mineralogical
Society of America and its journal is dedicated -- as its logo proclaims
-- to all three interactive disciplines.

Paul H. Ribbe
Series Editor
Blacksburg, VA

PREFACE and ACKNOWLEDGMENTS

Unlike sedimentation and volcanism, active metamorphism is not directly observable. Metamorphic petrologists therefore must infer what constitutes the process of metamorphism by examining the products of metamorphic events. The purpose of this volume is to review the use of a powerful probe into metamorphic process: mineral assemblages and the composition of minerals. Put very simply, this volume attempts to answer the question: "What can we learn about metamorphism through the study of minerals in metamorphic rocks?" It is not an encyclopedic summary of metamorphic mineral assemblages; instead it attempts to present basic research strategies and examples of their application. Moreover, in order to limit and unify the subject matter, it concentrates on the chemical aspects of metamorphism and regrettably ignores other important kinds of studies of metamorphic rocks and minerals conducted by structural geologists, structural petrologists, and geophysicists.

An overview of the chemical aspects of modern metamorphic petrology is timely because it brings together three areas of research which have reached maturity only in the last 25 years: (1) chemical analysis of minerals by microanalytical techniques; (2) application of reversible and irreversible thermodynamics to petrology; and (3) laboratory phase equilibrium experiments involving metamorphic minerals. Chemical thermodynamics is the formal mathematical framework which links measurable variables (i.e., mineral composition) to metamorphic variables which cannot be directly measured (i.e., chemical potential, pressure, temperature, fluid composition). Results of phase equilibrium studies involving metamorphic minerals at metamorphic pressures and temperatures (together with calorimetric and heat capacity data) permit these links to be *quantitative*. It is the union of analysis, theory, and laboratory experiment which allows the modern metamorphic petrologist to make sophisticated inferences about conditions of metamorphism and the factors which control these conditions. This union is the principal subject of the volume.

The volume is organized much in the same way that one might approach a research project involving metamorphic rocks. Initially those chemical components which characterize the composition of minerals in the assemblages under consideration must be identified. In addition, the reaction relationships among components must be systematically characterized. The reaction relationships rationalize the prograde changes in mineralogy which rocks experience during metamorphism and, furthermore, form the basis for extracting information about intensive variables during metamorphism. Chapters 1-3 summarize strategies for identifying components in metamorphic minerals and for formulating chemical reactions among them.

Chapter 4 develops, from classical thermodynamics, those equations which can be used to explicitly relate mineral composition to other variables of interest such as metamorphic pressure, temperature, and chemical potentials of volatile species in any metamorphic fluid phase. Chapter 5 is specifically devoted to geologic thermometry and barometry, and Chapter 6 reviews strategies for the determination of metamorphic fluid composition.

Petrologists should not be content with simply calculating and cataloguing values of metamorphic pressure, temperature, and fluid composition. In order to characterize the *process* of metamorphism, we must try to understand what controls these measured values and the manner in which they evolve during metamorphism both as rocks are heated and buried and as rocks are cooled and uplifted. Chapter 7 explores how two concepts -- buffering and infiltration -- can act as general controls on fluid composition, mineral composition, and temperature during metamorphic events. In addition, this chapter develops procedures which can be used to evaluate the relative importance of buffering versus infiltration in the evolution of specific rocks. Chapter 8 demonstrates how integrated petrologic and stable isotope studies may be used, in principle, to reconstruct the prograde pressure-temperature-infiltration history of metamorphic rocks. Chapter 9 discusses the use of mineral inclusions and compositional zoning in minerals in evaluating both prograde and post-peak P-T paths of certain mineral assemblages. In addition, compositional zoning is considered as an indicator of cooling rates during post-peak uplift. Thus between Chapter 1 and Chapter 9 we go from the first step of describing a metamorphic mineral assemblage through a reconstruction of the physical state in which it crystallized to an analysis of what factors controlled that state and how it evolved with time.

The contents of the volume reflect two themes which underlie modern research in metamorphic petrology. The first of these is an ever-increasing emphasis on the quantitative characterization of metamorphism. Current research less involves description and classification than calculation of intensive and extensive variables attained during metamorphism. This volume hopefully serves as a text in the quantitative study of the chemical aspects of metamorphism. As a corollary to the emphasis placed on quantitative methods, we can see increasing attention paid to analytical as opposed to graphical treatments of mineral equilibria. Graphical representations, while undeniably valuable, can consider two

(or at most three) independent variables. Analytical treatment of mineral
equilibria is attractive because it rigorously keeps track of all variables
pertinent to an equilibrium assemblage. The second theme is an increasing
interest in the dynamics of metamorphism. Metamorphism obviously is not a
static process -- it involves changes in pressure, temperature, mineral and
fluid composition, etc. The classical static approach to quantitative metamorphic petrology, though, searches for the physical conditions of a unique
pressure-temperature state which a rock or mineral assemblage records. Mineral equilibria are used to estimate single values of pressure, temperature,
and fluid composition -- a sort of snapshot of what conditions were like. If
mineral assemblages indeed represent a fossilized metamorphic state, then calculated P, T, X_i, however, simply represent a single point along the $P-T-X_i$-
time path which a rock followed during metamorphism. Chapters 2, 7, 8, and 9
reflect an increasing interest among petrologists in the entire $P-T-X_i$-time
path (or at least in more than one point along it). We can expect to see less
satisfaction in the future with the snapshot model of metamorphism and more effort devoted to characterizing metamorphism as a *dynamic process*. Thus the
volume not only summarizes time-honored current practices in quantitative metamorphic petrology, but hopefully also identifies some paths which may be followed
in the future.

I am grateful to a number of individuals and organizations for helping to make
possible the short course and the review volume. First and foremost, credit goes
to the authors whose thoughtful manuscripts were produced on a tight schedule.
As series editor, Paul H. Ribbe turned manuscripts, covered with red scribbles,
into neat, attractive copy. The volume could not have been prepared without his
dedicated effort. Many original figures were improved by the drafting skills of
Sue Selkirk at Arizona State University. Lou Fernandez served as a local organizer
for the short course, presented in New Orleans, LA, October 15-17, 1982; he ex-
pertly solved a number of difficult logistical problems. Preparation of the
volume was partially supported by NSF grant EAR 80-20567 and a Cottrell Grant from
Research Corporation to J.M.F.

John M. Ferry
Tempe, Arizona
August 1982

Characterization of Metamorphism
through Mineral Equilibria

TABLE of CONTENTS

Chapter 1

COMPOSITION SPACE: An ALGEBRAIC and GEOMETRIC APPROACH

J.B. Thompson, Jr.

Chapter 2

REACTION SPACE: An ALGEBRAIC and GEOMETRIC APPROACH

J.B. Thompson, Jr.

Chapter 3

LINEAR ALGEBRAIC MANIPULATION of N-DIMENSIONAL COMPOSITION SPACE

F.S. Spear, D. Rumble III, and J.M. Ferry

Chapter 4

ANALYTICAL FORMULATION of PHASE EQUILIBRIA: The GIBBS' METHOD

F.S. Spear, J.M. Ferry, and D. Rumble III

Chapter 5

GEOLOGIC THERMOMETRY and BAROMETRY

E.J. Essene

Chapter 6

CHARACTERIZATION of METAMORPHIC FLUID COMPOSITION
through MINERAL EQUILIBRIA

J.M. Ferry and D.M. Burt

Chapter 7

BUFFERING, INFILTRATION, and the CONTROL of INTENSIVE VARIABLES during METAMORPHISM

J.M. Rice and J.M. Ferry

Chapter 8

STABLE ISOTOPE FRACTIONATION during METAMORPHIC DEVOLATILIZATION REACTIONS

D. Rumble III

Chapter 9

COMPOSITIONAL ZONING and INCLUSIONS in METAMORPHIC MINERALS

R.J. Tracy

Chapter 1

COMPOSITION SPACE: An ALGEBRAIC and GEOMETRIC APPROACH

J.B. Thompson, Jr.

INTRODUCTION

Most petrologic and geochemical processes involve transfers or exchanges
of matter either between adjacent mineral grains, or between mineral grains
and natural fluids with which they may come in contact. A convenient system
of bookkeeping is thus essential to any adequate analysis of such natural
processes: it must include both algebraic and geometric methods for the
representation of the chemical composition of both homogeneous and heterogeneous systems. The representations for the homogeneous susbstances present
must, in turn, be readily adaptable to the algebraic and geometric treatment of
the reactions that take place between them. The problem, at its roots, is a
purely mathematical one. It is *not* directly concerned with the idea of thermodynamic equilibrium, although convenience of thermodynamic formulation may
guide our choice of path where the mathematics gives us an option. The problem
is also not directly related to the structural chemistry of matter, but here
again we shall find that structural considerations will guide our choice of
path. Consideration of structural chemistry is particularly important in dealing with a crystalline phase in which variations in chemical composition are
strongly constrained by the nature of its crystal structure.

Many readers will find what follows to be disturbing, perhaps painfully so.
I therefore urge that you be tolerant and patient with what may appear at first
to be needlessly unconventional. The unconventional aspects are introduced
partly to illustrate the full generality of composition space and its associated algebra, and partly to take best advantage of the choices of path referred
to above. There is little point in adhering to unnecessarily conventional modes
of thought if these impede progress. The ideas of zero as a number, and of
negative numbers, were not easy ones for early mathematicians to accept, but
algebra could not get off the ground until they did. This illustration has a
direct parallel in the algebra associated with composition space.

In classical treatments of chemical systems components are usually identified with discrete physical entities or molecular species. Most of us still
subconsciously think this way although we have known for years that such a
physical picture has little reality in dealing with many crystalline substances.

For example, there is no set of conventional end-member components which can be combined solely in positive amounts to represent the composition of all amphiboles. Vestiges of classical thought, however, make the idea of a negative value for the "content" of a component somehow abhorrent, even though we admit that if something can be put in, it can also be taken out again. In the classical calculation of a "norm" this leads to an unnecessary re-selection of components whenever a negative value appears, which leads in turn to unnecessarily complex computer programs and unnecessary expenditures of effort. If my use of "unnecessary" disturbs you it is probably because you are still associating "component" with some discrete physical entity rather than regarding it simply as an algebraic operator or a reference point needed to define a coordinate system in a composition space. If you are willing to divest the idea of a component from its classical physical trappings, however, *it may save you time and money*. Having thus appealed to your baser instincts, I urge a radical shift in scientific thinking as the best route to a conservative fiscal policy.

COMPONENTS AND COMPOSITION SPACES

We shall regard a chemical component as sufficiently defined if we can write a chemical formula for it (but not necessarily one with simple, rational subscripts, nor necessarily positive ones), *and* if we assign to it a unit weighting or unit quantity that specifies the number of formula units in one unit of the component. We shall therefore designate the unit quantity of component i as U_i, that of component j as U_j, and so on. It is essential that units of quantity be based on some conservative property of matter such as (for non-radioactive processes) number of atoms, numbers of atoms of a given kind or -- with care -- mol units. Negative subscripts, if any, count *against* the total.

If C represents a unit quantity of a composition that may be specified in terms of i and j we may then write:

$$aU_i + bU_j = (a + b)C \qquad (1)$$

where a and b are numerical coefficients. Volume is *not* an adequate measure of quantity, inasmuch as it is non-unique for even a single formula unit; hence (1) will not hold for volume units in other than very special circumstances. The volume occupied by a gram-formula unit of SiO_2, for example, is dependent not only on the state of aggregation, but also on temperature and pressure.

We will, for most purposes, use only sets of components that are linearly

2

independent in the algebraic sense. This constraint is met if no component of the set has a composition that can be expressed in terms of any of the others. A set of c (linearly) independent components is then sufficient to represent all possible compositions in a c-component chemical *system*, though not necessarily with positive values of all coefficients in equations such as (1). The algebraic representation has a corresponding geometric one in that c independent components may be regarded as occupying c linearly independent points in a (c-1)-dimensional *composition space*. A one-dimensional space is thus defined by two non-coincident points, a two-dimensional space by three non-colinear points, a three-dimensional space by four non-coplanar points and so on.

The above constraints allow considerable latitude, both in the selection and weighting of components, and in their geometric array. For a ternary system, for example, it is necessary only that the three points define a triangle of non-zero area. *Any* triangle will do. Equilateral triangles are aesthetically pleasing, but if the appropriate coordinate paper is not available any other kind will suffice. If, as is customary, we take all compositions within the triangle of reference as having positive coefficients in the analogue of equation (1), then the remaining regions of the composition plane follow the scheme of Figure 1. Points 1, 2, and 3 are not necessarily at the physical

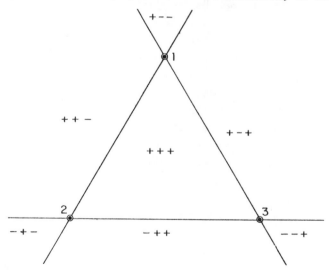

Figure 1. Triangle of reference for a ternary system (not necessarily equilateral). Unit compositions, C, in the plane are defined by relations of the form $aU_1 + bU_2 + cU_3 = (a + b + c)C$. If a,b, and c are taken as positive for C's within the triangle, then their signs in the remaining region follow (in order) the indicated pattern.

3

limits of the system, and compositions outside the reference triangle are as well represented as those within. A point in the ++- region of Figure 1 may be attained by *removing* some 3 from an additive mixture of 1 and 2. We have thus, I hope, removed any lingering reservations about negative coefficients. It is *not* necessary to proceed to a new set of, for example, normative components should some of the coefficients for the initial one be negative.

INDEPENDENTLY VARIABLE COMPONENTS

For most purposes we shall be concerned only with a special class of components known as *independently variable* (or I.V.) components. Let us suppose that a single homogeneous substance or phase may be separated into no more than c chemically distinct phases (the differences may be infinitesimal) whose compositions are linearly independent. If these c phases may then be connected to the initial one by further phases such that all (including the initial one) are connected by continuous variations in composition and physical properties, we may say that the initial phase has c independently variable components. Other phases related to the initial one by continuous variations in composition and physical properties are said to be in the same *phase region*. The chemical description of a phase region requires (c-1) independent compositional variables. The number of I.V. components, c_ϕ, for a given phase, ϕ, is in principle fixed but we have, as before, a great deal of latitude in their selection. The qualification "in principle" is necessary because in practice the assessment of the magnitude of c_ϕ may be constrained by the nature and precision of analytical techniques. The omitted components in such instances, however, usually represent only minor variations of little interest.

A heterogeneous assemblage, α, (of two or more phases) may also be regarded as having a fixed number c_α of independently variable (and linearly independent) components. Any component that is I.V. in one or more of the constituent phases is necessarily I.V. in the assemblage, as is any component that may be expressed in terms of these.

The above paragraphs may seem abstract and obscure. If so, their meaning may be clarified by some simple geometric illustrations. In Figure 2 the composition plane is defined by the oxide components (not necessarily I.V. in all phases) $MgO-FeO-SiO_2$. In Figure 2a the weighting of these components is in mass units, in 2b it is in mol units, and in 2c it is in oxy-equivalents. The relations shown are typical for assemblages of crystalline phases in that the phases taken individually have fewer I.V. components than do assemblages of two or more phases.

4

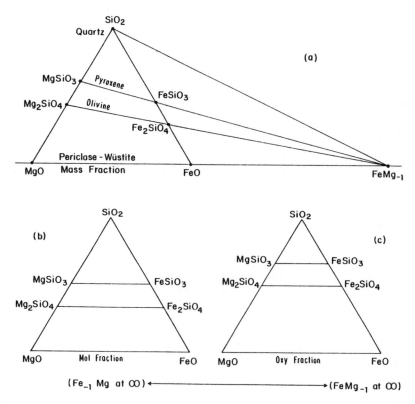

Figure 2. The composition space defined by $MgO-FeO-SiO_2$ with three different choices of unit quantity of these components. (a) may be regarded as in mass fraction (or weight percent); (b) is in mol fraction oxides, as written; and (c) is in oxygen fraction. The component $FeMg_{-1}$ has zero atoms (including oxygen atoms), but represents a positive mass of 31.53 gms and lies in Figure (a) at $X_{SiO_2} = 0$, $X_{MgO} \approx -1.28$, and $X_{FeO} \approx 2.28$.

Quartz in this system has, for all practical purposes, only one independently variable component: SiO_2. Mg-Fe olivines and pyroxenes, on the other hand, may be regarded as each having two I.V. components inasmuch as the chemical variations in these phases are essentially along straight lines in Figure 2. Either phase may be described chemically with reference to any two distinct points (components) along the line of variation. A pure Mg-olivine or Mg-pyroxene, however, has only one, because variations therefrom are physically possible in only one direction. We may speak, however, (in the sense of Gibbs, 1948) of other points on the lines of compositional variation as *possible* components of the magnesium end-member phases. Points to the right of the Mg end-members may be further regarded as positively possible, and those to the left as negatively possible.

5

EXCHANGE AND ADDITIVE COMPONENTS

We shall refer to points on and within the triangles $MgO-FeO-SiO_2$ in Figure 2 as belonging to the physically accessible region because only such points can correspond chemically to real physical substance. Our liberal definition of a component, however, is not so restricted and we should find that each point *outside* the triangle has, in fact, a real meaning in terms of some chemical variation. Let us consider, for example, the point labeled $FeMg_{-1}$. This point is clearly visible in Figure 2a but is approached only by moving infinitely to the right in 2b and 2c (its negative, $MgFe_{-1}$, is similarly approached in 2b and 2c by moving infinitely to the left). We shall show later that the three triangles of Figure 2 may be regarded as simple projections of each other through points outside their planes. In projective geometry points at infinity have no special significance. If the diagrams of Figure 2b and 2c are viewed in perspective from the left, with the eye as point of projection, we find that $MgFe_{-1}$ is again clearly visible!

$FeMg_{-1}$ does not represent a physically realizable chemical composition, but it does represent a physically realizable operation, specifically the substitution of an iron atom for a magnesium atom. We may therefore take the formula as representing an independently variable component of any substance (including an Fe-Mg alloy) in which this may occur. In Figure 2 it is the common point to the lines of variation of all three ferro-magnesian phases. This is a necessary consequence of the equations:

$$
\begin{aligned}
FeMg_{-1} &= Fe - Mg \\
&= Fe^{2+} - Mg^{2+} \\
&= FeO - MgO \\
&= FeSiO_3 - MgSiO_3 \\
&= \tfrac{1}{2}(Fe_2SiO_4 - Mg_2SiO_4)
\end{aligned}
\tag{2}
$$

$FeMg_{-1}$ is not at infinity in Figure 2a simply because an iron atom is heavier than a magnesium atom. We shall refer to formulas such as $FeMg_{-1}$ as *exchange* components, or as *exchange vectors* inasmuch as they operate to produce linear displacements in composition space. In crystalline phases they have the unique property of altering the composition of the crystal without altering its quantity as measured in unit cells. Because exchange components may be defined algebraically in terms of more conventional or *additive* components, they may be substituted for them without loss of information. In any case additive components may always be reintroduced, if desired, by means of equations such as (2). The classical stoichiometric equation for a simple exchange reaction may also

6

be greatly simplified as shown by the following sequence of equations:

$$\tfrac{1}{2}Mg_2SiO_4[Olivine] + FeSiO_3[Pyroxene] = \tfrac{1}{2}Fe_2SiO_4[Olivine] + MgSiO_3[Pyroxene]$$

$$(FeSiO_3 - MgSiO_3)[Pyroxene] = \tfrac{1}{2}(Fe_2SiO_4 - Mg_2SiO_4)[Olivine] \qquad (3)$$

$$FeMg_{-1}[Pyroxene] = FeMg_{-1}[Olivine]$$

Because of the simplicity of the latter formulation we shall find it convenient to adopt a highly unconventional choice of components for crystalline phases. At least one component, but no more, need be additive. All the rest may be exchange components. We shall therefore follow a suggestion of W.L. Bragg (1937; see also Bragg and Claringbull, 1965; Smith, 1959) and select one simple end-member composition as the additive component, taking the mol (gram-formula unit) or some simple multiple or submultiple thereof as the unit of quantity. The remaining components may then be exchange components or vectors, so designed that they do not alter the number of unit cells, and selected so that they are, insofar as possible, ones shared by other crystalline phases. The composition variables, X_{ex}, are then simply the numbers of mols of the exchange components per mol of additive component. The X_{ex} may then be regarded as the components of a vector in a composition space that connects the additive component to the actual composition of the phase.

TRANSFORMATION OF COMPONENTS

Many of the classical operations of mineralogy and petrology are, at their roots, transformations of components. The recalculation of a weight percent oxide analysis to atoms per unit cell is one example, and a normative calculation is another. The former goes from oxide components in mass units, normalized to total mass as weight percent, to atomic components normalized to a standard unit cell. The classical CIPW norm transforms weight percent oxide components to weight percent normative components. Normative calculations such as those prepared by Barth (1948) yield normative components in oxy-equivalents normalized to an arbitrary number of oxygens. Normative calculations were originally conceived as an attempt to calculate abundances of molecular species. Vestiges of this still affect our thinking though we know them to be unrealistic. What a normative calculation really does is to transform the original components to a new set localized in the part of composition space that is of interest. It is thus analogous to transforming latitude and longitude to a system based on local landmarks. An address in central Manhattan is not very readily located if given only with reference to the Equator and the Greenwich Observatory.

7

Attempts to calculate modal abundances are much like normative calculations but usually with more complex "new" components and a final conversion to volumes. The calculations needed to prepare most condensed and projected composition diagrams used in various aspects of petrology are also, at their roots, transformations of components as are many of the calculations relevant to the evaluation of mineral deposits. Brady (1975) and Brady and Stout (1980) have discussed some uses of transformation of components in petrology and geochemistry. Further applications of the transformation of components to the study of metamorphic rocks are summarized in Chapter 3 of this volume by Spear et al.

To pass from an "old" to a "new" set of components we must be able to write an equation giving the content of each new component in terms of the old. Thus if the old are in atoms or gram-atoms and the new are in formula units or mols we have equations such as:

$$1 \ U'_{Fe_2O_3} = 2 \ U_{Fe} + 3 \ U_O \tag{4}$$

where the new component is indicated by a prime. The simple formula, Fe_2O_3, is in fact a concise notation for equation (4). When both the old and new are in atom or mol units we shall therefore simply let the chemical formula itself represent the unit quantity of the component indicated by it. For mol units (transformation from oxide to normative components) we may thus write:

$$1 \ MgSiO_3 = 1 \ MgO + 1 \ SiO_2$$
$$1 \ Mg_2SiO_4 = 2 \ MgO + 1 \ SiO_2 \tag{5}$$

In atom units, for both old and new, we have, however:

$$1 \ U'_{MgSiO_3} = \frac{2}{5} \ U_{MgO} + \frac{3}{5} \ U_{SiO_2}$$
$$1 \ U'_{Mg_2SiO_4} = \frac{4}{7} \ U_{MgO} + \frac{3}{7} \ U_{SiO_2} \tag{6}$$

and with oxy-equivalents for both old and new:

$$1 \ U'_{MgSiO_3} = \frac{1}{3} \ U_{MgO} + \frac{2}{3} \ U_{SiO_2}$$
$$1 \ U'_{Mg_2SiO_4} = \frac{1}{2} \ U_{MgO} + \frac{1}{2} \ U_{SiO_2} \tag{7}$$

The corresponding equations for mass units, old and new, have coefficients on the right that are very nearly, but not precisely, those of (6). This is fortuitously because the gram-formula weight of MgO is ~ 40 and that of SiO_2 ~ 60. The mass coefficients for the ferrous analogue of (6) would be quite different.

Note that the coefficients on the right hand sides of (6) and (7) sum to unity, but that those of (4) and (5) do not. This is because both old and new in (6) and (7) are based on the same measure of quantity whereas in (4) and (5) they are not. If the unit quantities of old and new are inconsistent as in (4) or (5) the transformation is of a highly general or *projective* type. If unit quantities are consistent as in (6) or (7), or in mass-to-mass transformations, the transformation is *affine*, a special case of the more general projective transformation.

In either case we may obtain an array of equations involving unit quantities of old (unprimed) and new (primed) components of the form:

$$1\ U_{1'} = \nu_{11'}U_1 + \nu_{21'}U_2 + \cdots \nu_{i1'}U_i$$
$$1\ U_{2'} = \nu_{12'}U_1 + \nu_{22'}U_2 + \cdots \nu_{i2'}U_i$$
$$1\ U_{3'} = \nu_{13'}U_1 + \nu_{23'}U_2 + \cdots \nu_{i3'}U_i \qquad (8)$$
$$\vdots \qquad \vdots \qquad \vdots \qquad \vdots$$
$$1\ U_{i'} = \nu_{1i'}U_1 + \nu_{2i'}U_2 + \cdots \nu_{ii'}U_i$$

where the coefficients $\nu_{ii}{}'$ are to be understood as the number of units of i needed to make one unit of i'. These coefficients may be positive, negative, or zero. Because both old and new must describe the same range of composition, and must be linearly independent, they must also be equal in number. The coefficient matrix for the right hand sides of equation (8) is thus a square array, and represents a linear transformation that can be written in matrix notation as:

$$U' = AU \qquad (8a)$$

where U' represents the $U_i{}'$ (a column matrix), A the coefficient matrix of (8), and U the row matrix of the U_i. Equations (8), with the U_i regarded as unknowns, may be solved by Cramer's Rule obtaining:

$$1\ U_1 = \nu_{1'1}U_{1'} + \nu_{2'1}U_{2'} + \nu_{3'1}U_{3'} + \cdots \nu_{i'1}U_{i'}$$
$$1\ U_2 = \nu_{1'2}U_{1'} + \nu_{2'2}U_{2'} + \nu_{3'2}U_{3'} + \cdots \nu_{i'2}U_{i'}$$
$$1\ U_3 = \nu_{1'3}U_{1'} + \nu_{2'3}U_{2'} + \nu_{3'3}U_{3'} + \cdots \nu_{i'3}U_{i'} \qquad (9)$$
$$\vdots \qquad \vdots \qquad \vdots \qquad \vdots \qquad \vdots$$
$$1\ U_i = \nu_{1'i}U_{1'} + \nu_{2'i}U_{2'} + \nu_{3'i}U_{3'} + \cdots \nu_{i'i}U_{i'}$$

where the coefficients are now to be understood as the numbers of units of i' needed to make one unit of i. Equation (9) may be written in matrix notation as

$$U = A^{-1}U'$$

(9a)

Equation (9) or (9a) thus gives the compositions of the old components in terms of the new, and A^{-1}, the new coefficient matrix, is the *inverse* of A.

Cramer's Rule will not work and hence A will have no inverse, if the determinant of A is zero. If this happens it is usually because there has been a mistake in the selection of the new (and generally more complex) components so that there are one or more linear dependencies among them. If this occurs these can be identified by successively eliminating terms from the right-hand sides of equation (8) and discarding the equations used to perform each elimination. If the process comes to end with equations (residual equations) still remaining when *all* right-hand terms are eliminated, then these residual equations must be a relation among the U_i' that demonstrates a linear dependency among them. (See also discussion of this matter in Chapter 3 of this volume by Spear et al.)

Equations such as (8) or (4) suffice for some petrologic purposes but we are more often concerned with the transformation of the analysis of a sample from its expression in old components to its expression in the new. If we let the n_i be the numbers of units of the old components in the sample and the n_i' be the corresponding number for the analysis in terms of the new components we may write:

$$n_1 = v_{11'}n_{1'} + v_{12'}n_{2'} + v_{13'}n_{3'} + \cdots v_{1i'}n_{i'}$$
$$n_2 = v_{21'}n_{1'} + v_{22'}n_{2'} + v_{23'}n_{3'} + \cdots v_{2i'}n_{i'}$$
$$n_3 = v_{31'}n_{1'} + v_{32'}n_{2'} + v_{33'}n_{3'} + \cdots v_{3i'}n_{i'}$$
$$\vdots \qquad \vdots \qquad \vdots \qquad \vdots \qquad \vdots$$
$$n_i = v_{i1'}n_{1'} + v_{i2'}n_{2'} + v_{i3'}n_{3'} + \cdots v_{ii'}n_{i'}$$

(10)

and, conversely:

$$n_{1'} = v_{1'1}n_1 + v_{1'2}n_2 + v_{1'3}n_3 + \cdots v_{1'i}n_i$$
$$n_{2'} = v_{2'1}n_1 + v_{2'2}n_2 + v_{2'3}n_3 + \cdots v_{2'i}n_i$$
$$n_{3'} = v_{3'1}n_1 + v_{3'2}n_2 + v_{3'3}n_3 + \cdots v_{3'i}n_i$$
$$\vdots \qquad \vdots \qquad \vdots \qquad \vdots \qquad \vdots$$
$$n_{i'} = v_{i'1}n_1 + v_{i'2}n_2 + v_{i'3}n_3 + \cdots v_{i'i}n_i$$

(11)

where the coefficients have the same meaning as in equations (8) and (9). Note, however, that the coefficient matrix of (10) is that of (9), but with rows and columns interchanged (transposed), and that the matrix of (11) is related in the same way to that of (8). We thus have:

$$N = A^T N' \tag{10a}$$

$$N' = A^{-T} N \tag{11a}$$

and where N and N' denote the n_i and n_i' respectively. The superscript, T, indicates a transposed matrix. The coefficient matrices of (9) through (11) are thus all uniquely related. A or A^T is given when one has identified the new components; obtaining their inverses, however, requires work! Two special cases are worth considering:

(a) If the old and new components have identical compositions, hence differ only in selection of unit quantity as in mass to mols or mols to mass, and are numbered correspondingly, then the initial coefficient matrix can contain non-zero terms only in the leading diagonal. The inversion of the coefficient matrix is then trivial.

(b) The units of old and new may be consistent, and hence the transformation affine as in atom units to atom units or (as in the "direct" calculation of a CIPW norm) weight percent to weight percent. In this case the coefficients in each row of (8) or (9) and each column of (10) or (11) must sum to unity. Equations (10) and (11) may all be divided on both sides by the total number of units, converting the n_i and n_i' to X_i and X_i', their corresponding unit fractions. We then have $\sum_{i'} X_{i'} = 1$ and can simplify the transformation matrix by eliminating one row and one column, although in hand calculation it may be helpful to retain the full array and let the redundance serve as a built-in check.

In any case the tedious procedure of Cramer's Rule is often unnecessary. In practice many coefficients are zero. If molar or atom-equivalents of some kind are used for both old and new components, furthermore, the remaining coefficients will be simple rational numbers. The following strategy is therefore recommended for hand calculation:

(1) Convert the old components to molar or atomic units and calculate the new components, initially, in the same units (say oxy-equivalents).

(2) Number first the old components that have non-zero coefficients in only one of the equations (8). There will usually be several of this kind. Number the corresponding new components in the same order.

(3) Number next the old components that have non-zero coefficients in only one of the remaining equations (9), and number the corresponding new components in the same order.

(4) Repeat step (3) as many times as possible. Any old and new components that still remain may be numbered in any desired order.

TABLE 1. "Old" and new components for transformation example in text. Equations correspond to equations (8).

	Old		New		New in terms of old (in oxy-units)
(1)	K_2O	(1')	$KAlSi_3O_8$	(1') =	$[\frac{1}{16}]$ (1) + $[\frac{3}{16}]$ (6) + $[\frac{3}{4}]$ (8)
(2)	Na_2O	(2')	$NaAlSi_3O_8$	(2') =	$[\frac{1}{16}]$ (2) + $[\frac{3}{16}]$ (6) + $[\frac{3}{4}]$ (8)
(3)	MgO	(3')	$MgSiO_3$	(3') =	$[\frac{1}{3}]$ (3) + $[\frac{2}{3}]$ (8)
(4)	Fe_2O_3	(4')	Fe_3O_4	(4') =	$[\frac{3}{4}]$ (4) + $[\frac{1}{4}]$ (5)
(5)	FeO	(5')	$FeSiO_3$	(5') =	$[\frac{1}{3}]$ (5) + $[\frac{2}{3}]$ (8)
(6)	Al_2O_3	(6')	$CaAl_2Si_2O_8$	(6') =	$[\frac{3}{8}]$ (6) + $[\frac{1}{8}]$ (7) + $[\frac{1}{2}]$ (8)
(7)	CaO	(7')	$CaSiO_3$	(7') =	$[\frac{1}{3}]$ (7) + $[\frac{2}{3}]$ (8)
(8)	SiO_2	(8')	SiO_2	(8') =	1 x (8)

A simple example of the foregoing procedure is given in Table 1. The old components in Table 1 are standard oxide components and the new components are simple normative components. We may now proceed directly to equation (11), although beginners may find it helpful to write out equations (9) first. If both old and new are in oxy-equivalents we then have:

$$
\begin{array}{c|cccccccc}
 & n_1' & n_2' & n_3' & n_4' & n_5' & n_6' & n_7' & n_8' \\
\hline
n_1 = & \frac{1}{16} & 0 & 0 & 0 & 0 & 0 & 0 & 0 \\
n_2 = & 0 & \frac{1}{16} & 0 & 0 & 0 & 0 & 0 & 0 \\
n_3 = & 0 & 0 & \frac{1}{3} & 0 & 0 & 0 & 0 & 0 \\
n_4 = & 0 & 0 & 0 & \frac{3}{4} & 0 & 0 & 0 & 0 \\
n_5 = & 0 & 0 & 0 & \frac{1}{4} & \frac{1}{3} & 0 & 0 & 0 \\
n_6 = & \frac{3}{16} & \frac{3}{16} & 0 & 0 & 0 & \frac{3}{8} & 0 & 0 \\
n_7 = & 0 & 0 & 0 & 0 & 0 & \frac{1}{8} & \frac{1}{3} & 0 \\
n_8 = & \frac{3}{4} & \frac{3}{4} & \frac{2}{3} & 0 & \frac{2}{3} & \frac{1}{2} & \frac{2}{3} & 1 \\
\end{array}
\qquad (12)
$$

corresponding to equations (10) for the example of Table 1. We have retained the full array, even though the components are consistent and the transformation thus affine, in order to take advantage of the built-in check that the coefficient columns must sum to one. All terms not already in the leading diagonal may now be eliminated by using the equations in the order given until the only non-zero terms on the right-hand-side are in the leading diagonal. Multiplying or dividing so as to make all coefficients on the right unity, then interchanging right and left sides, we obtain:

	n_1	n_2	n_3	n_4	n_5	n_6	n_7	n_8	
n_1'	16	0	0	0	0	0	0	0	
n_2'	0	16	0	0	0	0	0	0	
n_3'	0	0	3	0	0	0	0	0	
n_4'	0	0	0	$\frac{4}{3}$	0	0	0	0	(13)
n_5'	0	0	0	-1	3	0	0	0	
n_6'	-8	-8	0	0	0	$\frac{8}{3}$	0	0	
n_7'	3	3	0	0	0	-1	3	0	
n_8'	-10	-10	-2	$\frac{2}{3}$	-2	$-\frac{2}{3}$	-2	1	

These correspond to equations (11) for the example of Table 1. Note that each column must again sum to unity, a useful check to avoid mistakes. As in the use of Cramer's Rule the procedure will not work if the new components are linearly dependent. In application to a rock-analysis, some of the n_i' may turn out to be negative. Classical procedure calls for a re-selection of normative components, but this is not necessary! Negative $CaAl_2Si_2O_8$, for example, simply means "peralkaline" and negative SiO_2 (in the new set) simply means a silica-undersaturated norm. If desired the new components may be converted to weight fraction or percent. Oxy-equivalents, however, correlate much more closely with modal abundances of the probable phases. The oxy-equivalent may readily be converted to oxy-fraction, oxy-percent, or normalized to a Barth Cell.

In Table 2a the first column gives components commonly reported (as weight percent) in rock and mineral analyses. [Can you explain the special handling of fluorine?] The second column gives formulas for a mixed set of oxide and exchange components in terms of the old oxide components. In Table 2b are

Table 2. A component transformation with exchange components.

(a) Equations for new components in terms of old (mol units). These correspond to equations (8) in text. The formulas here represent unit quantities.*

Old	New in Terms of Old
SiO_2	$SiO_2' = SiO_2$
Al_2O_3	$Al_2O_3' = Al_2O_3$
TiO_2	$TiFe_{-1} = TiO_2 + FeO - Fe_2O_3$
Fe_2O_3	$FeAl_{-1} = \frac{1}{2}Fe_2O_3 - \frac{1}{2}Al_2O_3$
Cr_2O_3	$CrFe_{-1} = \frac{1}{2}Cr_2O_3 - \frac{1}{2}Fe_2O_3$
•••	•••
FeO	$FeMg_{-1} = FeO - MgO$
MnO	$MnFe_{-1} = MnO - FeO$
•••	•••
MgO	$MgO' = MgO$
CaO	$CaO' = CaO$
SrO	$SrCa_{-1} = SrO - CaO$
•••	•••
Na_2O	$Na_2O' = Na_2O$
K_2O	$KNa_{-1} = \frac{1}{2}K_2O - \frac{1}{2}Na_2O$
Rb_2O	$RbK_{-1} = \frac{1}{2}Rb_2O - \frac{1}{2}K_2O$
•••	•••
H_2O	$H_2O' = H_2O$
F_2O_{-1}*	$F(OH)_{-1} = \frac{1}{2}P_2O_{-1} - \frac{1}{2}H_2O$

*F_2 as reported in gravimetric analyses should more properly be reported as F_2O_{-1} with a correspondingly lowered weight percent. To convert weight percent F_2 to mols F_2O_{-1} the "weight percent F_2" should be divided by the gram formula weight of F_2.

(b) Equations to transform an analysis (in mol units) from old to new. These correspond to equations (11) in text. The formulas here represent numbers of unit quantities(n'_i and n'_i).*

$SiO_2' = SiO_2$

$Al_2O_3' = Al_2O_3 + (Fe_2O_3 + Cr_2O_3 \cdots) + TiO_2$

$TiFe_{-1} = TiO_2$

$FeAl_{-1} = 2(Fe_2O_3 + Cr_2O_3 \cdots) + 2TiO_2$

$CrFe_{-1} = 2Cr_2O_3$

• • • •

$FeMg_{-1} = (FeO + MnO \cdots) - TiO_2$

$MnFe_{-1} = MnO$

• • •

$MgO' = MgO + (FeO + MnO \cdots) - TiO_2$

$CaO' = (CaO + SrO \cdots)$

$SrCa_{-1} = SrO$

• • • •

$Na_2O' = Na_2O + (K_2O + Rb_2O \cdots)$

$KNa_{-1} = 2(K_2O + Rb_2O \cdots)$

$RbK_{-1} = 2Rb_2O$

• • • •

$H_2O' = H_2O + F_2O_{-1}$

$F(OH)_{-1} = 2F_2O_{-1}$

*The different significance of the chemical formulas in (a) and (b) can lead to confusion although it is common practice.

given the equations that transform a sample analysis from mol units old to mol units new. Can you duplicate this result? Is this transformation affine?

<u>Note</u>: A possible source of trouble is that petrologists and chemists often use chemical formulas to denote very different things. In balanced stoichiometric equations such as (8) and (9), or in reaction equations, a formula usually represents a U_i in mols. In other contexts it may represent an n_i, again usually in mols, as in equations (10) or (11) that deal with analyses of samples (as in Table 3) or in giving plotting coordinates for certain diagrams. These different usages (the n_i are dimensionless, but the U_i are not) are usually clear in context but may also cause confusion. Beware!

Table 3.   A vector representation of amphibole composition.

(a) Equations to transform analysis (in mol units) from old to new. The chemical formulas here represent numbers of mols as in Table 2b:

$$Ca_2Mg_5Si_8O_{22}(OH)_2 = \frac{1}{24}(2RO_2 + 3R_2O_3 + RO + R_2O) \equiv A$$

$$H_2 = (H_2O + F_2O_{-1}) - A$$

$$FeMg_{-1} = (FeO + MnO+\cdots) - TiO_2 - 2(H_2O + F_2O_{-1}) + 2A$$

$$FeAl_{-1} = 2(Fe_2O_3 + Cr_2O_3+\cdots) + 2TiO_2 + 2(H_2O + F_2O_{-1}) - 2A$$

$$MgCa_{-1} = \tfrac{1}{2}(R_2O_3 + RO - R_2O) - (CaO + SrO+\cdots) - A$$

$$NaSiCa_{-1}Al_{-1} = -\tfrac{1}{2}(R_2O_3 + RO - R_2O) + 3A$$

$$NaAlSi_{-1} = \tfrac{1}{2}(R_2O_3 + RO + 3R_2O) - 2(H_2O + F_2O_{-1}) - 3A$$

$$Al_2Mg_{-1}Si_{-1} = -(R_2O_3 + RO + R_2O) + 2(H_2O + F_2O_{-1}) + 14A - SiO_2$$

Formulas for remaining exchange components are as in Table 2b.

$RO_2 \equiv (SiO_2 + TiO_2)$ $\qquad\qquad$ $RO \equiv (CaO + SrO+\cdots +MgO + FeO + MnO+\cdots)$

$R_2O_3 \equiv (Al_2O_3 + Fe_2O_3 + Cr_2O_3+\cdots)$ $\quad$ $R_2O \equiv (Na_2O + K_2O + Rb_2O+\cdots+H_2O + F_2O_{-1})$

A is unity if analysis has been normalized to 24 anions. In any case dividing by A yields the desired X_{ex}.

(b) Equations to obtain X_{ex} from a standard 24 anion recalculation: Element symbols represent numbers of atoms per 24 anions. M is total number of metal atoms; M^+ is total number of alkali atoms.

X_{H_2}	=	$\tfrac{1}{2}(H+F)-1$	$X_{TiFe_{-1}} = Ti$
$X_{FeMg_{-1}}$	=	$(Fe^{2+}+Mn^{2+}+\cdots)-Ti-(H+F)+2$	$X_{CrFe_{-1}} = Cr$
$X_{FeAl_{-1}}$	=	$(Fe^{3+}+Cr^{3+}+\cdots)+2Ti+(H+F)-2$	$\vdots \qquad \vdots$
$X_{MgCa_{-1}}$	=	$M-M^+-(Ca+Sr+\cdots)-13$	$X_{MnFe_{-1}} = Mn$
$X_{NaSiCa_{-1}Al_{-1}}$	=	$-M+M^++15$	$X_{SrCa_{-1}} = Sr$
$X_{NaAlSi_{-1}}$	=	$M-15$	$\vdots \qquad \vdots$
$X_{Al_2Mg_{-1}Si_{-1}}$	=	$8-Si+X_{NaSiCa_{-1}Al_{-1}}-X_{NaAlSi_{-1}}$	$X_{RbK_{-1}} = Rb$
$X_{F(OH)_{-1}}$	=	F	$\vdots \qquad \vdots$

Although our approach thus far has been mainly algebraic, such terms as "affine" and "projective" are geometric in origin. These describe geometries successively more general and hence less constrained than the Euclidean geometry with which most of us grew up. The affine transformation of a one-dimensional composition space is shown in Figure 3. In Figure 3a the composition line is

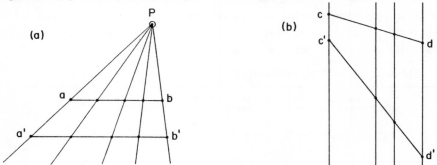

Figure 3. Affine transformation of a line: The primed and unprimed heavy lines are related by an affine transformation. In (a) the point of projection is at P and the lines ab and a'b' are parallel. In (b) the point P may be regarded as removed to infinity, hence the projection lines are parallel, though cd and c'd' need not be.

is transformed by projection through a point to a new line parallel to the initial one. In Figure 3b the composition line is transformed by projection on parallel lines (projection through a point at infinity). In affine transformations (those in which the units of old and new are consistent), there may be changes of scale, of origin, and of angles, but parallel lines remain parallel and compositions or points at infinity (where parallel lines meet) remain there. The proportional spacing of corresponding points on a line, moreover, is unaffected by affine transformation.

In Figure 4, however, we see five equally valid representations of the one dimensional space of the system iron-oxygen. Each of the five composition coordinates may be related to the others by a transformation of components as described previously. The transformations that relate them are clearly non-affine inasmuch as the proportional spacing of corresponding points is different in each and the compositions at infinity are different in each. In Figures 4a and 4b the compositions at infinity are not physically realizable, but in 4c, 4d, and 4e they are. A geometric construction that relates any one of these five representations to any other is presented in Figure 5. These relations are clearly a matter of perspective if the eye of the observer, in Figure 5, is placed at point P. Projective geometry is thus not so strange as it may at first seem. It is, in a real sense the geometry of space as we see it rather

16

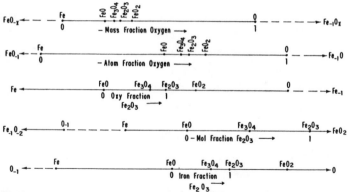

Figure 4. The composition coordinate defined by iron and oxygen: (a) is in mass fraction of the elemental components and (b) is in atom fraction of the elemental components. In (c), (d), and (e) the components are taken as FeO and Fe_2O_3 and the units of quantity are gram-equivalents of oxygen, mols of oxides, and gram-equivalents of iron, respectively. The physically accessible portions are indicated by solid lines, and the physically inaccessible ones by dashed lines. Arrows show formulas approached in moving indefinitely to the right ($+\infty$) or to the left ($-\infty$). In each case the formulas of the components at ∞ are formulas of zero quantity. The formulas for zero mass are $Fe_{+1}O_{+x}$ where x = 3.490625.

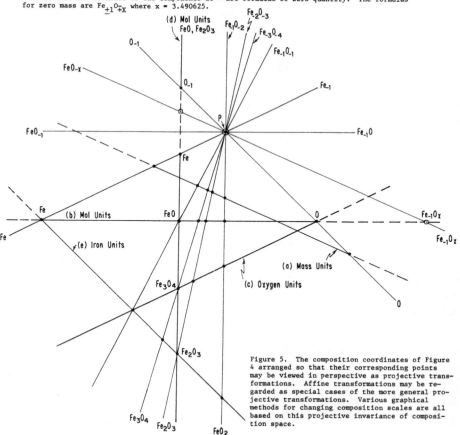

Figure 5. The composition coordinates of Figure 4 arranged so that their corresponding points may be viewed in perspective as projective transformations. Affine transformations may be regarded as special cases of the more general projective transformations. Various graphical methods for changing composition scales are all based on this projective invariance of composition space.

17

than as we imagine it to be, and its basic ideas were first explored by Renais-
sance artists, notably Leonardo da Vinci. The composition lines of Figure 5
really all look alike depending on how you tilt the page. Figure 5 may be a
bit difficult for most humans but should pose no problems for Kermit the Frog.
Note that points at infinity may also be regarded as points of zero quantity, if
quantity is measured in the units selected. The composition $Fe_{-1}O_x$ (where x =
3.490625) has zero mass.

A two-dimensional representation of the system Fe-Si-O is given in Figure 6
with reference to mol units of the components FeO, Fe_2O_3 and SiO_2. It may be
disturbing to find that the triangle Fe-Si-O appears to be in two parts. The
"split" triangle is split in appearance only. The two apparent parts meet on
the line (now at infinity) defined by SiO_3 and FeO_2. Many of the features of
ordinary Euclidean triangles are still there such as the three bounding straight
lines. Even the three interior angles still sum to a half-turn (180°) if due
care is taken to observe that the one at the oxygen vertex is negative. All
non-degenerate conic sections are projectively equivalent. This should be
readily apparent if one considers the vertex of a double cone as the point of
projection and allows the image plane to assume various orientations. The
triangle of Figure 6 is related to more conventional ones much as a hyperbola
is related to an ellipse or circle. Figure 6, after all, makes perfect sense
algebraically inasmuch as:

$$O_{-1} = 2FeO - Fe_2O_3 \tag{14}$$

and

$$O_{-1} + FeO = Fe \tag{15}$$

The projective correspondence of Figure 6 to Fe-Si-O in atom units is
shown in Figure 7. All of the essential features of composition space and of
phase diagrams that make use of compositive spaces are projectively invariant.
Chemographic rules are thus dependent only on the axioms of projective geometry
or its associated algebra, and are more neatly stated in such terms without
bringing in extraneous concepts. Some, in fact, are purely topologic, hence
even less constrained. There is thus even a "lever-law" for any projective
representation provided the weighting is given in the appropriate units of
quantity.

Projective transformation carries straight lines into straight lines,
planes into planes, and so on. Interestingly, all representations make algebraic sense if the signs of all subscripts in the formulas for all points are

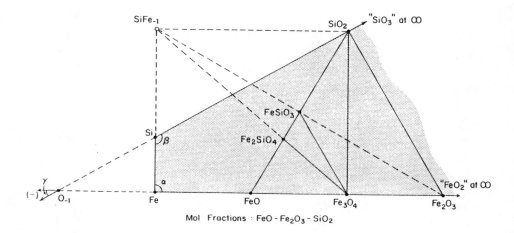

Mol Fractions : FeO - Fe₂O₃ - SiO₂

Figure 6. The plane Fe-Si-O as represented in full by the components FeO, Fe₂O₃ and SiO₂ in mol
units. The physically accessible triangle (shaded) is split by a line at infinity that connects
the compositions FeO₂ and SiO₃. (FeO₂ = Fe₂O₃ - FeO and SiO₃ = SiO₂ + Fe₂O₃ - 2FeO, hence both
are compositions of zero mol units). The diagram makes full algebraic sense: for example,
O₋₁ + FeO = Fe; and also makes geometric sense if tie-lines are taken as pointing away from
negative and toward positive. Tie-lines that cross the line at infinity are thus necessarily
split there. Note that α + β - γ = 180°.

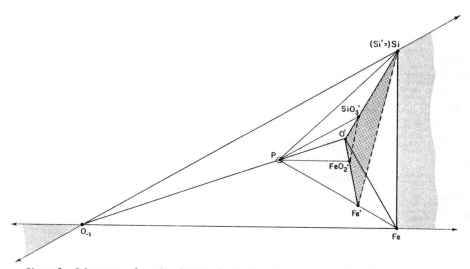

Figure 7. Enlargement of a part of Figure 6 with the tetrahedron, Fe-Si-O₋₁-O', erected on the
triangle Fe-Si-O₋₁ as base. The vertex O' may be regarded as toward the observer. The large
tetrahedron is divided by the plane triangle P-Fe-Si into two smaller ones. The tetrahedron
P-O'-Fe-Si is divided, in turn, by the triangle Fe'-Si'-O' that represents the space Fe-Si-O in
atom fractions of the elements. The triangle Fe'-Si-O' meets the base of the larger tetrahedron
at Si(=Si'). The plane P-FeO₂'-SiO₃' is parallel to Fe-Si-O₋₁. Projection of the triangle Fe'-
Si'-O' through P onto a plane parallel to P-FeO₂'-SiO₃' yields the triangle Fe-Si-O₋₁ of Figure
6. The line FeO₂-SiO₃ must therefore lie at ∞ in Figure 6.

19

reversed. This is a substitution of the usual notation for a point by its "antipodal" symbol. The antipodal points at infinity are really the same point, simply operating in the opposite sense (plus means "atom in" and minus means "atom out"). Antipodal pairs are linearly dependent, by virtue of relations such as:

$$FeMg_{-1} + Fe_{-1}Mg = 0 \tag{16}$$

and hence are really coincident points that cannot define a line any more than can the "same" point used twice, since:

$$FeO - FeO = 0 \tag{17}$$

Readers interested in learning more about affine and projective geometries may find the book by H.S.M. Coxeter (1961) to be helpful. A more algebraic approach can be found in Birkhoff and MacLane (1977, pp. 305-317). [Can you show that a diagram such as the strontium evolution (isochron) diagram of Lanphere et al. (1964, pp. 269-320, note also the Sniften reference therein) is a projective transformation of the triangle $Rb^{87}-Sr^{87}-Sr^{86}$ in atom units?]

CRYSTALLINE PHASES

It has been traditional to describe the composition of crystalline phases wholly in terms of additive or "end-member" components. Bragg (1937), however, pointed out that the number of "end-members" may greatly exceed the number of independent compositional variations, and suggested a different approach which we shall follow here. We shall describe each phase by selecting a *single* additive component, and let all variations therefrom be accounted for by exchange components. A composition variable, X_{ex}, may then be defined as the ratio of the number of mols of the exchange component to the number of mols of the additive component: $X_{ex} = n_{ex}/n_{ad}$, where "ad" represents the additive component. The compositions of the exchange components are then all points at infinity. The number of X_{ex} is $(c_\phi - 1)$. Unit values of the X_{ex} may then be regarded as a set of basis vectors that describe displacements in composition space from the additive component as origin. If we consider only component transformations that preserve the above features the resulting transformations are affine. Selection of a new additive component is simply a shift of origin in the vector space, and selection of a new set of independent exchange components is simply equivalent to a new set of basis vectors that spans the same space (see Venit and Bishop, 1981).

The above procedure has so much in common with the standard recalculation of a mineral analysis that, with practice, the numerical values of the X_{ex} can often be read directly from the recalculated mineral analysis. Thus if a simple K-Na-Ca ternary feldspar is recalculated to eight oxygens, with Ab selected as the additive component, the value of $X_{KNa_{-1}}$ is equal to the number of K-atoms (or mol fraction Or), and that of $X_{CaAlNa_{-1}Si_{-1}}$ is equal to the number of Ca-atoms (or mol fraction An). Had we chosen $NaSiCa_{-1}Al_{-1}$ rather than its inverse then $X_{NaSiCa_{-1}Al_{-1}}$ would have been *minus* the number of Ca-atoms (or *minus* the mol fraction An). The X_{ex}, unlike mol fractions (or other unit fractions) do not, however, sum to unity except in special cases. Thus if we take $Ca_2Mg_5Si_8-O_{22}(OH)_2$ as the additive amphibole component and consider the plane defined by it and the components $Al_2Mg_{-1}Si_{-1}$ and $NaSiCa_{-1}Al_{-1}$, then an idealized tschermakite has $X_{Al_2Mg_{-1}Si_{-1}}$ = 2 and $X_{NaSiCa_{-1}Al_{-1}}$ = 0. An idealized glaucophane has $X_{Al_2Mg_{-1}Si_{-1}}$ = $X_{NaSiCa_{-1}Al_{-1}}$ = 2.

To illustrate the transformation from an analysis in practical oxide components to such a vector space we shall use one of the most complex examples possible, a generalized amphibole. All of the components in Table 2, either old or new, may be regarded as I.V., at least within limits, in amphiboles. If we replace the six "new" additive components in Table 2a (SiO_2', Al_2O_3', MgO', CaO', Na_2O', H_2O') by the single additive component, $Ca_2Mg_5Si_8O_{22}(OH_2)$ and the five vectors, $MgCa_{-1}$, $Al_2Mg_{-1}Si_{-1}$, $NaSiCa_{-1}Al_{-1}$, $NaAlSi_{-1}$, and H_2 we have a new set that contains only one additive component. "H_2" may look like an additive component, but in amphibole (and in other ferrous silicates containing hydroxyl) it is not. This is because H_2 may be varied in content without altering the number of unit cells. The variation is achieved by the substitution of $Fe^{3+} + \square$ for $Fe^{2+} + H$, hence $FeHFe_{-1}\square_{-1}$ = H. The substitution may be verified by laboratory experiment. Certain ferrous amphiboles lose hydrogen on being heated in air, becoming oxyamphiboles and regain it on being heated (carefully) in hydrogen. The oxide equation (in mols) for H_2 is thus: H_2 = H_2O + 2FeO - Fe_2O_3. Exchange component formulas need not be encumbered with valencies or site coordination symbolism. Composition space is concerned only with bulk chemistry. The components $TiFe_{-1}$ (= TiO_2 + FeO - Fe_2O_3) and $MnFe_{-1}$ (= MnO - FeO or = ½Mn_2O_3 - ½Fe_2O_3, depending on the oxide components selected) are further examples. If you are upset by this apparent disregard for convention I urge a careful inspection of the diagrams of Figure 8. $FeMn_{-1}$ is as much I.V. in an Fe-Mn alloy as it is in bixbyite, $(Mn,Fe)_2O_3$. The exchange reaction between an ilmenite and a spinel in the system Fe-Ti-O can be written very simply as $TiFe_{-1}[Ilm] = TiFe_{-1}[Spl]$.

The transformation formulas for an amphibole analysis in mols oxides to

21

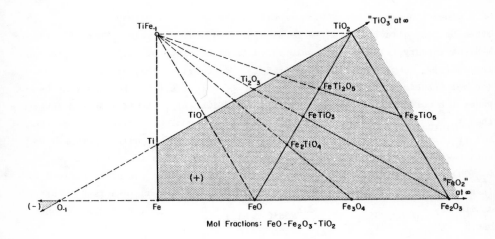

Mol Fractions: $FeO-Fe_2O_3-TiO_2$

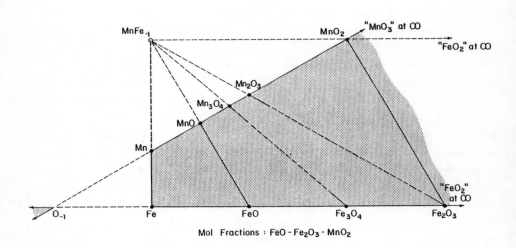

Mol Fractions : $FeO-Fe_2O_3-MnO_2$

Figure 8. Ternary composition planes containing iron and oxygen. (a) is Fe-Ti-O based on mol fractions of the components $FeO-Fe_2O_3-TiO_2$, and (b) is Fe-Mn-O based on mol fractions of the components $FeO-Fe_2O_3-MnO_2$. As in Figure 6 the physically accessible regions are shaded. The points $TiFe_{-1}$ and $FeMn_{-1}$ are common points to several lines of chemical variation of phases in each of the two systems, and hence are I.V. components of these phases. Although not physically accessible compositions, these components are physically realizable chemical displacements. The point Mn_2O_7 in (b) is off the diagram as $Mn_{-2}O_{-7}$ at the coordinates 6, -3, -2 for the mol fractions of FeO, Fe_2O_3 and MnO_2, respectively.

22

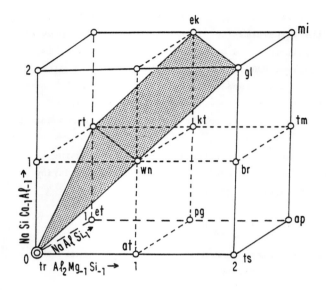

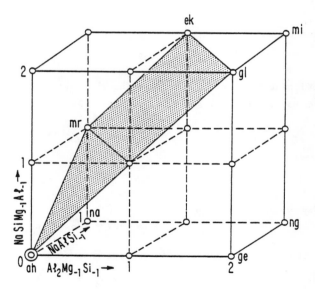

Figure 9. Amphibole composition spaces: (a) is a part of the space $Na_2O-CaO-MgO-Al_2O_3-SiO_2-H_2O$ (NCMASH). The additive component (at the origin) is taken as $Ca_2Mg_5Si_8O_{22}(OH)_2$. Displacements from the origin are indicated by a coordinate system defined by $X_{Al_2Mg_{-1}Si_{-1}}$, $X_{NaSiCa_{-1}Al_{-1}}$ and $X_{NaAlSi_{-1}}$. The standard formulas of points not at the origin may be obtained by applying the operation of each vector X_{ex} times to the additive component. (b) is a part of the space $Na_2O-MgO-Al_2O_3-SiO_2-H_2O$ (NMASH). The components of (b) may be derived from those of (a) by applying the operation $MgCa_{-1}$ to them. See Thompson (1981) for explanation of abbreviations.

23

mols new components are given in Table 3a. The formulas are readily derived
using the procedures for the transformation of components described previously.
If the oxides have been normalized to 24 anion units the quantity, A, is unity,
and the X_{ex} will be numerically equal to the n_{ex}. Table 3b shows how the X_{ex}
may be read directly from a 24-anion recalculation. The diagrams of Figure 9
show representations of two simple amphibole spaces. The components of Figure
9b are related to those of Figure 9a by transforming all calcic components by
the vector $MgCa_{-1}$.

The formulation of the thermodynamic properties of crystals using X_{ex}
is straightforward. Just as for wholly additive components we may define
μ_i as $[\partial E/\partial n_i]_{P, T, all\ n_j}$, whence, for reversible processes:

$$dE = -PdV + TdS + \sum_{ex} \mu_{ex} dn_{ex} + \mu_{ad} dn_{ad} \qquad (18)$$

where the subscript ex indicates exchange components and ad indicates the
additive component. By Gibbs' integration:

$$E = -PV + TS + \sum_{ex} \mu_{ex} n_{ex} + \mu_{ad} n_{ad} \qquad (19)$$

hence

$$0 = VdP - SdT - \sum_{ex} n_{ex} d\mu_{ex} - n_{ad} d\mu_{ad} \qquad (20)$$

Defining $X_{ex} \equiv n_{ex}/n_{ad}$ and letting the n_{ad} be the number of mols of the phase
we have, from (19) and (20), respectively:

$$\overline{E} = -P\overline{V} + T\overline{S} + \sum_{ex} \mu_{ex} X_{ex} + \mu_{ad} \qquad (21)$$

and:

$$0 = \overline{V}dP - \overline{S}dT - \sum_{ex} X_{ex} d\mu_{ex} - d\mu_{ad} \qquad (22)$$

hence

$$d\overline{E} = -Pd\overline{V} + Td\overline{S} + \sum_{ex} \mu_{ex} dX_{ex} \qquad (23)$$

and because $\overline{G} = \overline{E} + P\overline{V} - T\overline{S}$ we also have

$$\overline{G} = \mu_{ad} + \sum_{ex} \mu_{ex} X_{ex} \qquad (24)$$

and

$$d\overline{G} = \overline{V}dP - \overline{S}dT + \sum_{ex} \mu_{ex} dX_{ex} \qquad (25)$$

Equation (25) shows that $\mu_{ex} = [\partial \overline{G}/\partial X_{ex}]_{P, T, \text{all other } X_{ex}}$. The μ_{ex} are thus directly obtainable as first partial derivatives from a formulation for $\overline{G}$ as a function of P, T, and the X_{ex}. Only μ_{ad} requires further steps in that $\mu_{ad} = G - \Sigma_{ex} \mu_{ex} X_{ex}$.

The analogue to the above procedure, where the components are wholly additive involves several more steps, though the results, in the end, are equivalent. Comparison of the two procedures shows that the exchange potentials (μ_{ex}) are either numerically equal to *differences* between the chemical potentials of most standard additive components or simple submultiples of these differences. Thus:

$$\mu_{KAlSi_3O_8} - \mu_{NaAlSi_3O_8} = \mu_{KNa_{-1}}$$

and

$$\mu_{Ca_2Mg_3Al_4Si_6O_{22}(OH)_2} - \mu_{Ca_2Mg_5Si_8O_{22}(OH)_2} = 2\mu_{Al_2Mg_{-1}Si_{-1}}.$$

Chemical potential differences correspond to activity ratios.

Although the method suggested here is recommended primarily for crystals, it may, in fact, be adapted to non-crystalline phases. Silicate melts, for example, could be referred to with SiO_2 as the additive component, and aqueous fluids could be referred to with H_2O as the additive component. Many of the classical treatments of aqueous solutions indeed incorporate much of the above, but cloaked in different language. Activity ratios, after all, correspond in turn to exchange potentials. Many aspects of this have been explored by Burt (1974, 1975a,b, 1979, 1981). For certain petrologic purposes, however, it is desirable to take more than one component of natural fluids as additive (H_2O and CO_2, for example) although others may be carried as exchange components.

REDUCED COMPOSITION SPACES

A difficulty in dealing with polyphase, multicomponent systems such as those in rocks and mineral deposits is that the composition spaces required for their full representation require too many dimensions for their direct visualization. As a consequence many schemes have been devised for reducing the dimensionality so that at least some essential features may be graphically displayed. Despite superficial differences, all such schemes may be regarded as projections or mappings of spaces into subspaces of fewer dimensions. Any such reduction involves some loss of information, generality, or utility; hence the detailed procedure must be tailored to the problem at hand.

The projection or mapping of a many-component composition space into a simpler one may always be accomplished by regarding such an operation as fundamentally a transformation of components. The strategy is to select a set of "new" components that has two distinct subsets. One subset, often three or four in number for ease of pictorial portrayal, is the "image" subset that defines the compositional subspace that receives the mapped or projected image. The other subset must contain the compositions selected as projection points. We may call these the image components and projection components, respectively. At least one of the image components must be an additive component, otherwise there are no formal constraints other than those imposed by the problem at hand. We may, however, distinguish two limiting cases with respect to the nature of the projection components. If all of the projection components are exchange components (or vectors) the result is what may be called a *condensed* composition space. At the other extreme, all of the projection components may be additive. Most projected diagrams in current use are of mixed type, at least those dealing with natural mineral assemblages, but some (mainly in experimental studies) involve only additive components as projection points. Reduced composition spaces are usually referred to as "projections" but, as we have seen, *any* composition space may be regarded as a projection from some other choice of components.

The actual plotting (or, better, "mapping") of compositions into a reduced space is simple. Once the transformation of components has been accomplished the projection components are simply dropped so that the reduced representation is purely in terms of the image components. The actual act of projection is in the "neglect" of the new projection components. If these are still included the representation is complete, not reduced.

The idea of condensed spaces is far from new, but the procedure for obtaining them has often been regarded as more of a magical art than the simple algebraic operation it really is. An examination of Table 3 will show that the results are very nearly the same as the "Niggli-values" long in use among European petrologists (see Niggli, 1954). In Niggli's recalculation (a component transformation) the additive components si, al, fm, c, alk, and h correspond to the new additive components of Table 2, SiO_2', Al_2O_3', MgO', CaO', Na_2O' and H_2O' (NCMASH), respectively. The only essential difference is in Niggli's handling of the new components that replace old TiO_2 and Fe_2O_3. In Table 2 we have used $TiFe_{-1}$ and $FeAl_{-1}$. If these are replaced by new TiO_2 and 0 (or O_2), respectively, the Niggli values will be reproduced exactly. Although Niggli did not make explicit use of exchange vectors, his k and mg fulfill the roles of our KNa$_{-1}$ and FeMg$_{-1}$, respectively. Many of Niggli's graphical portrayals are further reductions of his condensed space by projections through related additive components.

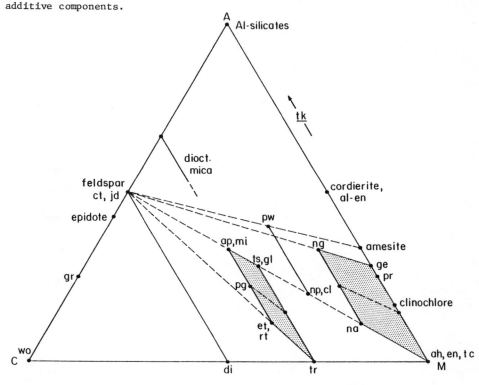

Figure 10. The condensed space CaO-MgO-Al$_2$O$_3$-SiO$_2$-H$_2$O(CMASH) as projected through the additive components SiO$_2$ and H$_2$O. Symbols for amphibole compositions correspond to those of Figure 9. The CMASH condensation corresponds to that of O'Hara (1968). The diagram presented here also has many features characteristic of the ACF diagram of Eskola (1939). The arrow tk corresponds to the operation Al$_2$Mg$_{-1}$Si$_{-1}$. See Thompson (1981) for explanation of abbreviations.

Other reduced composition spaces have been employed, notably those of Eskola (1939) with various modifications by subsequent authors. Most of these involve condensation on exchange vectors (projection through exchange components) as well as projection through additive components. If the new component Na$_2$O' in Table 2 is replaced by the exchange component NaSiCa$_{-1}$Al$_{-1}$ the transformation leaves only CaO', MgO', Al$_2$O$_3'$, SiO$_2'$ and H$_2$O' (CMASH) as additive components. Projections through new SiO$_2'$ and H$_2$O' yields the diagram of Figure 10. Figure 10 has much in common with the ACF diagram of Eskola, but it differs in the plotting of alkali-bearing phases. The condensation into CMASH corresponds

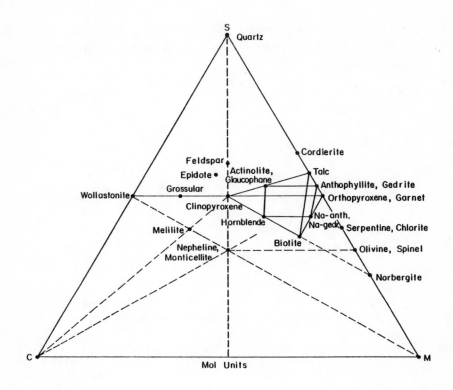

Figure 11. The condensed space CaO-MgO-SiO$_2$-H$_2$O(CMSH) as projected through the additive component H$_2$O. Condensation into CMSH differs from that into CMASH by additional condensation on the vector Al$_2$Mg$_{-1}$Si$_{-1}$. Further condensation on MgCa$_{-1}$ would permit a complete two-dimensional representation in MgO-SiO$_2$-H$_2$O(MSH) without projection though additive components.

very closely to the condensation scheme of O'Hara (1968). The process can be continued as long as at least one additive component is retained. Replacement of new Na$_2$O' and Al$_2$O$_3'$ by NaSiCa$_{-1}$Al$_{-1}$ and Al$_2$Mg$_{-1}$Si$_{-1}$ in Table 2 condenses the full space into CaO-MgO-SiO$_2$-H$_2$O (CMSH). A projection of this through new H$_2$O is shown in Figure 11. The diagrams of Figure 9 may also be used to represent more generalized amphiboles by condensation along the omitted exchange vectors. Figure 9a is a part of NCMASH and Figure 9b is a part of NMASH.

[In Chapter 3 of this volume, Spear et al. review a number of projections which are useful in the representation of mineral assemblages in metamorphic rocks.]

PROJECTED PHASE DIAGRAMS

Phase diagrams for the representations of heterogeneous systems at equilibrium often contain compositional coordinates. In complex systems it is not uncommon to utilize a reduced composition space. Condensation along exchange vectors, however, is not suitable, as a rule, for this purpose, particularly if the fractionation between coexisting phases is strong (as in many Fe-Mg systems) and the compositional range is large. For minor or trace component exchanges, on the other hand, some condensation of this type causes no difficulty. Extremely weak fractionation (as among isotopic species) is also of little consequence. In this last sense virtually *all* phase diagrams may be regarded as condensed along isotopic exchange vectors.

The most useful projected phase diagrams are those in which the dimensional reduction is by projection through major additive components. This may be either the additive component for some ubiquitous ("saturating" or "excess") phase of fixed composition (or very limited compositional range), or it may be some additive or exchange component for which the activity or chemical potential is buffered by some substance external to the phase assemblage that is regarded as within the system. Many of the best examples are in experimental petrology. In the pioneering studies of the "granite system" of Tuttle and Bowen (1958) and in subsequent work, all compositions are projected through H_2O because of the presence of a saturating aqueous phase (essentially high-density steam). Another early example is in the iron-crucible experiments on systems such as "FeO"-SiO_2 by Bowen and Schairer (1932). Here all phases coexist with essentially pure Fe-metal. Under such conditions the liquid phase, and also certain crystals such as wustite, are commonly slightly ferric so that compositions are projected into the purely ferrous subsystem for graphical presentation. Curiously the calculations involved in plotting these diagrams are not usually equivalent to a simple projection through Fe (which would be logically correct). The reason is probably that the Fe-content is negative in most projected compositions, a circumstance that classically would be taboo. Most projected compositions, however, are so close to the image plane that the precise point of projection makes little quantitative difference. Bowen and Schairer's calculation is equivalent to projection through $SiO_{1.755625}$ (parallel to the line Si-O, in weight percent). Allen and Snow (1955) project through oxygen! For highly ferric phases such as acmite (which can and do appear in such experiments) these varied modes of projection can yield quite different graphical results. Other projected phase diagrams appear in experiments with controlled atmospheres on iron-bearing systems, as for example the work of Muan and Osborn (1965), and in the results on oxygen-buffered systems by Eugster (1959) and subsequent workers.

Projected phase diagrams are also in common use for the study of natural assemblages in metamorphic rocks. These inevitably involve some condensation on exchange vectors for minor components (natural systems are "dirty"), but are for the most part projections through additive components of major ubiquitous phases of fixed composition or of limited chemical variability. Common projection points are SiO_2 for quartz-bearing rocks, $CaCO_3$ for calcite-bearing rocks, and components such as H_2O and CO_2, whose activities are taken as controlled by some external medium. Many such diagrams are multiply-projected with several projection components. One complex example is the AFM diagram (Thompson, 1957; see also Greenwood, 1975) for assemblages of phases in or near the system $SiO_2-Al_2O_3-MgO-FeO-K_2O-H_2O$. Al_2O_3, MgO and FeO are taken as image components and H_2O, SiO_2, KAl_3O_5 as projection components (Thompson, 1970, p. 550) so that assemblages coexisting with quartz and muscovite $(\sim KAl_3Si_3O_{10}(OH)_2)$ can be represented as a function of P, T, and a_{H_2O}. Minor amounts of other components may be eliminated by condensing on exchange vectors, and, for special purposes, a third dimension may be added (Froese, 1969; Thompson, 1972). Because muscovite can vary fairly widely in composition, a representation is also possible showing the response of the muscovite itself to the coexisting assemblage (Thompson, 1979). The last diagram is related to the AFM diagram much as a liquidus diagram is related to its corresponding solidus diagram in a ternary system.

Some projected phase diagrams appear in subsequent chapters of this volume. Can you identify the image components and a set of projection components for each?

ACKNOWLEDGMENTS

This presentation is the result of many years of exchanges with students on the chemical representation of rocks and minerals, and the calculations related thereto. To acknowledge a few, by name, would do injustice to many others. I do wish to thank Stephen K. Dobos and John Ferry, however, for their helpful comments on the manuscript for this paper on iron-bearing systems, as for example the work of Muan and Osborn (1965), and in the results on oxygen-buffered systems by Eugster (1959) and subsequent workers.

CHAPTER 1 REFERENCES

Allen, W.C. and R.B. Snow (1955) The orthosilicate-iron oxide portion of the system CaO-"FeO"-SiO$_2$. J. Am. Ceram. Soc. 38, 264-280.

Barth, T.F.W. (1948) Oxygen in rocks: A basis for petrographic calculations. J. Geol. 56, 50-60.

Birkhoff, G. and S. MacLane (1977) *A Survey of Modern Algebra, 4th Ed.* MacMillan, New York.

Bowen, N.L. and F. Schairer (1932) The system FeO-SiO$_2$. Am. J. Sci., 5th ser. 24, 177-213.

Brady, J.B. (1975) Chemical components and diffusion. Am. J. Sci. 275, 1073-1088.

_____ and J.H. Stout (1980) Normalization of thermodynamic properties and some complications for graphical and analytical problems in petrology. Am. J. Sci. 280, 173-189.

Bragg, W.L. (1937) *Atomic Structure of Minerals, 1st Ed.* Cornell Univ. Press., Ithaca, N.Y.

_____ and G.F. Claringbull (1965) *Crystal Structure of Minerals.* G. Bell and Sons, London.

Burt, D.M. (1974) Concepts of acidity and basicity in petrology The exchange operator approach. Geol. Soc. Am. Abstr. Prog. 6, 674-676.

_____ (1975a) Beryllium mineral stabilities in the model system CaO-BeO-SiO$_2$-P$_2$O$_5$-F$_2$O$_{-1}$, and the breakdown of beryl. Econ. Geol. 70, 1279-1292.

_____ (1975b) Hydrolysis equilibria in the system K$_2$O-Al$_2$O$_3$-SiO$_2$-H$_2$O-Cl$_2$O$_{-1}$; comments on topology. Econ. Geol. 71, 665-671.

_____ (1978) Tin silicate-borate-oxide equilibria in skarns and greisens: The system CaO-SnO$_2$-H$_2$O-B$_2$O$_3$-CO$_2$-F$_2$O$_{-1}$. Econ. Geol. 73, 269-282.

_____ (1981) Acidity-salinity diagrams--application to greisen and porphyry deposits. Econ. Geol 76, 832-843.

Coxeter, H.S.M. (1969) *Introduction to Geometry, 2nd Ed.* John Wiley and Sons, N.Y.

Eskola, P. (1939) Die metamorphen Gesteine. *In* Barth, T.F.W., C.W. Correns and P. Eskola, Eds., *Die Entstehung der Gesteine,* Springer-Verlag, Berlin, 263-407.

Eugster, H.P. (1959) Reduction and oxidation in metamorphism. *In* Abelson, P.H., Ed., *Researches in Geochemistry,* John Wiley and Sons, N.Y., 397-426.

Froese, E. (1969) Metamorphic rocks from the Coronation Mine and surrounding area. Geol. Surv. Canada Paper 68-5, 55-77.

Gibbs, J.W. (1948) *Collected Works.* Yale Univ. Press, New Haven.

Greenwood, H.J. (1975) Thermodynamically valid projections of extensive phase relations. Am. Mineral. 60, 1-8.

Lanphere, M.A., G.J.F. Wasserburg and A.L. Albee (1964) Redistribution of strontium and rubidium isotopes during metamorphism, World Beater Complex, Panamint Range, California. *In* Craig, H., S.L. Miller and G.J. Wasserburg, Eds., *Isotopic and Cosmic Chemistry,* North-Holland, Amsterdam, 269-312.

Muan, A. and E.F. Osborn (1956) Phase equilibria at liquidus temperatures in the system MgO-FeO-Fe$_2$O$_3$-SiO$_2$. J. Am. Ceram. Soc. 39, 121-140.

Niggli, P. (1954) *Rocks and Mineral Deposits.* W.H. Freeman and Co., San Francisco.

O'Hara, M.J. (1968) The bearing of phase equilibria studies on synthetic and natural systems on the origin and evolution of basic and ultra-basic rocks. Earth Sci. Rev. 4, 69-133.

Smith, J.V. (1959) Graphical representation of amphibole compositions. Am. Mineral. 59, 1069-1082.

Thompson, J.B., Jr. (1957) The graphical analysis of mineral assemblages in pelitic schists. Am. Mineral. 42, 842-858.

_____ (1970) Geochemical reaction and open systems. Geochim. Cosmochim. Acta 34, 842-858.

_____ (1972) Oxides and sulfides in regional metamorphism of pelitic schists. Proc. 24th Int'l Geol. Cong., 1972, Montreal, Sect. 10, 27-35.

_____ (1979) The Tschermak substitution and reactions in pelitic schists (in Russian). *In* Zharikov, V.A., W.I. Fonarev and S.P. Korikovskii, Eds., *Problems in Physicochemical Petrology, Vol. I,* Moscow. Acad. Sci.

_____ (1981) An introduction to the mineralogy and petrology of the biopyriboles. *In* Veblen, D.R., Ed., *Amphiboles and Other Hydrous Pyriboles-Mineralogy,* Reviews in Mineralogy 9A, 141-188.

Tuttle, O.F. and N.L. Bowen (1958) *Origin of Granite in the Light of Experimental Studies in the System NaAlSi$_3$O$_8$-KAlSi$_3$O$_8$-SiO$_2$-H$_2$O.* Geol. Soc. Am. Memoir 74.

Venit, S. and W. Bishop (1981) *Elementary Linear Algebra.* Prindle, Weber and Schmidt, Boston.

31

Chapter 2

REACTION SPACE: An ALGEBRAIC and GEOMETRIC APPROACH

J.B. Thompson, Jr.

INTRODUCTION

In Chapter 1 we considered the problems of characterizing the compositions of phases, one-by-one, though noting that a system of two or more phases
(a heterogeneous system) may also be regarded as having independently variable
(I.V.) components, and that any linearly independent set of these must be of
a definite number, c_s. We noted further that any component I.V. in one of
the phases is I.V. in the system and that any linear combination of such
components, though not necessarily I.V. in any of the constituent phases, is
also I.V. in the system. We have not, however, presented a systematic procedure for determining the number, c_s, of I.V. components in a heterogeneous
system, nor have we shown any simple means for obtaining a linearly independent set. It is *not*, however, safe to assume that all oxide or elemental
components may be taken as I.V. in the system even though this is indeed often
so in multiphase heterogeneous systems -- the more phases, in fact, the better
the chances.

A second problem not yet considered is the number of independent ways
(i.e., heterogeneous reactions) in which matter may be redistributed among
the phases in a system. If the redistributions are wholly among the phases
within the defined bounds of the system, these are *closed-system* reactions.
There may also be redistribution involving phases taken as external to the
system, in which case the reaction is an *open-system* reaction. This distinction, however, is not as fundamental as it may appear owing to its dependence
on the definition of the bounds of the system. Any heterogeneous reacting
system, after all, contains subsystems that are necessarily open ones. We
shall also find that the problem of identifying possible open-system reactions
is trivial once a closed-system set has been found.

PHASE COMPONENTS AND SYSTEM COMPONENTS

Let us select a set of practical components to describe a heterogeneous
system such that the set is sufficient to specify the compositions of all I.V.
components of all phases therein. The practical set need not be I.V. components of the system (though some or all may prove to be) and might, for

33

example, be the chemical elements themselves. A set of standard oxide components may also suffice for a rock-mineral assemblage.

Each phase, ϕ, has c_ϕ I.V. components. The heterogeneous system as a
whole thus has $c_p \equiv \Sigma c_\phi$ phase-components. Some of these may have identical
chemical formulas, but we shall here regard them as different if they are
components of different phases. If the system has c_s I.V. components (a number yet to be determined) these may always be selected as a set corresponding
one-for-one to a specific set selected from among the c_p phase components.
All remaining phase components have compositions that may be written in terms
of the c_s selected as system components. We thus have:

$$n_r = c_p - c_s \qquad (1)$$

where n_r represents the number of such equations. It is also, however, the
number of independent closed-system reactions that can take place in the given
system inasmuch as each represents a possible redistribution of matter among
the phases of the system. Any other possible redistribution is necessarily a
linear combination of the n_r, but such additional equations are not independent. There is thus considerable latitude in the specific set of n_r equations
selected, but their number is fixed. In this sense there is a close analogy
between a set of independent reactions and a set of I.V. components: We must
have the right number, and the set must be linearly independent, but this
leaves a great deal of latitude (and room for shrewd strategy) in the specific
forms selected.

A REDUCTION MATRIX

We still, however, have the problem of evaluating the number, c_s. This
may be obvious in simple systems but often is not. The number of practical
components, c_e, may (and often does) exceed c_s. In such circumstances it is
easy to *underestimate* the n_r if $c_e > c_s$ and these last were assumed equal.
We shall therefore write an equation giving the compositions of *each* of the
c_p phase components, on the *left*, in terms of the c_e ($\geq c_s$) practical components on the *right*. There will thus be c_p equations, each with c_e terms
(allowing some coefficients to be zero) on the right-hand-side. The coefficient matrix on the right thus has c_e columns and c_p rows. Now use the first
equation to eliminate, by linear combination, one non-zero term on the right
from each of the others. Repeat the process with the second equation, then
the third, and so on until the process comes to an end. There will then be
c_p new equations. If there are n_r possible closed-system reactions, there

will be n_r equations that have no non-zero terms on the right. These equations are then n_r stoichiometric equations among the components of the phases that represent n_r independent ways in which matter may be redistributed among the phases in the system. The remaining equations, c_s in number, will have no non-zero terms in common on the right, and the compositions of each right-hand-side, regrouped as a single component, may be taken as the c_s linearly independent I.V. components of the system inasmuch as all are expressed in terms of phase components.

In the language of linear algebra we have accomplished a Gauss-Jordan reduction of the original set of equations (see Hildebrand, 1952; Venit and Bishop, 1981, p. 59), determining in the process, the row nullity, n_r, and column nullity, n_c ($n_c = c_e - c_s$) of the coefficient matrix. With practice the process may be speeded considerably by a strategic rearrangement of rows and columns. Such rearrangement may also affect the precise form of the desired results, but, in any case, yields a sufficient set of c_s system components and a set of n_r independent reaction equations (if any are possible). Both sets may be simplified, if desired, by linear recombination, provided care is taken to preserve their numbers, and to avoid linear dependence. The procedure can be applied to whole-rock assemblages or to any sub-assemblages thereof, and can be applied to wholly or partly hypothetical assemblages. One may wish to include now-vanished precursors, or possible phases yet to come. The variance of the assemblage is irrelevant here, as is any consideration of whether or not the assemblage represents, or ever represented an equilibrium state of the system.

Closely analogous procedures may be used to deduce the possible independent reactions within a single phase (homogeneous reactions). For the reactions between species in a molecular fluid, for example, the formulas of the species play the role of the phase components and the chemical elements may be used as the practical components (see Prigogine and Defay, 1954, p. 468–469, and references therein). Homogeneous reactions (mainly order-disorder) in crystalline phases are a bit more complex, but again may be deduced by much the same method (see Thompson, 1969).

OPEN-SYSTEM REACTIONS

If there is reason to believe that there may be a redistribution of matter between the system and its surroundings (either an exchange or the transfer of an additive component), this may be taken into account by a single extension of the above procedure. Simply add an extra equation to

the initial set giving the composition of the suspect "environmental" component in terms of the practical set of components used initially to describe
the system. In petrologic examples these might include components such as
H_2O, CO_2, O_2 or KNa_{-1}. An open-system reaction involving the environmental
medium will result, however, only if the environmental component is also
independently variable in the closed system. In other words, it must be
possible to express its composition in terms of I.V. components of the phases
within the system. No more than one open-system reaction per environmental
component need be considered. If two arise these can always be linearly
recombined to provide one closed-system reaction and one open-system reaction.
It is preferable, in most applications, to keep the number of open-system
reactions in a set to a minimum. In thermodynamic treatments of equilibrium,
for example, the closed-system reactions are dependent only on temperature,
pressure, and the composition variables of the phases involved, whereas open-
system reactions must carry an explicit activity or chemical potential term
for each environmental component.

EXCHANGE REACTIONS

If two phases have a component of the same formula one of the possible
reactions $U_{ia} = U_{ib}$ is apparent by inspection. It is thus desirable to select
phase components so that, insofar as possible, there are components that are
I.V. in other phases in the system. In systems with crystalline phases we
may take advantage of this by selecting all but one of their components as
exchange components. Crystals with very different additive components may
vary in composition by the same exchanges. $FeMg_{-1}$, for example, is I.V. in
most ferromagnesian minerals, and KNa_{-1} is I.V. in most alkali-bearing minerals. $NaSiCa_{-1}Al_{-1}$ (or its inverse) is I.V. in most framework silicates,
and also in amphiboles, pyroxenes, some micas, and melilites. $Al_2Mg_{-1}Si_{-1}$ is
I.V. in most sheet and chain silicates, in melilite, and even, to a limited
degree, in high-temperature feldspars (Longhi, 1976; Longhi et al., 1976).
$NaAlCa_{-1}Mg_{-1}$ is I.V. in pyroxenes, amphiboles, and melilites, but it is the
linear sum of the two preceding ones ($NaAlCa_{-1}Mg_{-1}$ = $NaSiCa_{-1}Al_{-1}$ +
$Al_2Mg_{-1}Si_{-1}$) and hence is *not* independent. It is therefore unnecessary inasmuch as the other two are I.V. in the same mineral phases. Exchange components such as $TiFe_{-1}$, $MnFe_{-1}$, $CrFe_{-1}$ and the like are I.V. in the common
oxide phases as well as in most ferromagnesian silicates. Others commonly
encountered include $NaAlSi_{-1}$ (amphiboles, micas, and many framework silicates)

and $MgCa_{-1}$ (pyroxenes, amphiboles, garnets, and carbonates).

If an exchange component is I.V. in n phases then there are (n-1) independent exchange reactions relating them. In most applications a large fraction
of the n_r independent closed-system reactions can be accounted for immediately,
greatly reducing the work needed to identify the remainder. It is also evident that reactions involving minor and trace elements can be disposed of very
simply in this way.

A further advantage in taking as many as possible of the independent reactions as exchanges is that the equilibria corresponding to them are but
slightly affected by pressure; hence their petrologic significance is primarily one of geothermometry. The volume effect of exchange on a single phase may
be significant, but in most cases this is cancelled by a very nearly equal
effect, of opposite sign, in the other phase. The thermodynamic formulation
of a crystal-crystal exchange is also free of any activity or chemical potential terms of environmental components, though it may involve other compositional variables of the crystals themselves.

NET-TRANSFER REACTIONS

If as many reactions as possible have been taken as exchanges, any that
remain (often few in number) involve two or more additive components and
hence require a *net-transfer* of matter from one phase (or group of phases)
to another. Such reactions are petrographically conspicuous because they
lead to significant variations in modal abundances, whereas the modal effect
of exchanges is comparatively negligible. The equilibria corresponding to
net-transfer reactions may also have large volume effects leading to a strong
pressure-dependence. Thus if temperatures can be estimated from exchange
equilibria, then the closed-system net-transfer reactions provide the geobarometry. With T and P known, the open-system reactions may be used to
estimate H_2O activity ("geohygrometry") or the activities of other environmental components.

If we denote the number of exchange reactions as n_{ex} and the number of
closed-system net-transfer reactions as n_{nt} we have from equation (1):

$$n_r = n_{ex} + n_{nt} = c_p - c_s \tag{1a}$$

hence

$$n_{nt} = c_p - c_s - n_{ex} \tag{2}$$

37

Some net-transfer reactions are trivial and hence evident at sight (such as simple polymorphic changes). Others, such as feldspar-feldspar or amphibole-amphibole reactions, can be made equally apparent by assigning the same additive component to both feldspars or to both amphiboles. In these there is one simple net-transfer reaction and a set of exchanges.

Net-transfer reactions must contain at least two additive components but may also involve exchange components. For example, the reaction between Mg-Al orthopyroxene and a pyropic garnet may be written:

$$4MgSiO_3[Opx] + Al_2Mg_{-1}Si_{-1}[Opx] = Mg_3Al_2Si_3O_{12}[Gar] \qquad (3)$$

A net-transfer reaction for a similar pyroxene reacting with olivine and spinel can be written:

$$Mg_2SiO_4[Olv] + Al_2Mg_{-1}Si_{-1}[Opx] = MgAl_2O_4 [Spl] \qquad (4)$$

showing that the modal abundances of olivine and spinel may be altered by making use of the *exchange capacity* of the associated pyroxene. Similarly:

$$4SiO_2[Qtz] + Fe_3Al_2Si_3O_{12}[Gar] + 3Al_2Fe_{-1}Si_{-1}[Mica] = 4Al_2SiO_5[Sil] \qquad (5)$$

showing that the modal proportions of garnet and sillimanite may vary, in a quartz-bearing rock, by means of the exchange capacity of associated muscovite or biotite. We have also:

$$2Ca_2Al_3Si_3O_{12}(OH)[Epi] + 2MgCa_{-1}[Gar] =$$
$$SiO_2[Qtz] + [CaMg_5Si_8O_{22}(OH)_2 + 3Al_2Mg_{-1}Si_{-1}][Amp] \qquad (6)$$

as a more complex example. Reactions such as these often give surprisingly simple interpretations for otherwise baffling textural relations.

Two net-transfer reactions suffice for a generalized four-phase assemblage of olivine, orthopyroxene, olivine and spinel. These may be taken as equations (3) and (4). The remaining reactions are exchanges such as $FeMg_{-1}$, $FeAl_{-1}$ and so forth, to accommodate the other compositional variations. If, however, the phases are in the limiting Mg-Al-Si system, then the compositions of the phases themselves are linearly dependent, and the assemblage, if at equilibrium, is univariant. There are then two possible reactions, both net-transfers and these may be taken, again, as equations (3) and (4). *Both* are needed for a full thermodynamic treatment of the equilibrium. There is, however, another equation of the form:

$$\text{Garnet + Olivine = Orthopyroxene + Spinel} \qquad (7)$$

that is a formal statement of the *linear dependence of the actual phase compositions*. A balanced equation having the form of (7) must be a linear combination of (3) and (4). The coefficients of such a combination, however, are not unique owing to the variable composition of the pyroxene. The exact balancing will, in general vary from point to point along the univariant curve. Equations such as (7) are unique only for phases of fixed composition. In this case there is only one independent heterogeneous reaction in the indicated assemblage, and equation (7) is it. Linearly dependent phases, however, are a digression in the present context, although they may be treated by much the same algebraic methods.

SOME SHORTCUTS

We have already shown that exchange reactions and some net-transfer reactions are apparent by inspection provided a strategic selection of components has been made. There are also strategies that make the completion of the reaction set a relatively simple task. A major one is to select the additive components so that all lie in a simple subspace of the full system. Most major rock-minerals, for example, have at least one realizable end-member composition in $Na_2O-CaO-MgO-Al_2O_3-SiO_2-H_2O$ (NCMASH). The potassic analogue (KCMASH) would serve as well in many instances, but there are sodic phases that do not have potassic analogues, such as the sodic pyroxenes. The ferrous analogues (NCFASH and KCFASH) have some advantages, particularly in dealing with oxide-silicate cross-reactions, but the results in either subsystem are readily transposed with the aid of exchange vectors. It is even possible to select additive components, in many instances, in the simpler system CMASH, but some of the end-members selected as additive components might be a bit unnerving to a beginner.

Having selected a simple sub-system, say NCMASH, and having selected a set of additive components in it, we need only list all phase components that are in NCMASH. We must include, however, *both* the additive components *and* the exchange components that are in NCMASH. The principal independent exchange components in NCMASH are $Al_2Mg_{-1}Si_{-1}$, $CaAlNa_{-1}Si_{-1}$, $MgCa_{-1}$ and $NaAlSi_{-1}$. The matrix then contains terms only for NCMASH components, and there are but six practical components on the right-hand side. These may be taken as the six oxides of NCMASH (see Thompson et al., 1982 for a worked-out example). For more rapid results, the set SiO_2, $NaAlSi_{-1}$, $CaAlNa_{-1}Si_{-1}$, $MgCa_{-1}$, $Al_2Mg_{-1}Si_{-1}$ and H_2O is recommended as many if not all of them will also be I.V. in the assemblage. This last recommendation, however, may again be difficult for a beginner inasmuch as the formulas for the additive components in terms of them may not be obvious.

In any case the reduction will contain all net-transfer reactions, including an open-system H_2O-reaction, if one is possible. If carbonates are present CO_2 may be added. All other reactions of the set will be exchanges (and there may be many) involving non-NCMASH components. This presupposes, of course, that all phases of concern can be assigned additive components in NCMASH or one of its analogues. This may not always be possible in exotic silicate or in non-silicate assemblages, but adaptations can be devised as need arises.

Most minor elements that do not enter major phases by exchange are safely locked in unique phases that do not participate in most rock reactions. Zirconium, for example, is only in zircon in all but extreme peralkaline or silica-undersaturated assemblages. This very non-reactivity is, in fact, the key to much of the current interest in zircon.

Apatite is the only phosphate in a great many rocks and may thus be taken as non-reactive except for $(OH)F_{-1}$ and perhaps ClF_{-1} exchanges.

The strategies, in summary, are:

(1) Make maximum use of exchange components and exchange reactions.
(2) Select additive components in as highly condensed a subspace as possible.
(3) Ignore non-reactive (or nearly so) phases and their corresponding components.

The exact form for these strategies leaves a great deal of room for personal choice and esthetics, but there has to be some room for art in science!

Can you obtain a sufficient set of net-transfer reactions for the assemblages: (a) quartz-muscovite-biotite-almandine-sillimanite; or (b) quartz-muscovite-chlorite-chloritoid-kyanite? Both may be obtained in terms of KMASH or KFASH components. One, however, may be done easily in terms of MASH or FASH components. Can you tell why at a glance? Although both contain hydrous phases only one permits a hydration-dehydration reaction (open-system) without the introduction of at least one more phase. Even with allowance for H_2O gain or loss each requires only two net-transfer reactions.

REACTION SPACE

If the terms of a reaction equation are all gathered on one side, in such a way that the signs of the coefficients of components that one wishes to designate as the "products" (thus defining the positive direction of the reaction) are positive, then a unit advancement on the reaction coordinate can be defined as that which converts some arbitrary quantity of "reactant" components into "product" components. For exchange reactions we may simply take an exchange of one mole (one gram exchange-formula) as defining the unit advancement. For net-transfer reactions, however, we shall take the unit advancement as that which transfers one oxygen equivalent of reactant additive components into one oxygen equivalent of product additive components. We may then define, for each reaction, r, an extent of reaction, ξ_r, that represents the number of unit advancements in passing from some initial state to some final state of the system. If some arbitrary state of the system is selected as origin, then the n_r quantities ξ_r may be regarded as *reaction coordinates* or *extents of reaction* (Prigogine and Defay, 1954, Chapter 1). Geometrically the reactions may thus be regarded as defining a set of basis vectors spanning an n_r-dimensional *reaction space*. Any reaction of the designated assemblage is some linear combination of the n_r reactions; hence any point in a reaction space represents a state of the system that can be related to any other by quantities of the form $\Delta\xi_r$. The $\Delta\xi_r$ may therefore be regarded as the components of a vector joining the initial to the final state in reaction space.

A full reaction space may require too many dimensions for graphical portrayal, but it has a useful subspace defined by the n_{nt} net-transfer reactions. We shall therefore consider here only this subspace inasmuch as it contains all reactions that have an appreciable effect on modal abundances. Such a space may be regarded as a projection, along the coordinates of the exchange reactions into a simpler subspace. By writing as many as possible of the reactions as exchanges the dimensionality of a net-transfer reaction space may be kept to a minimum. If necessary, net-transfer space may be further simplified by omitting (projecting along) the coordinates for the more trivial net-transfer reactions such as polymorphic changes and those for reactions involving such reactions as pyroxene-pyroxene or feldspar-feldspar net-transfers. With this last simplification we here consider only total pyroxene or total feldspar. There is some loss of information in such simplifications, but not as much as one might expect.

Phase abundances in oxy-equivalents vary linearly with the ξ_r and are approximately proportional to modal abundances. The relative abundances of the phases, in oxy-equivalents of the additive components, are thus uniquely determined at each point in net-transfer space because we have for each additive component (and hence for each phase) an equation of the form:

$$\Delta n_{ad} = \sum_r \nu_{ar} \Delta \xi_r \tag{8}$$

where n_{ad} is the number of oxy-equivalents of the additive component in the assemblage and ν_{ar} is the coefficient of that component in reaction r as normalized to one oxy-equivalent. Equation (8) thus relates changes in quantities of additive components (n_{ad}, in oxy-equivalents) to changes in ξ_r ($\Delta\xi_r$) between any two *arbitrary* points in reaction space. Because certain exchange components (those in the condensed space containing the additive components) are also involved in the net-transfer reactions, we also have for each of these:

$$\Delta n_{ex} = \sum_r \nu_{xr} \, \Delta \, \xi_r \qquad (9)$$

where n_{ex} is the number of mols of the exchange component and ν_{xr} is the coefficient of that component (for its quantity in mols) in the normalized reaction equation. Equations such as (8) or (9) may also be written for "environmental" components, should an open-system reaction be included. Equations (8) and (9) thus correspond to equations (23), (24), and (25) of Thompson et al. (1982), and may be used to locate any state of the system relative to another, or to the origin.

REACTION POLYTOPES

A reaction space has, for any given system, certain bounds beyond which the reaction cannot proceed. If a reaction is proceeding in the positive direction it must cease when one of the "reactant" phases is exhausted, or, if proceeding in the negative direction, when one of the "product" phases is exhausted. A reaction must also cease if a necessary exchange capacity has reached its limit. In either case the remaining possibilities for further reaction are reduced and the assemblage has entered a reaction space of fewer dimensions. The physically accessible part of a two-dimensional reaction space is thus bounded by a polygon, that of a three-dimensional reaction space by a polyhedron, and that of an n-dimensional space by an n-dimensional polytope. The equations for the bounding faces of the reaction polytope may be found from equations (8) and (9) by setting the appropriate n_{ad}'s at zero (see Thompson et al., 1982, p. 16-17), or the n_{ex}'s at the appropriate limiting values in all phases in which the exchange components are I.V.

The reaction polytope thus bounds what is physically possible in the given assemblage. This does not mean, however, that all of a reaction polytope may actually be accessible in practice, either because of impossible conditions needed to reach certain regions (at least as equilibrium assemblages) or because some extra phase may appear that was not taken into account in

42

obtaining the reaction set. If this occurs, there must be one new net-transfer added for each such additional phase, and the reaction space has acquired an extra dimension.

Homogeneous reaction spaces may also be constructed by similar methods. In molecular fluids, isopleths of species abundance (within one phase), however, replace the phase-abundance isopleths of a heterogeneous reaction space. For this reason such a homogeneous reaction space may also be regarded as an expanded composition space which shows the site composition, species by species, for fixed bulk compositions. Homogeneous reactions in crystals are mainly site-to-site exchange or "order-disorder" reactions. The ordering parameters may be regarded here as the ξ_r, and a reaction space may also be regarded as a site-by-site composition space for a fixed bulk composition. In either case extra-dimensions may be added to allow for variations in bulk composition. Several examples of homogeneous reaction spaces for crystalline phases were given in an earlier paper (Thompson, 1969, Figs. 5-7, see also Barth, 1965).

USES OF REACTION SPACE

A reaction space is a graphical portrayal of what can happen in a given assemblage as a consequence of heterogeneous reaction among the phases considered. It thus has applications in experimental petrology for experiments using bulk compositions that are fixed (or fixed except for certain components for which the activities are externally buffered). Reaction spaces also have applications in metamorphic petrology, particularly in dealing with rocks that are of more or less constant composition except for gain or loss of volatile species (e.g., Ferry, 1980, 1981, 1982; Rumble et al., 1982; and especially Rice and Ferry, Chapter 7, this volume, Fig. 13).

Pressure-temperature grids are much in vogue, but in assemblages of high-variance (largely owing to the presence of major phases that vary widely in composition), these grids are not very revealing. In such assemblages the reactions are largely continuous and marked mainly by changes in modal abundances and in the compositions of a limited set of phases.

Mafic schists are a case in point. Low-grade mafic schists, encompassing greenschist, blueschist, epidote amphibolite, and certain amphibolites have been treated elsewhere (Thompson et al., 1982) by means of a three-dimensional net-transfer reaction space that permits correlation of the "whole-rock" reactions of Laird (1980) with the "whole-rock" experiments of Liou et al. (1974). Experiments on amphibole stabilities can be baffling because multicomponent systems are needed for even some of the simplest amphibole "end-members", and because the possibilities for unexpected reactions are many. The hoped-for amphibole may not be the one produced. An example of a three-dimensional net-transfer space for NMASH amphiboles has also been given elsewhere (Thompson, 1981). We shall be content here with a simple two-dimensional example.

REACTIONS IN ANHYDROUS MAFIC ROCKS

Anhydrous mafic rocks, particularly silica-saturated or oversaturated ones, may be treated very simply by choosing additive components in NCMAS. Most silica-undersaturated rocks require a three-dimensional representation, but for the more siliceous ones we may reduce this to two provided we consider only *total* pyroxene in obtaining the reaction set. In most metamorphic facies, such anhydrous mafic rocks contain assemblages involving plagioclase, pyroxene, garnet, and a silicic mineral, usually quartz. For these we may choose $NaAlSi_3O_8$(ab), $CaMgSi_2O_6$(di), $Mg_3Al_2Si_3O_{12}$(pp) and SiO_2(qz) as the additive components. We have also in NCMAS the exchange components $Al_2Mg_{-1}Si_{-1}$(tk), $NaSiCa_{-1}Al_{-1}$(pl) and $MgCa_{-1}$(mc). All three exchanges may be I.V. in pyroxene. The second, (pl), is also I.V. in plagioclase, and the third, (mc), is also I.V. in garnet. Matrix reduction shows that all five oxides are I.V. in the system, and hence that there are 9-5 = 4 heterogeneous reactions in the NCMAS assemblage. Two of these are exchanges:

$$pl\ [Pyx] = pl\ [Plg] \tag{10}$$

and:

$$mc\ [Pyx] = mc\ [Gar] \tag{11}$$

and two more must be taken as net-transfers. We shall select these as:

$$0 = pp + 2\ qz - 2\ ab + tk + 2\ pl - 2\ mc \tag{12}$$

and:

$$0 = 2\ di - pp + tk + 2\ mc \tag{13}$$

Equations (12) and (13) are written in mol units. They may be converted to oxy-units of additive components and to mol units of exchange component per oxy-unit transferred, by multiplying each additive component by the number of oxygens in its formula and then dividing by the number of oxy-units transferred in the equation as first written. The normalized equations then become:

$$0 = \frac{12}{16} \text{ pp} + \frac{4}{16} \text{ qz} - \frac{16}{16} \text{ ab} + \frac{1}{16} \text{ tk} + \frac{2}{16} \text{ pl} - \frac{2}{16} \text{ mc} \qquad (12a)$$

and

$$0 = \frac{12}{12} \text{ di} - \frac{12}{12} \text{ pp} + \frac{1}{12} \text{ tk} - \frac{2}{12} \text{ mc} \qquad (13a)$$

or, clearing fractions:

$$0 = \frac{3}{4} \text{ pp} + \frac{1}{4} \text{ qz} - \text{ ab} + \frac{1}{16} \text{ tk} + \frac{1}{8} \text{ pl} - \frac{1}{8} \text{ mc} \qquad (12a)$$

and

$$0 = \text{di} - \text{pp} + \frac{1}{12} \text{ tk} - \frac{1}{6} \text{ mc} \qquad (13a)$$

Equation (12) or (12a) represents a conversion (− to +) of plagioclase to garnet and quartz. It also leaves the amount of pyroxene unvaried. Isopleths on the abundance of pyroxene (in oxy-equivalents) must therefore be parallel to the vector in reaction space representing reaction (12). Reaction (12) also has a large ΔV. This means that the equilibrium corresponding to it must be markedly pressure-sensitive. It is also widely applicable to assemblages other than those of anhydrous mafic rocks and is a possible reaction in any assemblage containing plagioclase, garnet, quartz, and some phase, not necessarily pyroxene, that provides tk exchange capacity. In amphibolites the tk exchange capacity may be provided by amphiboles, in pelitic schists it may be provided by muscovite or biotite. A tk equivalent may also be expressed in terms of pp, qz, and a component such as Al_2SiO_5, obtaining the equation of Ghent (1976).

Equation (13) or (13a) represents a conversion (− to +) of garnet to pyroxene, but leaves the abundance of both plagioclase and quartz unvaried. Isopleths on the abundance (in oxy-equivalents) of plagioclase and quartz must therefore be parallel to the vector in reaction space corresponding to (13).

We shall use (12a) and (13a) to define the basis vectors for a two-dimensional net-transfer space. Certain other equations that are linear combinations of these, however, are of interest. We thus have, in mol units:

$$0 = \text{di} + \text{qz} - \text{ab} + \text{tk} + \text{pl} \qquad (14)$$

and

$$0 = \text{di} + \text{ab} - \text{pp} - \text{qz} + 2 \text{ mc} - \text{pl} \qquad (15)$$

Equation (14) leaves garnet unaltered and hence corresponds to vectors parallel to isopleths on the abundance of garnet, in oxy-equivalents. Equations (13) and (14) correspond to possible edges of the reaction polygon,

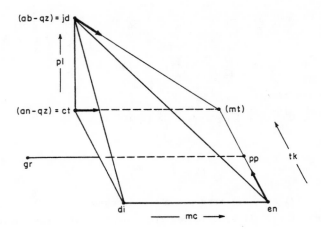

Figure 1. Composition space for NCMAS pyroxene, plagioclase and garnet. The diagram is on an oxy-fraction basis. The symbols di, en, mt, ct and jd, at the vertices of the pyroxene pyramid, correspond to the formulas $CaMgSi_2O_6$, $Mg_2Si_2O_6$, $MgAl_2SiO_6$, $CaAl_2SiO_6$ and $NaAlSi_2O_6$, respectively. Plagioclases, by projection through qz (SiO_2) lie along the line ct-jd. The garnet line is $gr(Ca_3Al_2Si_3O_{12})$ - $pp(Mg_3Al_2Si_3O_{12})$ and crosses the line di-ct at $Ca_2MgAl_2Si_3O_{12}$.

(13) corresponding to edges at the zero isopleths for feldspar or quartz, and (14) to the zero isopleth for garnet. Equation (12), however, cannot correspond to a possible edge of the reaction polygon because it requires a tk exchange capacity. This is possible only if the amount of additive component of pyroxene (di) is greater than zero. A "no-pyroxene" state of the system can thus occur only at a vertex of the reaction polygon.

Other edges of the reaction polygon may correspond to a loss of exchange capacity because certain phases have reached limiting compositions. A composition space for NCMAS pyroxene is shown in Figure 1. The pyroxene space in Figure 1 is the four-sided pyramid having the plane en-di-ct-mt as base and jd as vertex. The composition $MgAl_2SiO_6$(mt), though crystallochemically reasonable, is probably metastable under all conditions relative to other phases. The base also contains two-thirds of the garnet line, gr-pp. The plagioclase line, as projected through qz, coincides with the line jd-ct. Reactions (12) and (13), and any linear combination of them such as (14) or (15) can proceed freely if the pyroxenes are *within* the pyramid, or *within* its basal quadrilateral, or *within* the triangles jd-di-ct or jd-en-mt, or *in* the edge di-ct. All reactions are also possible with pyroxenes within the plane jd-di-en, or in the edges jd-di or jd-en, *provided* the associated plagioclase retains some pl exchange capacity. If, however, the plagioclase is pure ab then only reaction (14) is possible. Reaction (14) then corresponds to the familiar additive equation:

$$ab = jd + qz \qquad (14a)$$

Reaction (14) is also the only one possible for pyroxenes within the triangle jd-ct-mt, in its bounding edges, and at the vertices jd and ct. For pyroxenes at the vertex jd it corresponds to (14a) and for those in the line ct-mt, or at ct, it corresponds to (14) in the additive form:

$$an = ct + qz \qquad (14b)$$

Equations (14a) and (14b) are related to each other by pl-exchange which, owing to projection, is not shown. Equation (14) thus corresponds to a possible edge of the reaction polygon where all phases are at the limit [pl+] such that plagioclases are pure sodic ones and pyroxenes are constrained to jd-di-en. The opposite limit [pl-] is not a possible limit inasmuch as all the net-transfer reactions are possible between pure calcic plagioclase and pyroxenes in the base of the pyramid of Figure 1.

Pyroxenes in the line en-mt permit only reaction (13) which then corresponds to the additive equation:

$$en + mt = pp \qquad (13b)$$

Equation (13) may thus correspond to the limit [mc+] but this would not be encountered in a basaltic system inasmuch as all phases could not lie in the calcium-free subsystem NMAS. No reaction can be written for the limit [mc-] which would require that mc + 3 pp = 0 (i.e., the garnet is pure $Ca_3Al_2Si_3O_{12}$). The component tk is I.V. only in pyroxene. Pyroxenes in the line di-en (tk = 0) are thus at the limit [tk-] for which (15) is the only possible reaction. The limit [tk+] corresponds to jd-ct-mt pyroxenes and permits only reaction (14). The limit [tk+], however, requires a high-Al bulk composition beyond the range of normal basaltic rocks; hence is an unlikely bound to their reaction polygons.

THOLEIITE REACTION SPACE

A simple composition may be selected in NCMAS that can serve as a model for a tholeiite composition. In terms of C.I.P.W. components we shall use a composition corresponding to 96 oxy-equivalents, 48 of them as plagioclase (an_{50}) and 48 of them as pyroxene ($wo_{25}en_{75}$). This corresponds, in our components, to a plagioclase with $X_{pl} = -\frac{1}{2}$, and to a total pyroxene for which

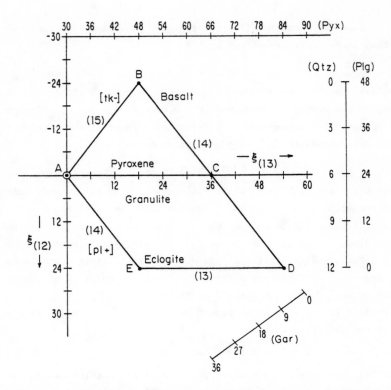

Figure 2. Reaction space for a model tholeiite. The bulk composition for which it is drawn is given in the text and contains 96 oxygen equivalents. The basis vectors are defined by equations (12a) and (13a). Edges of the polygons that are not parallel to the coordinate axes correspond to either equation (13) or equation (14) as indicated. Abundances of the mineral phases in oxy-equivalents are indicated by separate scales. The system is closed, hence the oxy-equivalents sum, at each point, to 96. Assemblages near D are probably metastable relative to a liquid phase.

$X_{mc} = \frac{1}{2}$ and all other X_{ex} are zero. Such a mineral content must lie at a vertex of the reaction polygon where the lines [Gar] (no garnet) and [tk-] meet. This would correspond to point B in Figure 2. B is connected to A in Figure 2 by reaction (15) at [tk-1] converting the an-component of plagioclase to garnet and quartz. At A only pure sodic plagioclase remains and the pyroxene is still on the line di-en of Figure 1, but very close to di considering the known mc exchange between pyroxene and garnet. An assemblage at A would thus contain 6 oxy-equivalents quartz, 24 of albite, 30 of a diopsidic pyroxene and 36 of a slightly calcic pyrope. This is an idealized pyroxene granulite.

Point A in Figure 2 lies at the intersection of [tk-] and [pl+]. Reaction (14), effectively in the form of (14a), transforms albite into quartz and pyroxene (as a jadeitic component). Loss of all feldspar occurs at E where the assemblage must consist of omphacite (48 oxy-equivalents) and quartz (12 oxy-equivalents). The garnet remains as at A. E clearly represents an idealized eclogite and lies at the intersection of the limits [pl+] and [Plg]. At the remaining vertex, D, the assemblage would contain 12 oxy-equivalents of quartz and 84 of highly aluminous pyroxene. Such an assemblage is probably not experimentally realizable because of melting.

The locations of the edges of the reaction polygon will be displaced if the properties of the initial additive components are altered. Such displacements, however, like those of crystal form, must be parallel ones. New edges may also appear or old ones disappear. In particular if we alter the example just considered solely by removal of silica, the vertex at 13 would be truncated by a [Qtz] limit parallel to D-E. To proceed beyond such a limit would require a new phase (olivine or spinel) and the assemblage would then be in a different reaction space. If, on the other hand, the initial additive components are unaltered, and the composition is varied only by exchange (as by the exchange $FeMg_{-1}$) then the reaction polygon remains the same. It is therefore possible, by exchanges such as $FeMg_{-1}$, to regard a more generalized (and more realistic) basaltic composition, than that with which we began, as sharing the same reaction space.

A reaction space may represent either equilibrium or disequilibrium assemblages. As it stands, the space of Figure 2 gives a great deal of modal information in either case. We can, however, go farther, if the assemblage is an equilibrated one, if we know something about the pyroxene miscibility gaps, or the nature of the exchange equilibrium between pairs of phases. We know, for example, that sodic plagioclases can coexist with highly calcic and aluminous pyroxenes. This means that in equilibrated assemblages the an-content of plagioclase must drop systematically from an_{50} at B to nearly zero along the line A-C, and that it must remain near zero in the field ACDE. Conversely the pyroxenes will be virtually free of Na in the field ABC and must have an increasing Na-component from the line AC to the line DE.

In addition to the isopleths related to modal abundance, as in Figure 2, there would be, for the equilibrium assemblages of a given composition, a set of isobars and isotherms. The quartz tholeiites studied by Green and Ringwood (1967) would have reaction spaces much like that of Figure 2. Their information on modal abundances and phase compositions is not quite sufficient to be precise, but is consistent with isotherms having the general form indicated in Figure 3. Isobars must be at a high angle to the isotherms, perhaps subparallel to a line AD. Even though two rocks might have nearly identical

49

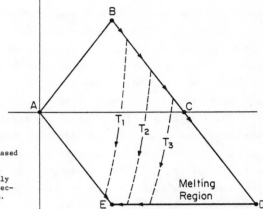

Figure 3. Schematic isotherms for quartz-tholeiites. The isotherms ($T_1 < T_2 < T_3$) are based on the experimental results of Green and Ringwood (1967). Isobars probably intersect these at a high angle, and are perhaps roughly parallel to a line A–D. Arrows show the direction of increasing pressure on each isotherm.

reaction spaces, however, the isotherms and isobars for equilibrium assemblages will be displaced by other composition variables. An increase in $X_{FeMg_{-1}}$ would, for example, both lower the pressure and raise the temperature at a given point owing to the stabilizing effect of $X_{FeMg_{-1}}$ on garnet.

Can you obtain an analogous reaction space to the one just presented for the assemblage: plagioclase–garnet–quartz–amphibole? Three dimensions are required (using total amphibole). One vector may be based on equation (10) of Thompson et al. (1982) and another on equation (12) above. Neither of these contains an additive amphibole component; hence there must be a third that does. The form selected, however, may be left to the ingenuity of the reader. Must it be an open-system reaction? Reaction (12), in this example, plays an equally powerful role and can transform a plagioclase–hornblende–cummingtonite (or orthoamphibole) rock into a garnet amphibolite or garnet glaucophanite.

SUMMARY

The example just considered and others given elsewhere should give the reader some idea of the utility of reaction space in both experimental investigation and in the analysis and interpretation of natural assemblages. The systematic procedures for obtaining a sufficient set of heterogeneous reactions, moreover, are useful in any thermodynamic analysis of either experimental results or natural occurrences. In equilibrium assemblages there must be an independent set of equilibrium conditions corresponding in number to the set of possible reactions. These are in a highly convenient form if they correspond directly to the reactions as obtained by the procedures presented above.

ACKNOWLEDGMENTS

Numerous discussions with Jo Laird and Alan Thompson have helped greatly in sharpening the concepts presented here. I am also grateful to Stephen K. Dobos and John Ferry for critical review of the manuscript.

CHAPTER 2 REFERENCES

Barth, T.F.W. (1965) On the constitution of the alkali feldspars. Tschermak's Mineral. Petrogr. Mitt. 10, 14-33.

Ferry, J.M. (1980) A case study of the amount and distribution of heat and fluid during a metamorphism. Contrib. Mineral. Petrol. 71, 373-385.

_____ (1981) Petrology of graphitic sulfide-rich schists from south-central Maine: An example of desulfidation during prograde regional metamorphism. Am. Mineral. 66, 908-930.

_____ (1982) Mineral reactions and element migration during metamorphism of calcareous sediments from the Vassalboro Formation, south-central Maine. Am. Mineral. 67, in press.

Ghent, E.D. (1976) Plagioclase-garnet-Al_2SiO_5-quartz: A potential geobarometer-geothermometer. Am. Mineral. 61, 710-714.

Green, D.H. and A.E. Ringwood (1967) An experimental investigation of the gabbro to eclogite transformation and its petrological applications. Geochim. Cosmochim. Acta 31, 767-833.

Hildebrand, F.B. (1952) *Methods of Applied Mathematics*. Prentice-Hall, Inc., N.Y.

Laird, J. (1980) Phase equilibria in mafic schist from Vermont. J. Petrol. 21, 1-37.

Liou, J.G., S. Kuniyoshi, and K. Ito (1974) Experimental studies of the phase relations between greenschist and amphibolite in a basaltic system. Amer. J. Sci. 274, 613-632.

Longhi, J. (1976) *Iron, Magnesium, and Silica in Plagioclase*. Ph.D. Thesis, Harvard University, Cambridge, Massachusetts.

_____, D. Walker, and J.F. Hays (1976) Fe and Mg in plagioclase. Proc. 7th Lunar Sci. Conf., 1281-1300.

Prigogine, I. and R. Defay (1954) *Chemical Thermodynamics*. (English translation by D.H. Everett) Longmans Green and Co., London.

Rumble, D., III., J.M. Ferry, T.C. Hoering, and A.J. Boucot (1982) Fluid flow during metamorphism at the Beaver Brook fossil locality, New Hampshire. Am. J. Sci. 282, 886-919.

Thompson, J.B., Jr. (1969) Chemical reactions in crystals. Am. Mineral. 54, 341-375.

_____ (1981) An introduction to the mineralogy and petrology of the biopyriboles. *In* D.R. Veblen, Ed., *Amphiboles and Other Hydrous Pyriboles - Mineralogy*, Reviews in Mineralogy 9A, 141-188.

_____, J. Laird, and A.B. Thompson (1982) Reactions in amphibolite, greenschist and blueschist. J. Petrol. 23, 1-27.

Venit, S. and W. Bishop (1981) *Elementary Linear Algebra*. Prindle, Weber and Schmidt, Boston.

Chapter 3

LINEAR ALGEBRAIC MANIPULATION of N-DIMENSIONAL COMPOSITION SPACE

F.S. Spear, D. Rumble III, and J.M. Ferry

INTRODUCTION

The purpose of this chapter is to provide a review of the application of
linear algebra to petrologic problems. Among the problems faced by petrologists is the graphical representation of the chemical compositions of rocks
and minerals. It is a familiar experience that once the number of chemical
components in a system exceeds three (or four, at best) visualization of
chemographic relations becomes difficult or impossible. Linear algebra, however, provides us with a formalism for dealing with, and extracting information from, n-component chemical analyses. The algebraic approach is completely
general and its power extends to systems consisting of any number of components.
Moreover, the techniques are not difficult to master, because they are merely
an extension of the algebra of a line ($y = mx + b$) to planes and surfaces of
n-dimensions.

The main thrust of this chapter will be to attempt to show the relation
between the graphical representation of rock and mineral compositions, with
which most petrologists are quite familiar, and the formalism of linear algebra. As will be seen, the mathematics are quite simple and, although a knowledge of linear algebra is helpful, it is not essential. Useful textbooks in
linear algebra include Aitken (1965), Anton (1973) and Strang (1980); books
and articles in the petrologic literature include Korzhinskii (1959), Perry
(1967a,b; 1968a,b), Davis (1973) and Greenwood (1967, 1975).

REPRESENTATION OF MINERAL (OR ROCK) COMPOSITIONS IN COMPOSITION SPACE

Composition space is defined as the space whose coordinate axes are the
chemical constituents of minerals or rocks (this chapter will deal principally
with minerals, but the approach is also applicable to rock compositions).
Composition space is completely analogous to physical space where the axes
represent units of distance (for example, the number of kilometers in a direction east, north and up), but in composition space the axes represent units
of chemical composition (for example, the number of moles of SiO_2, MgO, and
CaO). The units of quantity used for the axes of composition space are completely arbitrary and the choice should depend on the intended use. Molar
quantities of oxides are commonly used for minerals and rocks and in this

chapter they will be used exclusively. Other quantities such as grams, weight percent, atoms, and oxygen equivalents are also useful. Brady and Stout (1980) provide an informative discussion on the choice of components in petrologic applications.

There are two common types of coordinate systems to represent composition space in petrology. The most familiar to petrologists is the *barycentric* (center weighted) coordinate system where a two-component system is represented by a line, a three-component system as a triangle and a four-component system as a tetrahedron. Mineral assemblages and rock compositions are most typically represented graphically in barycentric coordinates. The other type of coordinate system is the *Cartesian* coordinate system, which is the conventional orthogonal x-y-z coordinate system. Most algebraic manipulations are done in Cartesian coordinates and the results converted to barycentric coordinates for graphical display.

The first step in applying algebraic methods to composition space is to define the coordinate system and to determine where minerals plot in this coordinate system. The Cartesian coordinate system will be discussed first, and then the relation between the Cartesian and barycentric coordinate systems will be shown.

Figure 1A represents the two-component system, SiO_2-MgO, in Cartesian coordinates. To determine where a mineral plot simply determine the amount of SiO_2 and MgO in the formula, e.g., quartz plots at (1,0) (i.e., 1.0 SiO_2 and 0.0 MgO), forsterite (Mg_2SiO_4) plots at (1,2), periclase (MgO) plots at (0,1), and enstatite ($MgSiO_3$) plots at (1,1). Readers familiar with vector notation will recognize that composition is being represented in this coordinate system as a vector, known as *composition vector* of the mineral. Each element of the vector represents the amount of a particular chemical species in the mineral of interest, i.e., $(n[SiO_2], n[MgO])$.

Note that the choice of how to write a mineral formula is somewhat arbitrary. For example, enstatite may be written as $MgSiO_3$, in which case its vector is (1,1), or as $Mg_2Si_2O_6$, in which case its vector is (2,2). Both of these points are shown in Figure 1A, and it can be seen that En_2 ($Mg_2Si_2O_6$) falls on a line from the origin (0,0) through En_1 ($MgSiO_3$). Since these are the same material, the conclusion is that in the Cartesian coordinate system it is the direction that the vector points in composition space, not its length, that defines a mineral composition.

The relation between Cartesian and barycentric coordinates is seen in Figure 1B. The latter have the property that the sum of the components

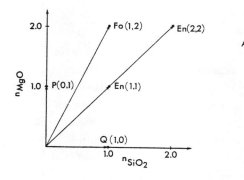

A.

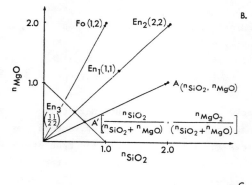

B.

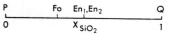

C.

Figure 1. (A) Cartesian coordinate system SiO_2-MgO showing plotting positions of periclase (P), forsterite (Fo), enstatite (En_1; En_2) and quartz (Q). (B) Diagram showing the relationship between the Cartesian and the barycentric coordinate system, SiO_2-MgO. The latter is shown as the line between MgO = 1.0 and SiO_2 = 1.0. The plotting positions of the minerals in the barycentric coordinate system are found as the intersection of the mineral vectors with this line (e.g., point En_3 or point A'). (C) Barycentric coordinate system for the system SiO-MgO.

equals 1.0: $\Sigma X_i = 1.0$. In this two-component system it means that $X(SiO_2) + (XMgO) = 1.0$ or $X(MgO) = 1.0 - X(SiO_2)$. In the Cartesian coordinate system of Figure 1B, this corresponds to the equation of a line with a slope of -1 between the points (0,1) and (1,0). Thus the barycentric coordinate system is simply the diagonal line between 1.0 MgO and 1.0 SiO_2.

The plotting position of any mineral in the barycentric coordinate system is the intersection of the vector representing the mineral composition in the Cartesian coordinate system with the line $X(SiO_2) + X(MgO) = 1.0$. Since $MgSiO_3$ and $Mg_2Si_2O_6$ have the same ratio of MgO and SiO_2, both are represented by vectors that point in the same direction, and both plot at the same point in the barycentric coordinate system. Thus, one property of the barycentric coordinate system is that it perserves the *ratios* of the chemical species in a substance.

The plotting position of a mineral in the barycentric coordinate system can be determined mathematically by solving for the intersection of the line representing the vector that points to the mineral with the line $\Sigma X_i = 1$. For an arbitrary mineral with a composition $n(SiO_2)$, $n(MgO)$ there is a vector that defines a line with the equation (of the form $y = mx + b$),

$$X(MgO) = \frac{n(MgO)}{n(SiO_2)} \cdot X(SiO_2) + 0 .$$

Solving this equation simultaneously with the equation $X(MgO) = 1 - X(SiO_2)$ yields

$$X(\text{MgO}) = \frac{n(\text{MgO})}{n(\text{MgO}) + n(\text{SiO}_2)} \quad \text{and} \quad X(\text{SiO}_2) = \frac{n(\text{SiO}_2)}{n(\text{MgO}) + n(\text{SiO}_2)} \ .$$

Thus the barycentric plotting coordinates of any mineral whose composition is known in Cartesian coordinates can be calculated by normalization.

There are two excellent reasons for using barycentric coordinates in petrology. First, intensive thermodynamic properties of minerals depend not on the absolute number of moles of a species in a substance, but on the mole fraction of a species, and barycentric coordinates deal only with mole fractions. Second, it is possible to graphically represent more composition axes on a piece of paper using barycentric coordinates than it is using Cartesian coordinates. For example, in Figure 1 it takes two dimensions to represent the system SiO_2-MgO with Cartesian coordinates, but only one dimension (i.e., a line) to represent the system with a barycentric coordinate system (Fig. 1C). Some information is lost in the process (specifically, the absolute amounts of MgO and SiO_2 in our mineral formula), but this is a small price to pay.

It is a simple extension of the above equations to include more components. The mineral or rock compositions are simply represented by vectors with the number of elements in the vectors equal to the number of components in the composition space, each element representing an axis in the Cartesian coordinate system. In this way, composition space of *any* dimension can easily be represented mathematically, even though only three Cartesian or four barycentric coordinates can actually be drawn.

TRANSFORMATION OF COORDINATE AXES

The transformation of coordinate axes is a mathematical technique of considerable importance to petrology for it can be used to calculate rock norms and end-member mineral components, balance chemical reactions, and compute the plotting coordinates of graphical projections. To transform coordinate axes simply means to compute the elements of a composition vector in terms of any one of several different sets of components that might be chosen. For example, an orthopyroxene solid solution may be expressed in terms of any of the following three sets of components: MgO-FeO-SiO_2; Mg_2SiO_4-Fe_2SiO_4-SiO_2; or MgSiO_3-FeSiO_3. Axis transformation, also called linear mapping from one set of coordinates to another, is a fundamental problem of linear algebra, and is treated in all textbooks on the subject (Aitken, 1965; Anton, 1973; Strang, 1980). This procedure is often referred to as the *transformation of components* from an "old" component set to a "new"

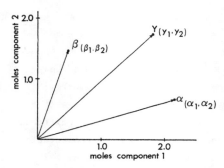

Figure 2. Coordinate system 1 and 2 (the "old" components) showing plotting positions of "new" components α and β and mineral Y, which is to be mapped into the new system α and β.

component set and has been discussed by Thompson (1957), Greenwood (1975), and Brady and Stout (1980).

The linear mapping of a rock or mineral composition vector from one set of coordinate axes into another set is performed in the following way. First, it is necessary to choose the "new" set of coordinate axes into which the mapping will be done. As an example, let there be a set of axes labeled 1 and 2 (the "old" coordinate system) and two minerals labeled α and β that will serve as "new" coordinate axes (see Fig. 2). The minerals α and β have composition vectors $\alpha = (\alpha_1, \alpha_2)$ and $\beta = (\beta_1, \beta_2)$ in the old coordinate system. Moreover, there is a third mineral Y where $Y = (y_1, y_2)$ that is to be mapped into the "new" coordinate axes α, β. In this and all examples in the chapter it is assumed that the units of quantity for the old and new coordinate systems are the same. But, as pointed out by Brady and Stout (1980, p. 181), this need not be the case, and the units in the old and new systems may be different.

The question asked in this example is: What is the composition vector of mineral Y relative to the coordinate axes α and β? Or, stated another way: How much of minerals α and β are needed to make mineral Y? Or, what linear combination of α and β will make mineral Y? Mathematically, this question is phrased

$$x_1 \cdot \alpha + x_2 \cdot \beta = Y$$

where Y, α, and β are known in the old system, and it is desired to find x_1 and x_2, which is the solution to where Y plots in the new system. Because this is a two-component system, the above equation actually represents two simultaneous equations:

$$x_1 \cdot \alpha_2 + x_2 \cdot \beta_1 = y_1$$

$$x_1 \cdot \alpha_2 + x_2 \cdot \beta_2 = y_2$$

or, in matrix notation

$$\begin{bmatrix} \alpha_1 & \beta_1 \\ \alpha_2 & \beta_2 \end{bmatrix} \cdot \begin{bmatrix} x_1 \\ x_2 \end{bmatrix} = \begin{bmatrix} y_1 \\ y_2 \end{bmatrix} \text{, or simply } A \cdot X = Y,$$

57

where A is the *coefficients matrix* and X is the *solution vector*.

Since the transformation is made from one set of coordinate axes to another, both of which have the same dimensionality, there will always be the same number of equations as unknowns and the system of equations can easily be solved for X. Note that the *columns* of the coefficient matrix, A, are the composition vectors of the minerals in the new coordinate axes (the "new" components). Also note that each equation has the physical significance of a *mass balance equation* -- that is, equation (1) describes the conservation of component 1 and equation (2) describes the conservation of component 2. Both these equations must be satisfied in the transformation from one component set to another. The solution vector, X, represents the composition vector for mineral Y relative to the new coordinate axes α, β.

To take a more specific example, consider the system SiO_2-MgO shown in Figure 1. Suppose that we wish to transform this coordinate system into the coordinate system enstatite-periclase, and to map both quartz and forsterite into this new coordinate system. To map quartz, we need to solve the two linear equations

$$x_1 \cdot [En] + x_2 \cdot [P] = [Qtz]$$

or

$$[SiO_2] x_1 \cdot 1 + x_2 \cdot 0 = 1$$
$$[MgO]\ x_1 \cdot 1 + x_2 \cdot 1 = 0$$

where the two equations represent mass balance for SiO_2 and MgO, respectively, as indicated by the square brackets. The two columns of the coefficients matrix are the composition vectors for the new components En and P. Solution of these two equations yields $x_1 = 1$, $x_2 = -1$, which is the composition vector for quartz in the new coordinate system (i.e., qtz' = (1,-1); the prime notation refers to the composition vector of a mineral in the transformed system). A similar solution to the equation

$$x_1 \cdot [En] + x_2 \cdot [P] = [Fo]$$

or

$$[SiO_2] x_1 \cdot 1 + x_2 \cdot 0 = 1$$
$$[MgO]\ x_1 \cdot 1 + x_2 \cdot 1 = 2$$

yields Fo' = (1,1). The minerals En and P could also be mapped from the old system to the new system with the result that En' = (1,0) and P' = (0,1) as they must, because they are the new coordinate axes.

A plot of the new system is shown in Figure 3A. Note the plotting position of Qtz' (1,-1). The new barycentric coordinate system can easily

TABLE 1.

Summary of the relationship between the composition vector of a mineral in two-component Cartesian and barycentric coordinate systems.

Sector (Fig. 5)	Sign of elements in Cartesian coordinates	Sum of elements in composition vector	Projects into Barycentric coordinates	Barycentric composition vector
I	(+,+)	+	directly	(+,+)
IIa	(+,-)	+	directly	(+,-)
IIb	(-,+)	+	directly	(-,+)
IIIa	(-,+)	-	through ∞	(+,-)
IIIb	(+,-)	-	through ∞	(-,+)
IIIc	(-,-)	-	through ∞	(+,+)
along line y=-x	(±,∓)	0	at ∞	(∞,∞)

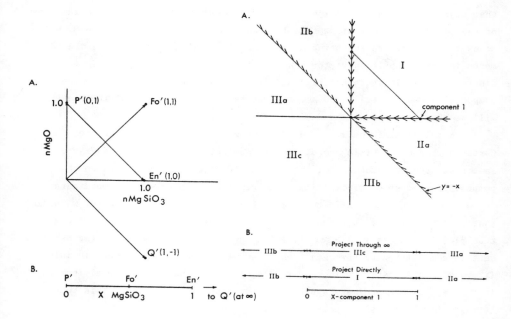

Figure 3, to the left. (A) Plot of the "new" Cartesian coordinate system En-P showing the plotting positions of quartz (Q') and forsterite (Fo') in the new system. Note the plotting position of Q' (1,-1).
(B) Barycentric coordinates for the new component system En-P. Note that Q' plots at infinity.

Figure 4, to the right. Cartesian (A) and barycentric (B) coordinate system showing relationship between plotting positions of minerals in the two systems (see text and Table 1).

be calculated by normalizing and is shown in Figure 3B. Thus, in barycentric coordinates, P' plots at (0,1), Fo' plots at $(\frac{1}{2},\frac{1}{2})$, En' plots at (1,0), but Q' plots at (∞,∞). The reason for this can be seen in Figure 3A: The composition vector for Q' (1,-1) is parallel to, and therefore does not intersect, the line for the barycentric coordinate system. Mathematically, this condition of parallelism between the composition vector of a mineral and the barycentric coordinate system arises when the sum of the elements in the composition vector = 0.

In fact, the Cartesian coordinate system can be divided into sectors showing how and where minerals will project into the barycentric coordinate system as shown in Figure 4. All minerals that plot in sectors I, IIa or IIb project directly into the barycentric coordinate system whereas all minerals that plot in sector IIIa, b or c project negatively (through ∞). The line that separates these two regions has the formula $y = -x$. Thus the coordinates of any point that plots along this line has the property that $x + y = 0$, which means that any mineral plotting along this line will have the sum of the elements in its composition vector = 0 and plots at ∞ in the barycentric coordinate system. In sectors I, IIa and IIb, the sum of the elements in the composition vector will be >0 and in sectors IIIa,b,c the sum of the elements will be <0.

The relationship between the plotting position of minerals in the Cartesian and barycentric coordinates is summarized in Table 1 and Figure 4. As can be seen from the table, there is a direct relationship between the composition vector of minerals in the Cartesian and barycentric coordinates. The *signs* of the elements in the composition vectors are the *same* for minerals that project directly into the barycentric coordinate system (i.e., minerals that plot in sectors I, IIa or IIb); the *signs* of the elements in the barycentric composition vector *change* when the mineral projects through infinity (i.e., minerals that plot in sector IIIa,b,c). If the sum of the elements in the Cartesian coordinate system = 0, the mineral plots at ∞. This result is valid for composition space of any dimension.

SOLUTION TO THE PROBLEM A·X = Y

The linear mapping described in the last section is an extremely powerful and versatile technique for manipulating composition space. In a later section, some examples and applications of this technique will be discussed. In this section, however, some techniques for solving the system of linear

60

equations will be explored. The purpose of this section is not to provide a discussion of numerical methods, but rather, to show the relationship and similarity among various methods that have been discussed in the literature. For details on the mathematics behind the various methods, the reader is referred to textbooks in linear algebra.

Matrix inversion

A general solution to the system of equations $A \cdot X = Y$ can be obtained by inversion of the coefficients matrix A and post-multiplication by the vector Y. Thus

$$A^{-1}Y = X .$$

Gauss-reduction

This method is a systematic procedure for reducing the system of equations to a form where all of the elements in the coefficients matrix below the diagonal are zero. This is achieved sequentially by adding and subtracting multiples of one equation from another. When all of the elements below the diagonal are zero, the solution can easily be calculated by back substitution.

Cramer's rule

This method utilizes the property that the ith element of the solution vector X (i.e., x_i) is equal to the determinant of the coefficient matrix with the Y vector substituted in for the ith column, divided by the determinant of the entire coefficient matrix. Thus:

$$x_i = \frac{\begin{vmatrix} a_1 & b_1 \cdots y_1 \cdots r_1 \\ a_2 & b_2 \cdots y_2 \cdots r_2 \\ a_3 & b_3 \cdots y_3 \cdots r_3 \\ \cdot & \cdot \quad\quad \cdot \quad\quad \cdot \\ \cdot & \cdot \quad\quad \cdot \quad\quad \cdot \\ a_r & b_r \cdots y_r \cdots r_r \end{vmatrix}}{\begin{vmatrix} a_1 & b_1 \cdots i_1 \cdots r_1 \\ a_2 & b_2 \cdots i_2 \cdots r_2 \\ a_3 & b_3 \cdots i_3 \cdots r_3 \\ \cdot & \cdot \quad\quad \cdot \quad\quad \cdot \\ \cdot & \cdot \quad\quad \cdot \quad\quad \cdot \\ a_r & b_r \cdots i_r \cdots r_r \end{vmatrix}} .$$

Each successive coefficient of the solution vector, x_i, can be obtained by successively substituting the vector Y for different columns in the

coefficient matrix and computing the determinant. Thus the entire solution
to the equation $x_1 \cdot \alpha + x_2 \cdot \beta + x_3 \cdot \gamma = Y$ becomes

$$\frac{\begin{vmatrix} y_1 & \beta_1 & \gamma_1 \\ y_2 & \beta_2 & \gamma_2 \\ y_3 & \beta_3 & \gamma_3 \end{vmatrix}}{\begin{vmatrix} \alpha_1 & \beta_1 & \gamma_1 \\ \alpha_2 & \beta_2 & \gamma_2 \\ \alpha_3 & \beta_3 & \gamma_3 \end{vmatrix}} \cdot \alpha + \frac{\begin{vmatrix} \alpha_1 & y_1 & \gamma_1 \\ \alpha_2 & y_2 & \gamma_2 \\ \alpha_3 & y_3 & \gamma_3 \end{vmatrix}}{\begin{vmatrix} \alpha_1 & \beta_1 & \gamma_1 \\ \alpha_2 & \beta_2 & \gamma_2 \\ \alpha_3 & \beta_3 & \gamma_3 \end{vmatrix}} \cdot \beta + \frac{\begin{vmatrix} \alpha_1 & \beta_1 & y_1 \\ \alpha_2 & \beta_2 & y_2 \\ \alpha_3 & \beta_3 & y_3 \end{vmatrix}}{\begin{vmatrix} \alpha_1 & \beta_1 & \gamma_1 \\ \alpha_2 & \beta_2 & \gamma_2 \\ \alpha_3 & \beta_3 & \gamma_3 \end{vmatrix}} \cdot \gamma = Y \ .$$

Korzhinskii's method

This method was designed specifically for balancing chemical reactions
(Korzhinskii, 1959) and is very similar to Cramer's rule except that the
entire equation is multiplied by the determinant of the coefficient matrix.
The above solution, therefore, becomes

$$\begin{vmatrix} y_1 & \beta_1 & \gamma_1 \\ y_2 & \beta_2 & \gamma_2 \\ y_3 & \beta_3 & \gamma_3 \end{vmatrix} \cdot \alpha + \begin{vmatrix} \alpha_1 & y_1 & \gamma_1 \\ \alpha_2 & y_2 & \gamma_2 \\ \alpha_3 & y_3 & \gamma_3 \end{vmatrix} \cdot \beta + \begin{vmatrix} \alpha_1 & \beta_1 & y_1 \\ \alpha_2 & \beta_2 & y_2 \\ \alpha_3 & \beta_3 & y_3 \end{vmatrix} \cdot \gamma = \begin{vmatrix} \alpha_1 & \beta_1 & \gamma_1 \\ \alpha_2 & \beta_2 & \gamma_2 \\ \alpha_3 & \beta_3 & \gamma_3 \end{vmatrix} \cdot Y \ .$$

The two solutions are equivalent except for a factor equal to the determinant of the coefficient matrix.

Pros and Cons

The advantages and disadvantages of the different methods principally
lie in ease of computation for hand versus computer calculation. Gauss-
reduction is particularly straightforward and amenable to hand computation,
but Cramer's rule or Korzhinskii's method are both rather simple if the
system is small or the coefficient matrix contains a large number of zeros.
With a computer the choice is arbitrary, but most computer systems contain
"canned" matrix inversion routines or routines to solve simultaneous linear
equations.

One advantage to Korzhinskii's method, expecially for balancing chemical
reactions, is in cases where there is a degeneracy in the system. Degenerate
systems will be discussed in more detail later, but briefly, if a system is
degenerate then the coefficient matrix, A, will be singular, which means
that it will not have an inverse and will have a determinant of zero.
Matrix inversion, Cramer's rule, or Gauss-reduction will not yield a solution if there is a degeneracy, but with Korzhinskii's method a solution will

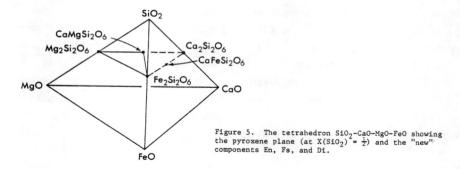

Figure 5. The tetrahedron SiO_2-CaO-MgO-FeO showing the pyroxene plane (at $X(SiO_2) = \frac{1}{2}$) and the "new" components En, Fs, and Di.

still be obtained, except that the coefficient of one of the minerals in the reaction (i.e., one of the "new components") will be zero because the determinant of a singular matrix is zero.

The matrix inversion approach is particularly useful for computer applications where a large number of minerals are to be mapped into a new coordinate system. In this case, the inverse of the coefficient matrix (A^{-1}) can be computed once, yielding a generalized set of equations to map from one coordinate set to another, as has been done by Greenwood (1975).

APPLICATIONS

There are a great number of applications for linear algebra in petrology, some of which will be discussed briefly below. All of the problems involve, in some way, the mapping from one coordinate set into another; the major difference is simply how the original question is formulated.

Calculation of end-member mineral components

This is a straightforward case of mapping a mineral Y, whose composition is known in an old component set (for example, moles of oxides) into a new component set, in this case end-member components. An example involving pyroxenes is shown in Figure 5. The old components in this example are moles of the oxides, SiO_2-CaO-MgO-FeO, and the new components are the end-members, enstatite ($Mg_2Si_2O_6$), ferrosilite ($Fe_2Si_2O_6$) and diopside ($CaMgSi_2O_6$). Note that there are only three linearly independent pyroxene components in the system SiO_2-CaO-MgO-FeO. (A fourth pyroxene component ($CaFeSi_2O_6$) can be calculated as Di + Fs − En). However, the choice of which three of the four to pick is completely arbitrary. Note also that there are four "old" components and only 3 "new" components. This is because the pyroxene stoichiometry (sum of the cations = 4) places an additional constraint on the system. The constraint can be seen in Figure 5: All of the pyroxenes must lie on the plane defined by $X(SiO_2) = 0.5$. In order to make the transformation, one of two things must be done. An arbitrary "new" component can be added that lies off the pyroxene plane and thus is not a pyroxene end-member (for example, SiO_2). By definition, this component will have a value of 0.0 for any stoichiometric pyroxene. Alternatively, it can be recognized that the only *independently variable* elements in the pyroxene are CaO, MgO and FeO, and the component SiO_2 can be ignored. Adopting the first procedure yields

$$x_1 Qt + x_2 En + x_3 Fs + x_4 Di = \text{"Px"}$$

$[SiO_2]$	$x_1 \cdot 1 + x_2 \cdot 2 + x_3 \cdot 2 + x_4 \cdot 2 = SiO_2$ in Px
$[MgO]$	$x_1 \cdot 0 + x_2 \cdot 2 + x_3 \cdot 0 + x_4 \cdot 1 = MgO$ in Px
$[FeO]$	$x_1 \cdot 0 + x_2 \cdot 0 + x_3 \cdot 2 + x_4 \cdot 0 = FeO$ in Px
$[CaO]$	$x_1 \cdot 0 + x_2 \cdot 0 + x_3 \cdot 0 + x_4 \cdot 1 = CaO$ in Px

where x_1, x_2, x_3 and x_4 are the amounts of the quartz, enstatite, ferrosilite and diopside components in the pyroxene, respectively. Back substitution quickly reveals that

$$x_4 = CaO \text{ in Px}$$

$$x_3 = FeO \text{ in Px}/2.0$$

$$x_2 = (MgO \text{ in Px} - x_4)/2 = (MgO - CaO)/2.0$$

$$x_1 = SiO_2 \text{ in Px} - 2x_4 - 2x_3 - 2x_2 \quad \text{or}$$

$$x_1 = SiO_2 - CaO - FeO - MgO \quad \text{or}$$

$$x_1 = 0 \text{ (since } SiO_2 = 2 \text{ and } CaO + MgO + FeO = 2 \text{ in a stoichiometric}$$
$$\text{pyroxene) .}$$

A slightly more complicated example involves the calculation of pyroxene end-member components in the system $SiO_2-Al_2O_3-MgO-FeO-CaO-Na_2O$. An arbitrary choice of linearly independent pyroxene end-member components for this system is enstatite ($Mg_2Si_2O_6$), ferrosilite ($Fe_2Si_2O_6$), diopside ($CaMgSi_2O_6$), Catschermak's ($CaAlAlSiO_6$), and jadeite ($NaAlSi_2O_6$). Here, the only independently variable cations in the pyroxene are Al(VI), Mg, Fe, Ca and Na, since Al(IV) and Si must be coupled to these cations to maintain electroneutrality and stoichiometry. Hence, there are five equations in five unknowns:

$$x_1En + x_2Fs + x_3Di + x_4Ct + x_5Jd = Px$$

[Mg]	$x_1 \cdot 2 + x_2 \cdot 0 + x_3 \cdot 1 + x_4 \cdot 0 + x_5 \cdot 0 = Mg$ in Px
[Fe]	$x_1 \cdot 0 + x_2 \cdot 2 + x_3 \cdot 0 + x_4 \cdot 0 + x_5 \cdot 0 = Fe$ in Px
[Ca]	$x_1 \cdot 0 + x_2 \cdot 0 + x_3 \cdot 1 + x_4 \cdot 1 + x_5 \cdot 0 = Ca$ in Px
[Al^{VI}]	$x_1 \cdot 0 + x_2 \cdot 0 + x_3 \cdot 0 + x_4 \cdot 1 + x_5 \cdot 1 = Al(VI)$ in Px
[Na]	$x_1 \cdot 0 + x_2 \cdot 0 + x_3 \cdot 0 + x_4 \cdot 0 + x_5 \cdot 1 = Na$ in Px

or

$$\begin{bmatrix} 2 & 0 & 1 & 0 & 0 \\ 0 & 2 & 0 & 0 & 0 \\ 0 & 0 & 1 & 1 & 0 \\ 0 & 0 & 0 & 1 & 1 \\ 0 & 0 & 0 & 0 & 1 \end{bmatrix} \cdot \begin{bmatrix} x_1 \\ x_2 \\ x_3 \\ x_4 \\ x_5 \end{bmatrix} = \begin{bmatrix} Mg \\ Fe \\ Ca \\ Al(VI) \\ Na \end{bmatrix}$$

where x_1-x_5 are the amounts of the respective components in the pyroxene. Back substitution yields

$$x_5 = Na$$
$$x_4 = Al(VI) - Na$$
$$x_3 = Ca + Na - Al(VI)$$
$$x_2 = Fe/2$$
$$x_1 = (Mg - Ca - Na + Al(VI))/2 .$$

An analogous procedure can be followed to calculate end-member components in amphiboles.

Perry (1967a; 1968a,b) has written extensively on the use of algebraic techniques in the classification of minerals and calculating end-member components of minerals. Perry has shown that, in general, 11-dimensional linearly-independent component space is necessary for the description of the chemical variability in feldspars and amphiboles. Moreover, for the amphiboles, there is generally not a set of components whereby all the coefficients of the component set for a particular amphibole have positive values. Negative values of components simply means that the physically accessible composition space of amphiboles cannot be encompassed by a single set of linearly independent end-members. For thermodynamic calculations, however, negative components are perfectly acceptable (*cf.* Brady, 1975).

65

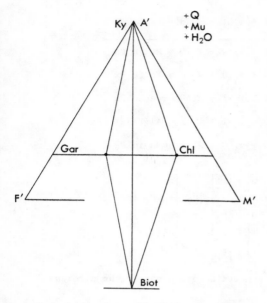

Figure 6. An AFM diagram showing the relationship among the minerals kyanite (Ky), biotite (Biot), chlorite (Chl) and garnet (Gar).

Calculation of rock norms

This problem is exactly analogous to calculating components in minerals except that the total set of normative minerals ("new" components) is not linearly independent. Sequential sets of normative minerals are calculated until all of the coefficients of the composition vector in the new coordinate system are positive.

Balancing chemical reactions

The problem of balancing a chemical reaction among minerals in a system can be formulated as follows: Given the composition of n minerals in an n-component system, what linear combination of these minerals is needed to make mineral Y? Or, mathematically, what is the solution (X) to the system of equations:

$$x_1 \cdot \alpha + x_2 \cdot \beta + \ldots x_n \cdot \gamma = Y?$$

Note that this is exactly the problem of mapping Y into the "new" coordinate system α, $\beta \ldots \gamma$, and the coefficients of the solution vector X are the stoichiometric coefficients of the reaction. Note also that the above algebraic formulation is consistent with the phase rule, that is, in an n-component system it takes, in general, n+1 phases (e.g., α, $\beta \ldots \gamma$ and Y) to balance a chemical reaction among the phases (provided there are no degeneracies in the system).

As an example, consider the crossing tie-line relation on the AFM diagram shown in Figure 6, which represents the reaction

garnet + chlorite + muscovite = kyanite + biotite + quartz + H_2O .

The composition of the garnet, chlorite and biotite are

garnet: $(Fe_2Mg_1)Al_2Si_3O_{12}$

chlorite: $(Fe_2Mg_7)Al_6Si_5O_{20}(OH)_{16}$

biotite: $K_2(Fe_3Mg_3)Al_2Si_6O_{20}(OH)_4$

66

Balancing this chemical reaction is equivalent to answering the question: What linear combination of quartz, muscovite, kyanite, chlorite, garnet and H_2O are necessary to make biotite? (Actually, the choice of which mineral to be mapped into the "new" coordinate system is completely arbitrary. Here biotite was chosen, but it could have been any of the six minerals or H_2O.) Mathematically, this is represented by the system of equations:

$$x_1 \cdot Qt + x_2 \cdot Mu + x_3 \cdot Ky + x_4 \cdot Ch + x_5 \cdot Ga + x_6 \cdot H_2O = Biot$$

$[SiO_2]$	$x_1 \cdot 1 + x_2 \cdot 3 + x_3 \cdot 1 + x_4 \cdot 5 + x_5 \cdot 3 + x_6 \cdot 0 = 6.0$
$[AlO_{3/2}]$	$x_1 \cdot 0 + x_2 \cdot 3 + x_3 \cdot 2 + x_4 \cdot 6 + x_5 \cdot 2 + x_6 \cdot 0 = 2.0$
$[FeO]$	$x_1 \cdot 0 + x_2 \cdot 0 + x_3 \cdot 0 + x_4 \cdot 2 + x_5 \cdot 2 + x_6 \cdot 0 = 3.0$
$[MgO]$	$x_1 \cdot 0 + x_2 \cdot 0 + x_3 \cdot 0 + x_4 \cdot 7 + x_5 \cdot 1 + x_6 \cdot 0 = 3.0$
$[KO_{1/2}]$	$x_1 \cdot 0 + x_2 \cdot 1 + x_3 \cdot 0 + x_4 \cdot 0 + x_5 \cdot 0 + x_6 \cdot 0 = 2.0$
$[H_2O]$	$x_1 \cdot 0 + x_2 \cdot 1 + x_3 \cdot 0 + x_4 \cdot 8 + x_5 \cdot 0 + x_6 \cdot 1 = 2.0$

or, in matrix notation

$$
\begin{bmatrix}
1 & 3 & 1 & 5 & 3 & 0 \\
0 & 3 & 2 & 6 & 2 & 0 \\
0 & 0 & 0 & 2 & 2 & 0 \\
0 & 0 & 0 & 7 & 1 & 0 \\
0 & 1 & 0 & 0 & 0 & 0 \\
0 & 1 & 0 & 8 & 0 & 1
\end{bmatrix}
\cdot
\begin{bmatrix}
x_1 \\ x_2 \\ x_3 \\ x_4 \\ x_5 \\ x_6
\end{bmatrix}
=
\begin{bmatrix}
6.0 \\ 2.0 \\ 3.0 \\ 3.0 \\ 2.0 \\ 2.0
\end{bmatrix}
$$

Solution of these equations yields the reaction

$$1.25 \ Gar + 0.25 \ Chl + 2.0 \ Mus = 1.0 \ Biot + 4.0 \ Ky + 1.0 \ Qt + 2.0 \ H_2O$$

where the minerals with negative coefficients have been moved over to the side of the reaction with biotite.

Linear algebra has one of its widest applications in metamorphic petrology in balancing chemical reactions. Korzhinskii (1959) pointed out the utility of linear algebra in balancing petrologic reactions, and the approach has been expanded on by Perry (1967b). Applications can be divided into two types. The first is where the number of equations is equal to the number of unknowns (that is, the number of phases in the reaction = the number of components + 1) and the solution is exact. This type of problem has greatest application in idealized systems involving end-member components (e.g.,

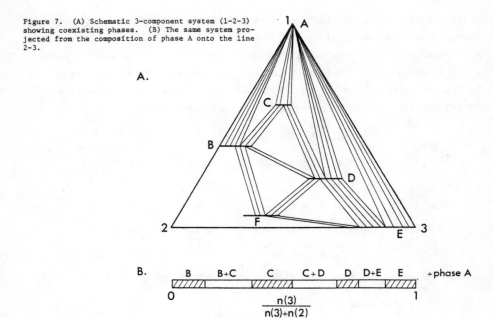

Figure 7. (A) Schematic 3-component system (1-2-3) showing coexisting phases. (B) The same system projected from the composition of phase A onto the line 2-3.

Perry, 1967a,b; 1968a,b). The second application is where the number of equations exceeds the number of unknowns (the number of phases is less than the number of components + 1) in which case there is no exact solution. In these cases least squares techniques have been employed, as will be discussed in a later section.

Petrologic mixing problems

Mixing problems are exactly analogous to balancing chemical reactions except that these problems usually deal with rock compositions rather than mineral compositions. The problem is usually phrased: How much of the minerals α, β, γ, etc. are necessary to make a rock of composition Y? Clearly, petrologic mixing problems can be formulated in many different ways.

Projective analysis of composition space

Projective analysis is a very powerful technique for the visualization of composition space when the number of components is too large to represent graphically. The goal of projective analysis is to reduce the dimensionality of the composition space by viewing only portions of it at a time -- but to do so in such a way that the projected composition space preserves the properties of the full composition space and is thus a rigorous phase diagram in a thermodynamic sense. Greenwood (1975) has presented an excellent discussion of the mathematics of projective analysis, and the reader is referred to that paper for further discussion (see also Brady and Stout, 1980).

The basis for projective analysis can be seen in diagrammatic form in Figure 7. This figure shows a phase diagram for the three-component system (1-2-3) at some arbitrary but fixed value of pressure (P) and temperature (T). Suppose it was desired to examine the phase relations of only those assemblages in Figure 7 that contained the phase A. Imagine placing one's eye at point A in Figure 7 and sighting down the tie-lines connecting phase A with B, C, and D and E. By continuing these tie-lines down to the join 2-3, the two-component projected phase diagram shown in Figure 7B is obtained. Figure 7B is thus a valid phase diagram for the three-component system 1-2-3 at T and P for all those assemblages that contain phase A.

Mathematically, the plotting positions of minerals B, C and D in the projected composition space 2-3 are easy to calculate by recognizing that the projected plotting coordinates preserve the ratio of components 2 and 3 in the phases (that is, they are projected radially from point A along lines of constant ratio of components 2 and 3). The plotting positions are calculated by simply deleting the element or elements in the transformed composition vector that correspond to the projection point or points (in this case A). The remaining components are then renormalized.

In order to make a projection, it is necessary to have both a projection point and a projection plane (in much the same way that an artist needs a perspective point and a perspective plane to render an accurate perspective drawing). In the above example, the projection point is one of the components of the system (i.e., phase A = component 1) and the projection "plane" is defined by two other components of the system (i.e., the line 2-3). Often, however, the projection point and/or the projection plane are not components of the system to start with, but are the compositions of mineral phases. In order mathematically to make the projection (which simply involves "dropping" the component(s) which represent the projection point(s) and renormalizing), it is first necessary to transform the coordinate system such that the projection point(s) and projection plane are all components of the new coordinate system. This is accomplished quite simply by linear mapping described earlier.

The complete procedure for projective analysis of composition space is:

(1) Transform the coordinate system so that the axes of the new coordinate system are all either projection points or part of

the projection plane and map all minerals of the system into this new coordinate system.

(2) Perform the projection by "dropping" the components representing the projection points from the transformed composition vectors and renormalize.

It can be seen from Figure 7 that there is a graphical basis for projective analysis. Furthermore, any projected phase diagram will be a valid representation of the phase relations as long as either the projection points are fixed compositions (i.e., not solid solutions) or that all of the projected minerals (e.g., B, C, D, and E) coexist with a phase of the same composition as the projection point. There is also a thermodynamic basis for the validity of the projected composition space. One of the conditions of heterogeneous equilibrium is that the chemical potentials of all components in all phases in an equilibrium assemblage are equal. If one envisions a chemical potential axis coming out of the page in Figure 7A, it will be noticed that all of the minerals that coexist with mineral A (i.e., B, C, D, and E) have the same value for the chemical potential of component 1. In other words,

$$\mu_1^A = \mu_1^B = \mu_1^C = \mu_1^D = \mu_1^E .$$

Hence, the projected phase diagram, Figure 7B, can be seen as a diagram drawn at constant chemical potential of component 1. Chemical potential is an intensive parameter of the system (as are T and P); therefore, a projected phase diagram is valid because it represents mineral coexistences at a *constant value of intensive parameters* (e.g., T, P and μ_i). This is a requirement of all thermodynamically "legal" projections (Greenwood, 1975).

As an example of the procedure, consider the AFM projection of Thompson (1957). The "old" components are those of the model chemical system for pelitic schists, $SiO_2-Al_2O_3-FeO-MgO-K_2O-H_2O$, and the "new" components include the projection points SiO_2 (quartz), $KAl_3Si_3O_{10}(OH)_2$ (muscovite), and H_2O as well as the projection plane $Al_2O_3-FeO-MgO$. Suppose a plotting position of a particular biotite with the composition $KFe_{2.5}Mg_{0.5}AlSi_3O_{10}(OH)_2$ on the projection plane $Al_2O_3-FeO-MgO$ is desired. The system of equations looks like:

$$x_1 \cdot Qt + x_2 \cdot Mus + x_3 \cdot H_2O + x_4 \cdot Al_2O_3 + x_5 \cdot FeO + x_6 \cdot MgO = Biot$$

$[SiO_2]$	$x_1 \cdot 1$	$+ x_2 \cdot 3$	$+ x_3 \cdot 0$	$+ x_4 \cdot 0$	$+ x_5 \cdot 0$	$+ x_6 \cdot 0$	$= 3.0$
$[Al_2O_3]$	$x_1 \cdot 0$	$+ x_2 \cdot 3/2$	$+ x_3 \cdot 0$	$+ x_4 \cdot 1$	$+ x_5 \cdot 0$	$+ x_6 \cdot 0$	$= 0.5$
$[MgO]$	$x_1 \cdot 0$	$+ x_2 \cdot 0$	$+ x_3 \cdot 0$	$+ x_4 \cdot 0$	$+ x_5 \cdot 0$	$+ x_6 \cdot 1$	$= 0.5$
$[FeO]$	$x_1 \cdot 0$	$+ x_2 \cdot 0$	$+ x_3 \cdot 0$	$+ x_4 \cdot 0$	$+ x_5 \cdot 1$	$+ x_6 \cdot 0$	$= 2.5$
$[K_2O]$	$x_1 \cdot 0$	$+ x_2 \cdot 1/2$	$+ x_3 \cdot 0$	$+ x_4 \cdot 0$	$+ x_5 \cdot 0$	$+ x_6 \cdot 0$	$= 0.5$
$[H_2O]$	$x_1 \cdot 0$	$+ x_2 \cdot 1$	$+ x_3 \cdot 1$	$+ x_4 \cdot 0$	$+ x_5 \cdot 0$	$+ x_6 \cdot 0$	$= 1.0$

or, in matrix notation $(A \cdot X = Y)$

$$\begin{bmatrix} 1 & 3 & 0 & 0 & 0 & 0 \\ 0 & \frac{3}{2} & 0 & 1 & 0 & 0 \\ 0 & 0 & 0 & 0 & 0 & 1 \\ 0 & 0 & 0 & 0 & 1 & 0 \\ 0 & \frac{1}{2} & 0 & 0 & 0 & 0 \\ 0 & 1 & 1 & 0 & 0 & 0 \end{bmatrix} \cdot \begin{bmatrix} x_1 \\ x_2 \\ x_3 \\ x_4 \\ x_5 \\ x_6 \end{bmatrix} = \begin{bmatrix} 3.0 \\ 0.5 \\ 0.5 \\ 2.5 \\ 0.5 \\ 1.0 \end{bmatrix}$$

The solution $(A^{-1} \cdot Y) = X)$ is

$[Qt]$													

$$\begin{matrix} [Qt] \\ [Mus] \\ [H_2O] \\ [Al_2O_3{}'] \\ [MgO'] \\ [FeO'] \end{matrix} \begin{bmatrix} 1 & 0 & 0 & 0 & -6 & 0 \\ 0 & 0 & 0 & 0 & 2 & 0 \\ 0 & 0 & 0 & 0 & -2 & 1 \\ 0 & 1 & 0 & 0 & -3 & 0 \\ 0 & 0 & 0 & 1 & 0 & 0 \\ 0 & 0 & 1 & 0 & 0 & 0 \end{bmatrix} \begin{bmatrix} 3.0 \\ 0.5 \\ 0.5 \\ 2.5 \\ 0.5 \\ 1.0 \end{bmatrix} = \begin{bmatrix} 0 \\ 1.0 \\ 0 \\ -1.0 \\ 0.5 \\ 2.5 \end{bmatrix}$$

or

$$0 \cdot Qt + 1 \cdot Mu + 0 \cdot H_2O - 1 \cdot Al_2O_3{}' + 0.5 \cdot MgO' + 2.5 \cdot FeO' = 1 \cdot Biot \ .$$

Dropping quartz, muscovite, and H_2O as the projection points and renormalizing yields:

$$Biotite = (-0.5 \cdot Al_2O_3{}', \ 0.25 \cdot MgO', \ 1.25 \cdot FeO') \ .$$

One can easily generalize these results to any mineral:

$$Al_2O_3{}' = (Al_2O_3 - 3K_2O)/(Al_2O_3 - 3K_2O + FeO + MgO)$$

$$FeO' \ = FeO/(Al_2O_3 - 3K_2O + FeO + MgO)$$

$$MgO' \ = MgO/(Al_2O_3 - 3K_2O + FeO + MgO) \ .$$

71

Examples of projective analysis in metamorphic petrology

Since Thompson (1957) presented his AFM projection for muscovite-bearing pelitic schists, graphical analysis of mineral assemblages using thermodynamically legal projections has become an essential tool of the metamorphic petrologist, and application of these techniques has contributed greatly to the understanding of the petrogenesis of metamorphic rocks. Projected phase diagrams are usually three-component triangles, simply because these are easy to draw and visualize. Tetrahedral phase diagrams are also extremely useful because they permit visualization of one more component than a triangle. Tetrahedral phase diagrams can easily be constructed in either perspective or stereoscopic pairs (e.g., Spear, 1980). A computer program for drawing such diagrams is contained in the appendix.

This section will review some of the more commonly used graphical projections and the types of information that have been gained through their application.

Projection involving predominantly pelitic assemblages. Certainly the most widely used graphical projection in metamorphic petrology is Thompson's (1957) AFM projection through the composition of muscovite, quartz, and H_2O. A similar projection for high-grade rocks that contain K-feldspar rather than muscovite was introduced by Barker (1961).

Greenwood (1975) presents three alternative projections of the AFM system that show different portions of composition space. Assuming that it is desirable to retain FeO and MgO as points on the projection plane, three-component phase diagrams can be constructed by projection through one of the following sets of phases:

projection points	projection plane
Quartz, muscovite, H_2O	Al_2O_3'-FeO'-MgO' (AFM)
Al_2SiO_5, muscovite, H_2O	SiO_2'-FeO'-MgO'
Al_2SiO_5, quartz, H_2O	Muscovite'-FeO'-MgO'
Quartz, Al_2SiO_5, muscovite	H_2O'-FeO'-MgO'

An example of the last of these projections is shown in Figure 8B (from Greenwood, 1975) along with the conventional AFM diagram (Fig. 8A). The projection in Figure 8B, which is valid for any pelite containing kyanite + quartz + muscovite, depicts the variable H_2O contents of different mineral assemblages at a particular P-T condition. For example, the assemblage

72

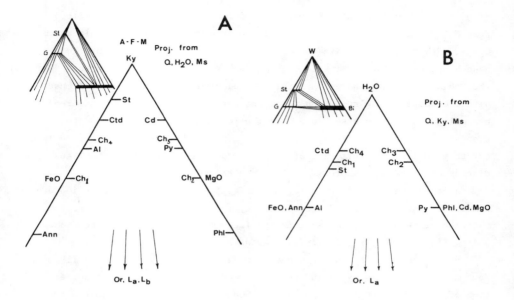

Figure 8. Projections of the six-component pelite system after Greenwood (1975). (A) The AFM projection of J. B. Thompson. The diagram in the upper left corner represents a possible topology at approximately 4 kb and 700°C. (B) Projection of the model pelite system onto the plane H_2O-FeO-MgO from quartz, kyanite and muscovite. Water is shown as an extensive parameter in this diagram in contrast to its role as an intensive variable in Figure 8A. The diagram in the upper left represents a possible topology at approximately 4 kb and 700°C, where Gar-St-Bi are stable with H_2O and Ms, but not with kyanite. In (B) Gar-St-Bi are shown as stable with Q, Ms and Ky but not with pure H_2O. Abbreviations: Ky = kyanite; St = staurolite; Ctd = chloritoid; Cd = cordierite; Ch = chlorite; Al = almandine; Py = pyrope; Ann = annite; Phl = phlogopite; Or = K-feldspar; La and Lb = assumed liquids near the Or-Q-H_2O "eutectic".

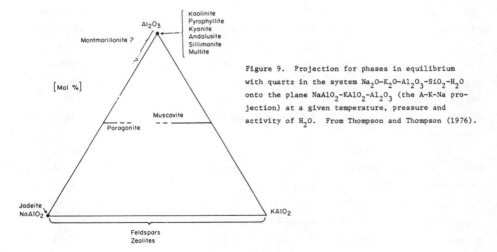

Figure 9. Projection for phases in equilibrium with quartz in the system Na_2O-K_2O-Al_2O_3-SiO_2-H_2O onto the plane $NaAlO_2$-$KAlO_2$-Al_2O_3 (the A-K-Na projection) at a given temperature, pressure and activity of H_2O. From Thompson and Thompson (1976).

staurolite + biotite + H_2O (+ quartz + kyanite + muscovite) exists with a pure H_2O fluid, and thus $\mu(H_2O)$ is defined at a maximum value by this assemblage. Staurolite + biotite (no H_2O) has a lower water content and the assemblage staurolite + biotite + garnet has a still lower H_2O content. Note that this latter assemblage, staurolite + biotite + garnet + quartz + muscovite + kyanite, would appear on a conventional AFM projection as a four-phase assemblage, and is an assemblage which buffers $\mu(H_2O)$. Projection onto planes containing volatile species, for example, H_2O-FeO-MgO, is an excellent way to visualize the buffering of volatile species by mineral assemblages (see discussion below and Chapter 7 by Rice and Ferry of this volume).

Another useful projection for medium-grade pelitic rocks, which is very similar to the projection discussed above, is a projection through the composition of staurolite, rather than kyanite. This projection has been used quite effectively by Rumble (1977) (see discussion below).

Pelitic schists often contain sodium, and Thompson and Thompson (1976) have discussed facies types in sodium-bearing pelitic schists by consideration of the subsystem SiO_2-Al_2O_3-K_2O-Na_2O-H_2O. Projecting from quartz and H_2O in this system gives the A-K-Na phase diagram shown in Figure 9, which is an adequate representation of this subsystem except in cases where plagioclase contains considerable Ca or muscovite or paragonite contains considerable phengite component. By consideration of naturally-occurring A-K-Na facies types, Thompson and Thompson (1976) derived a petrogenetic grid for this subsystem.

Projections in calcareous bulk compositions. Metamorphosed siliceous dolomitic limestones may be modeled by the system CaO-MgO-SiO_2-H_2O-CO_2. Phase equilibria for this rock type are commonly represented on a CaO-MgO-SiO_2 diagram which is projected through H_2O and CO_2 (Bowen, 1940). Resulting diagrams portray mineral equilibria at constant P, T, $\mu(H_2O)$, and $\mu(CO_2)$. Skippen (1974) derived quantitative petrogenetic grids in T-X_{CO_2} space at several pressures for metamorphosed siliceous dolomitic limestones using the projected CaO-MgO-SiO_2 diagram.

Calc-mica schists and marls can be considerably complex chemically and thus amenable to projective graphical analysis. The premetamorphic protolith of calc-mica schists can be thought of as a calcareous pelite or an alkali-bearing, aluminous carbonate, and it can be represented by the model system SiO_2-Al_2O_3-FeO-MgO-CaO-Na_2O-K_2O-H_2O-CO_2. Thompson (1975), in a study of mineral reactions in a calc-mica schist from Gassetts, Vermont, depicted a

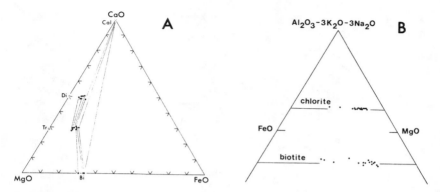

Figure 10. Projections of assemblages in calc-mica schists.
(A) From Thompson (1975). Compositions of clinopyroxenes, actinolite, biotite, and calcite (mole percent CaO-MgO-FeO) projected from SiO_2, $KAlO_2$, H_2O and CO_2. Clinozoisite would plot at CaO if projected from $CaAl_2O_4$ in addition. (B) From Ferry (1979). Compositions of biotite and chlorite plotted on a diagram that projects through SiO_2, H_2O, KAl_3O_5, and $NaAl_3O_5$ [biotite and chlorite coexisting with quartz and muscovite of fixed K/Na under conditions of constant P, T, and $\mu(H_2O)$].

portion of this system by projecting from quartz, $KAlO_2$, H_2O and CO_2 (Fig. 10A). The diagram in Figure 10A thus refers to mineral assemblages with quartz and K-feldspar at constant P, T, $\mu(H_2O)$, and $\mu(CO_2)$. Ferry (1979), in a study of metamorphosed argillaceous carbonate rocks from south-central Maine, utilized a projection through SiO_2, H_2O, KAl_3O_5 and $NaAl_3O_5$ onto Al_2O_3'-FeO'-MgO' in order to depict composition of coexisting biotite and chlorite (Fig. 10B). The diagram in Figure 10B thus refers to minerals coexisting with quartz and muscovite (with constant K/Na) at conditions of constant P, T, and $\mu(H_2O)$.

One problem with any graphical representation of phase relations in systems which contain both carbonate and hydrous phases is that most assemblages in these types of rocks buffer $\mu(H_2O)$ and $\mu(CO_2)$ rather than vice versa (see Rice and Ferry, this volume). Hence, it is usually impossible to draw a single phase diagram for a wide variety of assemblages.

Projective analysis of amphibolites. These rocks have long defied rigorous projective analysis because of their chemical complexity. The classic ACF diagram of Eskola (1915), while showing the major facies changes, cannot depict the importance of Fe-Mg solution, nor does it adequately account for phase relations involving plagioclase.

Amphibolites can be represented, to a first approximation, by the system SiO_2-Al_2O_3-FeO-MgO-CaO-Na_2O-H_2O. One approach to graphic representation of these rocks has been to project from quartz and H_2O into the tetrahedron Al_2O_3-CaO-Na_2O-$(FeO+MgO)$. This projection obscures the effect of Fe-Mg substitution but is useful for portraying phase relations

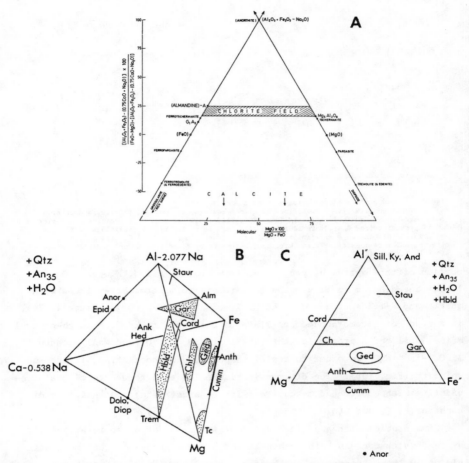

Figure 11. Projections for amphibolite assemblages. (A) From Harte and Graham (1975). Projection from epidote and albite onto the plane Al_2O_3-FeO-MgO. This diagram is valid for greenschists and albite-bearing epidote amphibolites. (B) From Spear (1982). Plotting positions for end-member minerals projected from quartz, plagioclase (An_{35}), H_2O, CO_2 into the tetrahedron Fe-Ca-Mg-Al. Abbreviations are: Ged = gedrite; Anth = anthophyllite; Cumm = cummingtonite; Chl = chlorite; Tc = talc; Cord = cordierite; Gar = garnet solid solutions; Alm = almandine; Hbld = hornblende; Trem = tremolite; Dolo = dolomite; Diop = diopside; Ank = ankerite; Hed = hedenbergite; Epid = epidote (or zoisite); Anor = anorthite; Stau = staurolite. Wonesite, which has not been shown, would plot between talc, chlorite, and cummingtonite but negatively with respect to Ca'. (C) From Spear (1982). Generalized plotting positions of phases projected from quartz, plagioclase (An_{35}), H_2O and 'average' hornblende [$Na_{0.5}Ca_{1.7}Fe_{1.85}Mg_{1.85}Al_{1.4}(Al_{1.5}Si_{6.5})O_{22}(OH)_2$]. Abbreviations as in (B). This diagram is valid for quartz-, plagioclase-, hornblende-bearing amphibolites.

involving plagioclase. Laird (1980) has made very effective use of this projection in representing phase relations in mafic schist from Vermont. A more rigorous projection was devised by Harte and Graham (1975), who depicted the phase relations of low-grade metabasites from the Dalradian on a type of A'F'M' projection that projects through the compositions of quartz, albite, epidote and H_2O (see Fig. 11A).

Robinson and Jaffe (1969) and Stout (1972) projected orthoamphibole-bearing assemblages from quartz, H_2O, and plagioclase of fixed composition onto the Al_2O_3-FeO-MgO plane. This projection adequately depicts phase relations in low-Ca amphibolites from New Hampshire and southern Norway. Because the projection has the effect of lumping Na and Ca, it does not work well when hornblende coexists with aluminous phases such as staurolite or kyanite, as pointed out by Spear (1982) and Robinson et al. (1981). Spear (1982; see also Robinson et al., 1981) presented an alternative projection from quartz, H_2O, and andesine (An_{35}) into the tetrahedron Al_2O_3-FeO-MgO-CaO (see Fig. 11B). Figure 11B adequately represents phase relations in amphibolites from the Post Pond Volcanics, Vermont, provided only assemblages that crystallized at similar values of $\mu(H_2O)$ are plotted on a single diagram (see below). Spear (1982) has made an additional projection of this system through the composition of hornblende onto the AFM plane (Fig. 11C). Figure 11C can be used as a thermodynamically valid representation of the phase relations in medium-grade, hornblende + andesine + quartz-bearing amphibolites.

Types of information to be gained from graphical analysis. The most obvious benefit of graphical analysis is that it provides a visual representation of mineral phase relations that is considerably easier to comprehend, as well as remember, than mineral analyses or lists of mineral assemblages. But a thermodynamically legal projection is much more than just a diagram. It can represent a phase diagram for complex chemical systems, as discussed in the above examples. It can also provide the basis for a petrogenetic grid for a particular rock type. For example, the grid of Thompson and Thompson (1976) for the A-K-Na facies types or the grids of Albee (1965b), Hess (1969), and Harte (1975) for pelites are all based on projected phase relations.

Perhaps the most extensive use to which projections are put is the testing of validity of the assumptions that go into making the projection. Following Greenwood (1975), the requirement of a thermodynamically legal projection is that "the equilibrium state of the system be uniquely determined by specifying pressure, temperature, and the proportions of the components represented in the diagram. If the diagram is a projection, then the phases or components from which the projection is made must either be present, pure, or have their chemical potentials fixed at a constant value over the whole of the diagram" (Greenwood, 1975, p. 1). If the projected

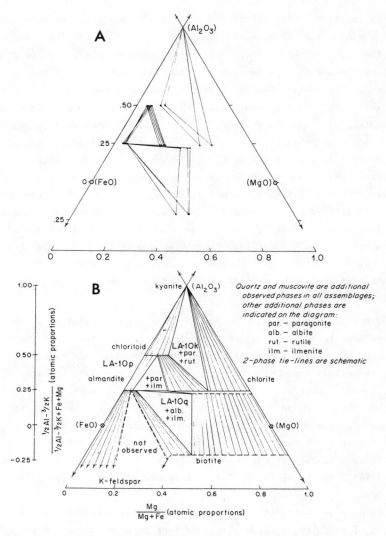

Figure 12. From Albee (1965). Pelitic schist assemblages from Mt. Grant, Vermont. (A) Compositions of coexisting mineral phases shown on a Thompson projection. (B) Observed mineral assemblages and compositions of coexisting phases, shown on a Thompson projection of the K_2O-Al_2O_3-FeO-MgO tetrahedron. Additional accessory phases may include combinations of: hematite, magnetite, apatite, tourmaline, graphite, clinozoisite-allanite, and pyrite.

phase relations of a suite of samples are regular and systematic (with consistent element partitioning and no crossing tie lines or overlapping phase volumes), then it can be concluded that the suite of assemblages are consistent with the above criteria (i.e., that all assemblages equilibrated at the same P, T and chemical potential of projected components). If, however,

78

crossing tie lines are present, then it can be concluded that a reaction relation existed among the participating assemblages, and one or more of these criteria must be violated.

This type of application has been utilized by many researchers. For example, Albee (1965a), in a classic study of pelitic schists near Mt. Grant, Vermont, used the AFM projection of Thompson (1957) to analyze phase relations in three different assemblages. From the compositions of coexisting minerals (Fig. 12A), Albee deduced the generalized phase relations shown in Figure 12B, and concluded that all assemblages crystallized at the same, P, T, and $\mu(H_2O)$.

On the other hand, Rumble (1977), in a study of metamorphosed pelitic layers from the Clough Formation, New Hampshire, found that phase relations projected onto an AFM diagram (Fig. 13A) were inconsistent with the crystallization of all rocks at the same conditions of P, T, and $\mu(H_2O)$. On the basis of additional projections through the compositions of muscovite, quartz, and either kyanite (Fig. 13B) or staurolite (Fig. 13C) onto the H_2O-FeO-MgO plane, Rumble concluded that assemblages in different beds crystallized at different chemical potentials of H_2O (see also Fig. 13D). A similar conclusion was reached based on phase relations projected through the compositions of quartz, muscovite, staurolite, and magnetite into the H_2O-FeO-MgO-TiO_2 tetrahedron. The conclusion was also confirmed by algebraic analysis (see Chapter 4 by Spear et al., this volume).

Spear (1982) presented a graphical analysis of low-variance amphibolite assemblages from the Post Pond Volcanics, Vermont. Figure 14A is a projection of quartz + plagioclase + hornblende-bearing assemblages through the composition of these phases and H_2O on to the plane $AlO_{3/2}$-FeO-MgO. The abundance of four-phase assemblages as well as crossing tie lines suggested that many of these assemblages buffered $\mu(H_2O)$, and that $\mu(H_2O)$ differed from one assemblage to another. Further projection from staurolite (Fig. 14B) as well as algebraic analysis of the phase relations confirmed chemical potential differences in H_2O, as depicted in Figure 14C.

Pitfalls of graphical analysis. Graphical analysis is a very powerful tool, and it is usually the starting point of most phase equilibria studies. It is not, however, without its pitfalls and shortcomings. There are many chemical systems that simply cannot be reduced by projective analysis to a dimension that will permit graphical portrayal. Moreover, it is almost always necessary to ignore or lump "extra" components, which can sometimes lead to erroneous conclusions regarding possible reaction relations. A very

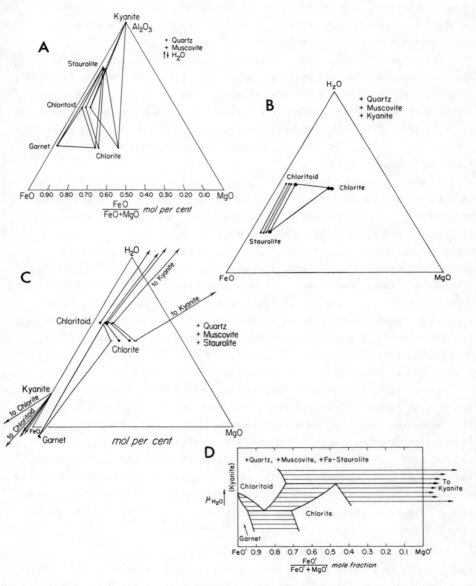

Figure 13. From Rumble (1977). Plotting positions of projected mineral assemblages from Black Mountain, New Hampshire. (A) Kyanite, staurolite, chloritoid, garnet, and chlorite compositions plotted on the AFM projection. The 4-phase assemblage staurolite-chloritoid-garnet-chlorite appears to be an assemblage which buffered μ(H₂O). (B) Projection of silicate mineral compositions from quartz, muscovite, and kyanite onto the plane FeO-MgO-H₂O. (C) Projection of silicate mineral compositions from quartz, muscovite, and staurolite onto the plane FeO-MgO-H₂O. (D) Isobaric, isothermal variations of μ(H₂O) with mineral assemblage and mineral composition for the system $SiO_2-Al_2O_3-FeO-MgO-K_2O-H_2O$. Constructed from (C) using Korzhinskii's (1959) method of equipotential lines.

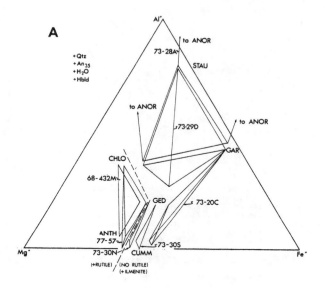

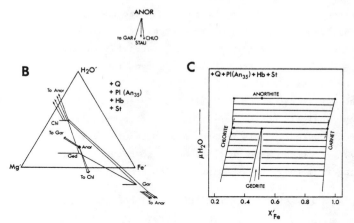

Figure 14. From Spear (1982). Projections of amphibolite assemblages from the Post Pond Volcanics, Vermont. (A) Projection of hornblende-bearing (carbonate-absent) assemblages from
quartz, H_2O, An_{35}, and 'average' hornblende onto the plane Al-Fe-Mg. Anorthite projects through
infinity; thus tie-lines between anorthite and either chlorite, staurolite, or garnet must project through infinity. Abbreviations as in Figure 11B. The dashed line separates assemblages
that contain rutile (the Mg-rich assemblages) from those that contain ilmenite (the Fe-rich assemblages). (B) Projections of the phase relations in (A) through staurolite onto the plane
FeO-MgO-H_2O, depicting the different water contents of the assemblages. (C) $\mu(H_2O)$ versus X_{Fe}
plot of the assemblages shown in (B).

81

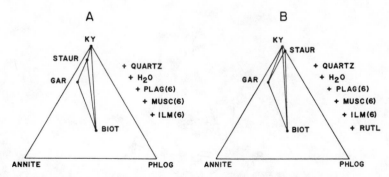

Figure 15. From Fletcher and Greenwood (1979). Projections of phase relations in a specimen of pelitic schist from the Penfold Creek area, British Columbia. (A) Projection from SiO_2, H_2O, plagioclase, muscovite and ilmenite. Points of projection are those mineral compositions present in the specimen. (B) Projection from SiO_2, H_2O, plagioclase, muscovite, ilmenite, and rutile. Note that the compositional degeneracy (reaction) inferred from the two projections, which are of the same assemblage, are different. The difference indicates how neglecting to project from a single phase (in this case rutile) can strongly influence resulting conclusions.

revealing example of this type of pitfall is given by Fletcher and Greenwood (1979). Figure 15 shows two projections of phase relations for a sample of pelitic schist; the only difference between the two diagrams is that the diagram on the right is projected from rutile whereas the diagram on the left is not. As can be seen by comparing the two figures, the reaction relation deduced from the two projections is different: Figure 15A suggests staurolite = kyanite + garnet whereas Figure 15B suggests staurolite + garnet = kyanite + biotite. Only the second reaction is correct, as confirmed by linear algebra.

The way around these pitfalls is, of course, not to rely entirely on graphical analysis but to use it as an aid to visualization. The algebraic methods discussed in this chapter and in Chapter 4 by Spear et al. (this volume), permit the same conclusions to be deduced without any simplification resulting from the process of projection.

DEGENERACIES AND LINEAR DEPENDENCIES

It is important to note that the situation sometimes arises in which components chosen for the set of "new" coordinate axes are not all linearly independent. That is, it is possible to make one of these new components out of a linear combination of the other new components. Phrased another way, there is a reaction relationship among the new components. In this case the new coordinate system is said to be degenerate and the coefficients matrix, A, will be singular (i.e., its determinant will be zero and it will not invert). The way to test for a degeneracy is simply to calculate the determinant of the coefficient matrix A. A zero determinant indicates a singular matrix. A degeneracy may provide important information about the system being studied or it may simply mean that a poor choice of new components has been made.

Linear dependence of component sets

As an example, consider the earlier problem of calculating end-member pyroxene components. It was already stated that in the system SiO_2-MgO-FeO-CaO, only three pyroxene components are linearly independent (see Fig. 6). However, if we had chosen the four components En, Fs, Di and Hd the transformation matrix would appear as

$$
\begin{array}{cccc}
 & [En] & [Fs] & [Di] & [Hd] \\
[SiO_2] & \begin{bmatrix} 2 & 2 & 2 & 2 \\ [MgO] & 2 & 0 & 1 & 0 \\ [FeO] & 0 & 2 & 0 & 1 \\ [CaO] & 0 & 0 & 1 & 1 \end{bmatrix}
\end{array}
$$

The determinant of this matrix is zero, indicating a linear dependence among the new components (i.e., En + Hd = Fs + Di).

Linear dependence in over- and underdetermined systems

In the above example the number of equations equals the number of unknowns and thus the coefficient matrix is square. However, there are situations where it is desirable to test for linear dependence when the coefficient matrix is not square. For example, many mineral assemblages have a variance greater than two and thus contain more components than phases. Systems such as this where the number of equations exceeds the number of unknowns are called overdetermined. In other situations, the system may contain more unknowns that equations. Systems such as this are called underdetermined. An example of an underdetermined system will be given below.

Testing for linear dependence in overdetermined or underdetermined systems can be done by consideration of a theorem of linear algebra which states that if a matrix is singular, then the matrix obtained by multiplying the singular matrix by any other matrix, will also be singular. A convenient matrix to multiply the singular matrix by is the transpose of the singular matrix (the transpose of a matrix is simply the matrix with the columns and rows exchanged and is denoted as A^T). Therefore, if A is singular, then A^TA will also be singular. Because A is an n x m matrix, the transpose of A is an m x n matrix and A^TA is an m x m square matrix. We can easily test for linear dependence in A^TA by calculating the determinant.

As an example, consider the situation described in "Algebraic Formulation of Phase Equilibria" (Spear et al., Chapter 4, this volume) in which we wish to know whether or not a system of equilibrium relations are linearly independent. The three equilibrium relations are

$$SiO_2 + Mg_2SiO_4 = 2MgSiO_3$$
$$SiO_2 + Fe_2SiO_4 = 2FeSiO_3$$
$$Fe_2SiO_4 + 2MgSiO_3 = Mg_2SiO_4 + 2FeSiO_3 \ .$$

In this example, it is clear that there *is* a linear dependence, but it is not always so obvious. The problem is formulated algebraically by first writing the three equations with all phases on one side of the equals sign:

$$0 = SiO_2 + Mg_2SiO_4 - 2MgSiO_3$$
$$0 = SiO_2 + Fe_2SiO_4 - 2FeSiO_3$$
$$0 = Fe_2SiO_4 - Mg_2SiO_4 - 2FeSiO_3 + 2MgSiO_3 \ .$$

The coefficient matrix, A, contains the coefficient of each mineral in the three reactions. A is therefore a 3 x 5 matrix and A^TA is computed as:

$$
\begin{array}{ccc}
A^T & A & A^TA
\end{array}
$$

$$
\begin{vmatrix}
1 & 1 & 0 \\
0 & 1 & 1 \\
1 & 0 & -1 \\
0 & -2 & -2 \\
-2 & 0 & 2
\end{vmatrix}
\cdot
\begin{array}{ccccc}
Q & Fa & Fo & Fs & En
\end{array}
\begin{vmatrix}
1 & 0 & 1 & 0 & -2 \\
1 & 1 & 0 & -2 & 0 \\
0 & 1 & -1 & -2 & 2
\end{vmatrix}
=
\begin{vmatrix}
2 & 1 & 1 & -2 & -2 \\
1 & 2 & -1 & -4 & 2 \\
1 & -1 & 2 & 2 & -4 \\
-2 & -4 & 2 & 8 & -4 \\
-2 & 2 & -4 & -4 & 8
\end{vmatrix}
$$

The determinant of A^TA is zero and thus a linear dependence exists among these three equations. Specifically, the third equation can be obtained by subtracting the second equation from the first.

Linear dependence and the variance of a mineral assemblage

One particularly useful application for the above test for linear independence is in consideration of the variance of a mineral assemblage. The variance, as calculated from the phase rule ($F = C + 2 - P$), is only valid if there are no linear dependencies (i.e., compositional degeneracies) in the system. Linear algebra provides a way to test for linear dependencies in a particular mineral assemblage and thus examine whether the phase rule variance applies. The necessary criterion for no linear dependencies in a mineral assemblage is either (a) that the determinant of the coefficient matrix not be zero or (b) for assemblages where there are more components than phases, that the determinant of A^TA not be zero.

This procedure tests for exact compositional degeneracies (that is, whether a chemical reaction written among phases of the assemblage balance exactly). It is also possible to test for the existence of compositional degeneracies within the error of an electron microprobe analysis by the method of least squares, as suggested by Greenwood (1968) (see also Pigage, 1976, and discussion below).

TREATMENT OF "EXTRA" COMPONENTS

Quite often it is found that a measured mineral or rock composition contains elements that are not considered as part of the "model" rock system. For example, a model pelitic schist contains the components SiO_2-Al_2O_3-FeO-MgO-K_2O-H_2O, but the components CaO, MnO, TiO_2, ZnO and Na_2O are commonly present in significant quantities in certain phases (e.g., Zn in staurolite; Mn or Ca in garnet; Na in white micas).

When there is a sufficient quantity of an "extra" component to stabilize an extra phase, for example TiO_2 stabilizing ilmenite or rutile or Na_2O stabilizing paragonite or albite, then this phase may be used as one of the "new" coordinate axes and either projected from, if graphical analysis is the goal, or used in balancing a chemical reaction. If, however, no new phase is stabilized, then the problem is how to treat the "extra" components. Mathematically, the problem is equivalent to transforming from an "old" set of coordinate axes with dimensionality n to a "new" set of coordinate axes with dimensionality less than n. That is, we are transforming from a higher order space into a lower order space. Thus, when the transformation matrix is set up, it will be seen that there are a greater number of equations than unknowns and that the transformation matrix is overdetermined. (Singularities in overdetermined systems were discussed briefly in the previous section.)

There are three common ways to handle the overdetermined transformation:

(1) Ignore the component entirely and recalculate the analysis as if the component were not present. This is effectively the same as projecting through the component.

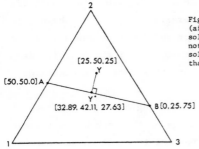

Figure 16. Schematic 3-component composition diagram (after Greenwood, 1968) showing the least-squares solution to the equation $x_1 \cdot A + x_2 \cdot B = Y$ where Y does not fall on the line A-B. Y* is the least-squares solution and represents the point along the line A-B that is closest to Y.

(2) Lump the extra component together with another component, especially one that behaves in a similar crystal chemical fashion. For example, lump MnO with FeO or lump Fe_2O_3 with Al_2O_3.

(3) Model the measured mineral composition by least squares approximation. Least squares provides a solution to the problem of linear mapping from a higher order space into a lower order space. For example, in Figure 16, the mineral Y is contained in the three-component system (1-2-3) and we wish to map Y into the two-component "new" system consisting of A-B (A and B might be end-member mineral formulas or perhaps we wish to balance the chemical reaction A + B = Y). As can be seen in Figure 16, mineral Y lies outside of the two-component space defined by A-B; thus, to perform the mapping we must somehow reduce the dimensionality of the space.

The least squares solution is the unique linear combination of the "new" components (A and B in this example) that *most closely approximates* the composition of the mineral that is being mapped. In Figure 16, Y is the mineral being mapped and Y* is the least squares solution. Y* falls along the line that joins A and B at a point that is closest to point Y. Thus, Y* defines the point where the length of the vector between Y and the join A-B is shortest (i.e., the shortest residual vector). The angle A-Y*-Y must be 90°. Some computational aspects of least-squares modeling are discussed in the next section.

The graphical effect of discarding, lumping and least-squares modeling can be seen in Figure 17. In this example, the "model" system consists of components 1 and 2, and the "extra" component is 3. Discarding component 3 from phase A is equivalent to projecting from component 3 and results in point A plotting at A_1 along the join 1-2. Note that this procedure preserves

86

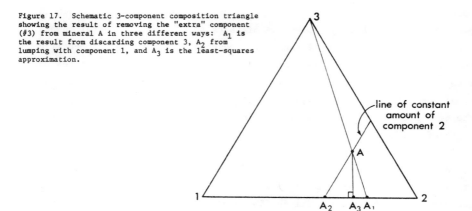

Figure 17. Schematic 3-component composition triangle
showing the result of removing the "extra" component
(#3) from mineral A in three different ways: A_1 is
the result from discarding component 3, A_2 from
lumping with component 1, and A_3 is the least-squares
approximation.

the ratio of components 1:2.   Lumping component 3 with another component
(say, with component 1) does not change the absolute amounts of the other
components; hence, point A_2 is found by moving along a line of constant
amount of component 2 to the join 1-2. Note that this procedure destroys
the original ratio of 1:2. The least-squares approximation is point A_3 and
is found by dropping a normal from point A to the 1-2 join.

    The three methods quite obviously do not result in the same projected
plotting position for point A.   Point A_3 is closest in absolute magnitude to
point A, point A_1 preserves the original ratio of 1:2, but point A_2 may be
desirable for crystal-chemical reasons. Note that as the amount of component 3 in phase A increases, the discrepancy between the points A_1, A_2 and
A_3 increases. Unfortunately, there is no "correct" way to handle extra
components; all three methods result in simplifications that will alter the
appearance of the phase diagram of the sub-system relative to the model
system.

LEAST-SQUARES MODELING IN PETROLOGY

    Least-squares solutions to petrologic problems have had considerable
application in petrology and deserve some discussion. A detailed discussion
of the theory of least-squares modeling is beyond the scope of this paper,
however, and the reader is referred to excellent textbooks on the subject
for further information (e.g., Davis, 1973; Draper and Smith, 1966).

    As outlined in the previous section, the least-squares approach is ap-
plicable in overdetermined systems where a mineral or rock is to be mapped
from a higher order composition space into a lower order space. The least-
squares solution is a "model" composition contained wholly in the lower

87

order space. This "model" composition is defined as the composition in the lower order space that is closest in distance to the actual mineral composition. That is, the length of the vector connecting the "model" composition and the actual composition is minimized. This vector is called the residual vector, R, and is defined as $R = Y* - Y$ where Y* is the model and Y is the actual composition. Formally, the least-squares solution minimizes ΣR_i^2 (the sum of the squares of the residuals), which is also equivalent to the length of R (the dot product of R on itself). Hence, both are minimized by the least-squares solution.

As an illustration of the application of least-squares modeling to petrology, consider the example of Greenwood (1968) (see Fig. 16). The two minerals A and B have compositions (50,50,0) and (0,25,75), respectively, and the mineral Y has a composition (25,50,25). The problem is to balance the equation $X_1A + X_2B = Y$, which is equivalent to performing a linear mapping of Y into the new coordinate system A,B. Clearly, this cannot be done exactly because Y does not fall on the line connecting A and B. The least-squares solution to this problem is the model composition Y*, which can be found graphically by finding the normal to the line A-B that passes through Y.

Mathematically, this problem can be expressed as follows. The mass balance equation to be solved is

$$X_1 \cdot A + X_2 \cdot B = Y + R$$

$$X_1 \cdot 50 + X_2 \cdot 0 = 25 + R_1$$
$$X_1 \cdot 50 + X_2 \cdot 25 = 50 + R_2$$
$$X_1 \cdot 0 + X_2 \cdot 75 = 25 + R_3$$

where R is the residual vector Y* - Y, or in matrix notation,

$$A \cdot X = Y + R .$$

The vector R is required to make this equation exact. The least-squares solution is given by

$$X = (A^T \cdot A)^{-1} A^T \cdot Y$$

where A^T is the transpose of A. For the above example, X = (0.658, 0.368). The model composition, Y*, is then calculated simply as

$$A \cdot X = Y*$$

or Y* = (32.89, 42.11, 27.63) with a residual vector R = (7.89, -7.89, 2.63)

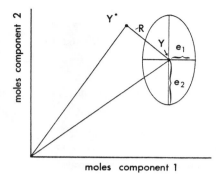

Figure 18. Schematic representation of the least-squares solution (in Cartesian coordinates) showing the model composition, Y^*, the actual mineral composition, Y, the residual vector, R, and the error ellipsoid defined by the error vector, E.

(note that this result is slightly different from that obtained by Greenwood, 1968).

One particular advantage to the least-squares approach is that it permits evaluation of the statistical significance of the solution obtained. This is particularly important when dealing with real mineral compositions. For example, if the mineral Y is analyzed by conventional techniques such as the electron microprobe, the composition of Y will have an error associated with each component, which can be denoted by an error vector, $E_Y = (e_1, e_2, e_3 \ldots e_i)$. Each element in the error vector is the standard deviation of the measurement of the ith element of the vector Y. If the minerals A and B are also measured by chemical analysis, as opposed to being fixed, end-member components, then there is also an error vector associated with each of these vectors, E_A and E_B. Hence, there is also an error associated with the model composition, Y^*. These error vectors can be combined into a single error vector that is the sum of the error on Y and the weighted sum of the error on A and B (that is, weighted by the least-squares solution, X). In other words,

$$E = E_Y + X_1 \cdot E_A + X_2 \cdot E_B .$$

This error vector then represents the total error in measurement associated with Y and Y^*, and it can be compared directly with R. This error vector defines an error ellipsoid around Y (or Y^*) where each element of E is the length of one of the axes of the ellipsoid (see Fig. 18) (note that an alternative approach is to draw two error ellipsoids, one around Y (= E_Y) and one around Y^* (= $X_1 \cdot E_A + X_2 \cdot E_B$), as has been done by Greenwood, 1968). This error ellipsoid can then be compared directly with R to determine whether or not Y^* falls within the error ellipsoid of Y. This can be determined by computing the length of the vector R, which is the distance between Y^* and Y, and comparing it with the size of the error ellipsoid in the direction of R. If this distance is greater than the length of R, then Y* is inside the error ellipsoid; otherwise, it is outside. For example, in Figure 18 the model Y* is outside the error ellipsoid.

The procedure of determining whether or not Y* is within the error ellipsoid permits evaluation of the quality of the least-squares model Y* as a statistically significant model of Y. If Y* is within the ellipsoid, then Y* is a significant model of Y within one standard deviation of the measurement on Y, A, and B. Quite obviously, the quality of Y* as a significant model of Y will depend directly on the errors associated with measurement of Y, A, and B.

The above example is an unweighted, unconstrained least-square problem, but it is also possible to weight the mass balance equations in the least-squares solution or to impose any number of constraints on the solution. Reid et al. (1973) have presented a generalized model for constrained least-squares problems. They stress the importance of careful examination of the physical meaning of the mixing model and state that meaningful models can only be obtained by properly weighting the equations, by constraining the solution, and by analyzing the uncertainties in the values of the parameters.

As pointed out by Reid et al., a rigorous solution to the weighted least-squares problem requires knowledge of the variance-covariance matrix. The variance-covariance matrix, C, is a matrix that contains the variance (standard deviations) of the measurements of each component as diagonal elements of the matrix and the covariance of the ith measurement with the jth measurement as the off diagonal elements. For example,

$$
C = \begin{vmatrix}
\sigma_1^2 & \sigma_{12}^2 & \sigma_{13}^2 & \sigma_{14}^2 & \cdots & \sigma_{1n}^2 \\
\sigma_{21}^2 & \sigma_2^2 & \cdots\cdots\cdots\cdots \\
\cdot & \cdot & & & & \cdot \\
\cdot & \cdot & & & & \cdot \\
\cdot & \cdot & & & & \cdot \\
\sigma_{n1}^2 & \sigma_{n2}^2 & & & & \sigma_n^2
\end{vmatrix}
$$

The solution to the weighted least-squares problem when the variance-covariance matrix is known is

$$
X = (A^T \cdot C^{-1} \cdot A)^{-1} \cdot A^T \cdot C^{-1} \cdot Y .
$$

Note that if the variance-covariance matrix equals the identity matrix, the weighted least-squares solution reduces to the unweighted solution.

Knowledge of the covariance of measurements used in petrologic mixing problems is usually not available. If one is dealing with mineral composi- tions determined from electron microprobe analysis, it is usually assumed that the measurement of one element is unaffected by the amounts of other elements, but this is clearly not true because absorption and fluorescence corrections (e.g., Bence and Albee, 1968) are dependent to a first order on the concentration of all other elements, as is the normalization of formula units. Moreover, in real mineral systems, elemental abundances are con- strained by the crystal chemistry of the phase. For example, in plagioclase, as Na and Si increase, Ca and Al must decrease. To our knowledge, however, a rigorous analysis of covariance for mineral analyses has never been pre- sented.

Reid et al. (1975) suggest that, for analytical techniques such as electron microprobe analysis, the covariance is probably small and the variance of the measurements can be used as weighting factors. In this case, the variance-covariance matrix becomes a diagonal matrix (off diag- onal elements = 0) with the diagonal elements equal to the square of the standard deviation of the measurements.

Reid et al. stress the importance of proper weighting. If no weight- ing is used, equal absolute errors are assumed and the solution is dominated by the mass balance equation for the most abundant elements. Weighting has the effect of increasing the importance of minor components in the regres- sion.

One additional aspect of weighted least squares needs to be discussed. As pointed out earlier, in most petrologic mixing problems there are not only errors in the mineral composition to be modelled (Y) but also in the mineral compositions in the coefficients matrix (A). These errors should also be considered in the weighting procedure and an algorithm for doing so was suggested by Reid et al. and Albarede and Provost (1977). The algorithm involves an iterative solution whereby in each iteration the weighting fac- tors are recomputed as the error on Y plus the sum of the errors on the coef- ficients matrix, weighted by the previous least-squares solution. Pigage (1982) has employed this algorithm in calculations and states that it gives results similar to the standard weighted least-squares technique.

It is often necessary to impose constraints on least-squares solutions in order that the solution have physical meaning, as discussed by Reid et al., Albarede and Provost, and LeMaitre (1979). LeMaitre, for example, has dis- cussed the advisibility of imposing the constraint $\Sigma X_i = 1.0$ on any

least-squares solution involving mineral or rock compositions. For further
discussions of constrained least-squares problems, the reader is referred
to these papers.

Application of least-squares modeling to petrology

Least-squares modeling in metamorphic petrology has had principal application in solving mass balance equations with specific application to
problems such as balancing chemical reactions, determining compositional
overlap between mineral assemblages (the n-dimensional tie-line problem),
and modal analysis. Other applications include derivation of thermochemical
data from reversed experimental studies.

Balancing chemical reactions and the n-dimensional tie-line problem.
Greenwood (1967,1968) discussed the problem of whether or not tie lines
cross in n-dimensional space and presented an algorithm to solve this based
on least-squares modeling (Greenwood, 1968; Greenwood, 1967, used a linear
programming technique to solve the n-dimensional tie-line problem). The
essence of Greenwood's argument is that if tie lines between two assemblages
cross (or n-dimensional phase volumes intersect) then there will be a reaction relationship between the two assemblages of the form

$$X_1A_1 + X_2A_2 + X_3A_3 = Y_1B_1 + Y_2B_2 + Y_3B_3$$

where the minerals A_1, A_2, A_3 and B_1, B_2, B_3 belong to the two different
assemblages. This equation can be expressed as

$$B_1 = Z_1A_1 + Z_2A_2 + Z_3A_3 + Z_4B_2 + Z_5B_3$$

where coefficients Z_1, Z_2 and Z_3 are positive and Z_4 and X_5 are negative.
This equation is a system of mass balance equations of the form $A \cdot X = Y$ and
is directly amenable to solution by least-squares techniques. The model
composition $Y*$ can be compared directly with Y to determine if the difference
(i.e., R) is within the errors estimated from electron microprobe analysis.

This approach is, of course, not simply restricted to the n-dimensional
tie line problem, but is applicable to the modeling of any type of metamorphic reaction. Greenwood (1968) tested his approach through an analysis of
the data of Engel and Engel (1962a,b) pertaining to the amphibolite to granulite facies transition in the Northwest Adirondacks of New York State.
Greenwood found that it was impossible to relate the mineral assemblages of
the Emmeryville Area (amphibolite facies) to those of the Colton Area (granulite facies) by simple mass balance equations. The difference between the
mineral assemblages in hornblende amphibolite and in pyroxene granulite

must be due, at least in part, to changes in bulk rock composition, as well as changes in P and T. In other words, the metamorphism cannot be considered to be isochemical, at least for the samples analyzed by Greenwood.

Fletcher (1976) examined possible reaction relations in a finely banded, two-pyroxene amphibolite from southeastern Ontario. All of the samples studied by Fletcher were from a very small outcrop where T and P could be assumed to have been constant. The effects of bulk composition on mineral assemblage therefore could be analyzed separately from the effects of P and T. No reaction relations could be found between many of the samples, and the mineralogical differences were attributed to differences in bulk compositions. Between several of the assemblages, however, it was found that reaction relations did, in fact, exist, which could only be explained by differences in the H_2O-content between assemblages. It was therefore concluded that several of the amphibole-two pyroxene assemblages acted as H_2O-buffers during metamorphism.

Pigage (1976), in a study of pelitic schists from southern British Columbia, was able to document through his regression analysis that the assemblage staurolite + kyanite + garnet + biotite + muscovite + quartz + plagioclase + ilmenite + rutile, which appears as a four-phase assemblage on an AFM projection, acted as an H_2O-buffer. This assemblage could have existed over the P-T interval observed in the field only if $P(H_2O) <$ Ptotal with the composition of the fluid buffered by the reaction. Pigage was also able to model prograde reactions responsible for the production of kyanite, the first appearance of sillimanite, and the disappearance of staurolite. In a separate study, Pigage (1982) used the linear regression technique to model probable sillimanite-forming reactions in pelitic schist from the Shuswap complex, British Columbia. The regression analysis, which is supported by textural criteria, indicates that fibrolite formed initially by the breakdown of staurolite and/or garnet with rutile as a reactant phase. When rutile was exhausted as a reactant phase, however, garnet became a product phase and continued growth of fibrolite was accompanied by the growth of a second generation of garnets. This is an excellent example of where regression analysis has aided in the interpretation of the prograde development of metamorphic minerals; in this case, the sequence was first generation garnet, which was resorbed and then followed by growth of second generation garnets.

Fletcher and Greenwood (1979) applied regression analysis to progressively metamorphosed pelitic schists from the Penfold Creek Area,

93

British Columbia. Their approach was to use regression techniques to detect reaction relations within and between mineral assemblages. Only three assemblages analyzed by Fletcher and Greenwood contained evidence of linear dependence, and thus the capacity of buffering H_2O, whereas the remainder of the samples studied did not have the capability of buffering H_2O. Their results contrast with the findings of Pigage (1976) who found evidence that pelitic schists commonly have a capability to buffer $\mu(H_2O)$. In contrast with Pigage (1976), but in accord with Pigage (1982), Fletcher and Greenwood found rutile to be a reactant phase during growth of Al_2SiO_5 in the progressive metamorphism of pelitic schist.

Laird (1980) used regression techniques in a somewhat different way to model whole-rock reactions during the progressive metamorphism of mafic schist from Vermont. Measured mineral compositions in the "common" assemblage amphibole + chlorite + epidote + plagioclase + quartz + Ti phase $\pm$ Fe^{3+} oxide + carbonate $\pm$ K-mica from different metamorphic grades were used to calculate the weight percent of minerals in two "average" bulk compositions. The results of these calculations were plots of calculated weight percent of minerals against metamorphic grade, and they graphically depict the change in modal mineralogy with metamorphic grade. From these calculations, generalized continuous "whole-rock" reactions between different metamorphic zones were deduced. For example, between the garnet-albite and garnet-oligoclase zones the whole-rock reaction was

$$amphibole_1 + chlorite_1 + epidote + albite \rightarrow$$
$$amphibole_2 + chlorite_2 + quartz + H_2O \ .$$

Whole-rock reactions were determined for low-, medium-, and high-pressure P-T paths. Thompson et al. (1982; also see Thompson, this volume) have factored these reactions into component parts.

In summary, least-squares modeling of metamorphic mineral assemblages has been used in two major ways. (1) The evaluation of linear dependencies *within* mineral assemblages has been used to confirm or disprove the validity of graphical projections, to determine the true variance of a mineral assemblage, and to evaluate the potential of an assemblage to buffer the chemical potential of volatile species. (2) Examination of linear dependencies *between* assemblages has led to determination of whether mineral assemblages overlap in composition space (the n-dimensional tie-line problem). This information is a necessary prerequisite to determining whether two mineral assemblages at different metamorphic grade can be related by an isochemical prograde mineral reaction, or whether essential differences in bulk composition exist between the two assemblages. With this information as a starting point, linear regression analysis has been used to deduce overall prograde metamorphic reactions.

Linear regression analysis is an extremely powerful tool and is ideally suited for petrologic problems where measured mineral compositions are involved. However, there are pitfalls. For example, in order to deduce a whole-rock reaction, the compositions of the phases participating in the reaction must be specified. In the above-mentioned studies, rim compositions were always used, but these may not necessarily be the composition of the phases that participated in the reaction. Moreover, it is not always clear that a particular phase should even be included in a reaction. Consequently, careful textural analysis and mineral zoning studies are indispensible to ensure that the reaction which evolves from the regression analysis is consistent with observed sequences of mineral growth.

Pigage (1976,1982) has also emphasized the difficulty in evaluating the significance of a reaction derived from regression analysis, especially where minor components are concerned. Regression errors (residuals) for minor components such as Zn in staurolite or Mn or Ca in garnet are often within the analytical error of the measurements. Because of this, Pigage concluded that a particular assemblage was univariant, and was a potential buffer of $\mu(H_2O)$, at least within analytical error of measurement. In this application, regression becomes a problem of determining the true variance of a mineral assemblage. As Pigage pointed out, the implication of the regression analysis is that if more precise analyses were possible (i.e., the error ellipsoid were smaller), then the assemblage might be of higher variance. Thus we have the concept, which follows directly from the statistical analysis of the quality of the regression, that the thermodynamic variance of a mineral assemblage is dependent on the analytical precision of the measured mineral compositions.

CONCLUSIONS

Linear algebra is an extremely powerful tool that can be applied to problems of n-dimensional analysis of composition space. Linear algebra has none of the limitations of graphical analysis, but it is not always easy to see how to formulate a problem algebraically. The goal of this paper has been to provide some insight into how petrologic problems can be formulated. The one most significant conclusion is that almost all petrologic applications of linear algebra can be described as linear mapping of a mineral composition vector from one set of coordinate axes into another. The mathematics necessary to solve diverse problems such as component transformations, balancing chemical reactions and petrologic mixing problems are, therefore, fundamentally identical. In the future, new applications of linear algebra to petrology beyond those discussed here should further enhance our insights into the complexities of the composition space of metamorphic mineral assemblages.

ACKNOWLEDGMENTS

The authors would like to acknowledge the helpful reviews of J. Selverstone, K. Kimball, J. Rice and H. Lang as well as the able typing of D. Frank and drafting of D. Hall. Partial support for this work was funded by National Science Foundation Grant EAR-8108617 (Spear) and a Joseph H. Defrees grant of the Research Corporation (Spear).

CHAPTER 3 REFERENCES

Aitken, A.C. (1956) *Determinants and Matricies.* 9th ed., Oliver and Boyd, Ltd., Edinburgh.

Albarede, F. and A. Provost (1977) Petrological and geochemical mass-balance equations: an algorithm for least-square fitting and general error analysis. Computers and Geosciences 3, 309–326.

Albee, A.L. (1965a) Phase equilibria in three assemblages of kyanite-zone pelitic schists, Lincoln Mountain Quadrangle, Central Vermont. J. Petrol. 6, 246–301.

———— (1965b) A petrogenetic grid for the Fe-Mg silicates of pelitic schists. Am. J. Sci. 263, 512–536.

Anton, H. (1973) *Elementary Linear Algebra.* John Wiley & Sons, Inc., New York.

Banks, R. (1979) The use of linear programming in the analysis of petrological mixing problems. Contrib. Mineral. Petrol. 70, 237–244.

Barker, F. (1961) Phase relations in cordierite-garnet-bearing Kinsman quartz monzonite and the enclosing schist, Lovewell Mountain Quadrangle, New Hampshire. Am. Mineral. 46, 1166–1176.

Bence, A.E. and Albee, A.L. (1968) Empirical correction factors for the electron microanalysis of silicates and oxides. J. Geol. 76, 382–403.

Bowen, N.L. (1940) Progressive metamorphism of siliceous limestone and dolomite. J. Geol. 48, 225–274.

Brady, J.B. (1975) Reference frames and diffusion coefficients. Am. J. Sci. 275, 954–983.

———— and J.H. Stout (1980) Normalizations of thermodynamic properties and some implications for graphical and analytical problems in petrology. Am. J. Sci. 280, 173–189.

Bryan, W.B., L.W. Finger, and F. Chayes (1969) Estimating proportions in petrographic mixing equations by least-squares approximation. Science 163, 926–927.

Davis, J.C. (1973) *Statistics and Data Analysis in Geology.* John Wiley & Sons, Inc., New York.

Draper, N. and H. Smith (1966) *Applied Regression Analysis.* John Wiley & Sons, Inc., New York.

Engel, A.E.J. and G.G. Engel (1962a) Hornblendes formed during progressive metamorphism of amphibolites, Northwest Adirondack Mountains, New York. Bull. Geol. Soc. Am. 73, 1499–1514.

————, ————, and R.G. Havens (1964) Mineralogy of amphibolite interlayers in the Gneiss Complex, Northwest Adirondack Mountains, New York. J. Geol. 72, 131–156.

Eskola, P. (1915) On the relations between the chemical and mineralogical composition in the metamorphic rocks of the Orijarvi region. Bull. Comm. Geol. Finlande 44.

Ferry, J.M. (1979) A map of chemical potential differences within an outcrop. Am. Mineral. 64, 966–985.

Fletcher, C.J.N. (1971) Local equilibrium in a two-pyroxene amphibolite. Canadian J. Earth Sci. 8, 1065–1080.

———— and H.J. Greenwood (1979) Metamorphism and structure of Penfold Creek area, near Quesnel Lake, British Columbia. J. Petrol. 20, 743–794.

Gray, N.H. (1973) Estimation of parameters in petrologic materials balance equations. J. Math. Geol. 5, 225–236.

Greenwood, H.J. (1967) The n-dimensional tie-line problem. Geochim. Cosmochim. Acta 31, 465–490.

———— (1968) Matrix methods and the phase rule in petrology. XXII Internat. Geol. Cong. 6, 267–279.

———— (1975) Thermodynamically valid projections of extensive phase relationships. Am. Mineral. 60, 1–8.

Harte, B. (1975) Determination of a pelitic petrogenetic grid for the eastern Scottish Dalradian. Carnegie Inst. Wash. Year Book 74, 438–446.

———— and C.M. Graham (1975) The graphical analysis of greenschist to amphibolite facies mineral assemblages in metabasites. J. Petrol. 16, 347–370.

Hess, P.C. (1969) The metamorphic paragenesis of cordierite in pelitic rocks. Contrib. Mineral. Petrol. 24, 191–207.

Korzhinskii, D.S. (1959) *Physiochemical Basis of Analysis of the Paragenesis of Minerals.* Consultants Bureau, Inc., New York.

Laird, J. (1980) Phase equilibria in mafic schist from Vermont. J. Petrol. 21, 1–37.

97

LeMaitre, R.W. (1979) A new generalized petrological mixing model. Contrib. Mineral. Petrol. 71, 133–137.

Palatnik, L.S. and A.I. Landau (1964) *Phase Equilibria in Multicomponent Systems*. Holt, Rinehart and Winston.

Perry, K.L., Jr. (1967a) An application of linear algebra to petrologic problems: Part 1. Mineral classification. Geochim. Cosmochim. Acta 31, 1043–1078.

_____ (1967b) Methods of petrologic calculation and the relationship between mineral and bulk chemical composition. Contrib. Geol. 6, 5–38.

_____ (1968a) Representation of mineral chemical analyses in 11-dimensional space. Part 1, Feldspars. Lithos 1, 201–218.

_____ (1968b) Representation of mineral chemical analyses in 11-dimensional space. Part 2, Amphiboles. Lithos 1, 307–321.

Pigage, L.C. (1976) Metamorphism of the Settler Schist, southwest of Yale, British Columbia. Canadian J. Earth Sci. 3, 405–421.

_____ (1982) Linear regression analysis of sillimanite-forming reactions. Canadian Mineral., in press.

Reid, M.J., A.J. Gancarz, and A.L. Albee (1973) Constrained least-squares analysis of petrologic problems with an application to lunar sample 12040. Earth Planet. Sci. Lett. 17, 443–445.

Robinson, P. and H. Jaffe (1969) Chemographic exploration of amphibole assemblages from central Massachusetts and southwestern New Hampshire. Spec. Pap. Mineral. Soc. Am. 2, 251–274.

_____, F.S. Spear, J.C. Schumacher, J. Laird, C. Klein, B.W. Evans, and B.L. Doolan (1982) Phase relations of metamorphic amphiboles: natural occurrence and theory. *In* D.R. Veblen, Ed., *Reviews in Mineralogy 9B*, Mineralogical Society of America, Washington, D.C.

Rumble, D., III (1977) Mineralogy, petrology and oxygen isotopic geochemistry of the Clough Formation, Black Mountain, Western New Hampshire, U.S.A. J. Petrol. 19, 317–340.

Skippen, G.B. (1974) An experimental model for low pressure metamorphism of siliceous dolomitic marble. Am. J. Sci. 274, 487–509.

Spear, F.S. (1982) Phase equilibria of amphibolites from the Post Pond Volcanics, Mt. Cube Quadrangle, Vermont. J. Petrol. 23, in press.

Stout, J. (1972) Phase petrology and mineral chemistry of coexisting amphiboles from Telemark, Norway. J. Petrol. 13, 99–145.

Strang, G. (1980) *Linear Algebra and Its Applications*, 2nd ed. Academic Press, New York.

Thompson, A. B. (1975) Mineral reactions in a calc-mica schist from Gassetts, Vermont, U.S.A. Contrib. Mineral. Petrol. 53, 105–127.

Thompson, J.B., Jr. (1957) The graphical analysis of mineral assemblages in pelitic schists. Am. Mineral. 42, 842–858.

_____ and A.B. Thompson (1976) A model system for mineral facies in pelitic schists. Contrib. Mineral. Petrol. 58, 243–277.

Wright, T.L. and P.C. Doherty (1970) A linear programming and least squares computer method for solving petrologic mixing problems. Bull. Geol. Soc. Am. 81, 1995–2008.

APPENDIX A

PROGRAM TETPLT: A FORTRAN Program to Plot Stereoscopic Phase Diagrams

Visualization of metamorphic mineral assemblages is greatly aided by graphical projections. Three-component phase diagrams are simple to draw, but accurate drawings of four-component tetrahedral phase diagrams are difficult without computer assistance. Spear (1980) presented an algorithm for converting four-component barycentric coordinates into spatial Cartesian coordinates (x-y-z) so that the tetrahedron may be plotted in stereoscopic perspective.

The FORTRAN program TETPLT contained in this appendix performs these calculations and generates stereoscopic drawings of tetrahedral phase diagrams from any perspective desired. The program will also plot triangular phase diagrams. The program is written to run under Dec's RT-11 operating system, but should be easily convertible to any operating system. It is an interactive program and will have to be modified to run under batch. The plotting subroutines are designed for a Hewlett-Packard HP 7221B 4-pen plotter, and are largely self-explanatory. The only information necessary to know about the plotting subroutines is that the scaling of the plot assumes an x-y plotting pattern of 15,000-10,000 (0.025 units/mm).

The only input necessary to run the program is a file containing the four component barycentric plotting coordinates of the minerals to be plotted. If a triangular plot is desired, then the fourth barycentric coordinate should be set equal to 0. The file must be structured like so:

```
TITLE OF PLOT (80 characters)
LABEL COMPONENT A
LABEL COMPONENT B
LABEL COMPONENT C
LABEL COMPONENT D
TITLE MINERAL 1
PLOTTING COORDINATES MINERAL 1
TITLE MINERAL 2
PLOTTING COORDINATES MINERAL 2
etc.
```

Running the program. Most commands are self-explanatory. However, the relationship between the input angles and the stereoscopic view that is drawn needs discussion. When the command to draw the tetrahedron is specified, the program asks

INPUT ALPHA, THETA, GAMMA, EYE DISTANCE and SCALE

The angles ALPHA and THETA control the rotation of the tetrahedron as shown in Figure A1 and can be any number ±360°. GAMMA determines the amount of stereo shift and EYE DISTANCE determines the amount of foreshortening. Both these numbers should be about the same magnitude (e.g., 5 and 5 or 20 and 20) but can be adjusted to suit individual preference. SCALE is the parameter that determines the size of the tetrahedron. It corresponds to unit distance in plotter units, usually 4000 or 5000 on a plotter that has x-y limits of 15,000-10,000. If in doubt about what to type in, try ... 0, 0, 5, 5, 4000.

Figure A1. Relationship between the 4 barycentric coordinates A, B, C, and D and the 3 Cartesian coordinates X, Y, and Z used in the program TETPLT. Shown in the original orientation of the tetrahedron. The angle α is a ± rotation around the Z axis and the angle θ is a ± rotation around the X axis. γ, which determines the amount of stereo shift, is a ± rotation around Y.

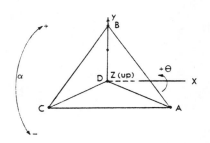

```
0001  C     PROGRAM TETPLT
      C     CODED BY F. SPEAR
      C     UPDATE 2-FEB-82 TO INCLUDE TRIANGLE PLOTS
      C     UPDATE MAY-82 FOR MSA SHORTCOURSE
      C
      C     ---DESCRIPTION OF PLOTTER ROUTINES--
      C     THE PLOTTER PLATTEN IS DIVIDED INTO A 15000X10000 GRID IN THE
      C     X AND Y DIRECTIONS. EACH PLOTTER UNIT = 0.025 MM.
      C
      C     THE PLOTTER SUBROUTINES ARE AS FOLLOWS:
      C
      C     CALL PLOTS       -- INITIALIZE PLOTTER
      C     CALL POFF        -- TURN PLOTTER OFF
      C     CALL PON         -- TURN PLOTTER ON
      C     CALL MOVE(X,Y,N)  -- MOVES TO X AND Y IN PLOTTER UNITS.
      C                         N=0 MOVES WITH PEN UP
      C                         N=1 MOVES WITH PEN DOWN
      C     CALL SYMB(N,W,H)  -- DRAWS A CENTERED SYMBOL OF WIDTH W AND HEIGHT H
      C                         (W AND H ARE IN PLOTTER UNITS)
      C                         N IS THE SYMBOL # (FROM 1-12)
      C     CALL LABSIZ(W,H)  -- SETS WIDTH AND HEIGHT OF LABEL IN PLOTTER UNITS
      C     CALL NEWPEN(N)    -- PICKS ONE OF 4 COLORED PENS (N=1 TO 4)
      C     CALL PENUP        -- LIFTS PEN
      C
      C     STATEMENTS SUCH AS
      C  100 WRITE(NPOUT,100) 126,39,.............3
      C  100 FORMAT(40A1)
      C     WRITE A LABEL ONTO THE PLOT. ASCII CODE 126,39 TURNS THE LABEL
      C     MODE ON AND 3 TURNS LABEL MODE OFF. EVERYTHING INBETWEEN IS A LABEL.
      C
      C     THESE SUBROUTINES ARE NOT HP PLOT/21 SUBROUTINES BUT WERE CODED TO
      C     RUN AN HP-7221B PLOTTER ON A SMALL COMPUTER. THESE DESCRIPTIONS
      C     ARE GIVEN TO FACILITATE ADAPTATION OF THE PROGRAM TO ANOTHER PLOTTING
      C     SYSTEM. IF DESIRED, A COPY OF THE PLOTTING SUBROUTINES CAN BE
      C     OBTAINED BY WRITING THE AUTHOR.
      C
0002        DIMENSION FLNAME(4)
0003        DIMENSION MINNO(12),PTITLE(40),APX(4,8),C1(4,4),LAB(8)
0004        DIMENSION BARY(4),TITLE(8,300)
0005        DIMENSION C(4,300),XLA(10),XRA(10),YLA(10),YRA(10)
0006        COMMON NPOUT
0007        NPOUT=5
      C     NPOUT IS THE PLOTTER OUTPUT DEVICE NUMBER
      C     CALL PLOTS INITIALIZES THE PLOTTER
0008        CALL PLOTS
0009        CALL POFF
      C     SET DEFAULT STEREO VALUES
0010        ISHIFT=50
0011        ALPHA=0.
0012        THETA=0.
      C     ALL ANGLES ARE IN RADIANS.  57.29578 CONVERTS DEGS. TO RADS.
0013        GAMMA=5./57.29578
0014        EYE=5.
0015        SCALE=5000.
      C     SET BARYCENTRIS COORDINATES FOR APICIES OF TETRAHEDRON
0016        DO 20 LJ=1,4
0017        DO 20 I=1,4
0018        C1(I,LJ)=0.
0019   20   C1(LJ,LJ)=1.
0020   22   CONTINUE
      C     LJ COUNTS THE # OF DATA POINTS IN THE DATA FILE
0021        LJ=1
0022        TYPE *,'INPUT FILE NAME FOR DATA'
0023        ACCEPT 92,FLNAME
0024   92   FORMAT(4A4)
0025        NFILE=1
0026        CALL ASSIGN (NFILE,FLNAME)
      C     READ TITLE OF PLOT
0027 2564   READ (NFILE,1012)(PTITLE(I),I=1,40)
0028 1012   FORMAT(40A1)
      C     READ LABELS OF 4 APICIES
0029    6   DO 6 I=1,4
0030        READ(NFILE,1012)(APX(I,J),J=1,8)
0031        TYPE 2012,(PTITLE(I),I=1,40)
0032 2012   FORMAT(' ',40A1)
0033        TYPE 2013,((APX(I,J),J=1,8),I=1,4)
0034 2013   FORMAT(' ',4(8A1,1X))
0035  100   READ(NFILE,1002,END=99)(TITLE(I,LJ),I=1,8)
0036 1002   FORMAT(20A4)
0037        READ(NFILE,*)(C(I,LJ),I=1,4)
      C     RENORMALIZE BARYCENTRIC COORDINATES TO 1.0
0038        SUM=0
0039   33   DO 10 I=1,4
0040   10   SUM=SUM+C(I,LJ)
0041        IF(SUM.EQ.0.)GO TO 16
0042        DO 15 I=1,4
0043   15   C(I,LJ)=C(I,LJ)/SUM
0044   16   CONTINUE
0045        TYPE 2015,(C(I,LJ),I=1,4),LJ,(TITLE(I,LJ),I=1,8)
0046 2015   FORMAT(4F9.3,I5,5X20A4)
0047        LJ=LJ+1
0048        GO TO 100
0049   99   CONTINUE
0050        LJ=LJ-1
      C     GENERATE HARD COPY OF DATA FILE IF DESIRED
0051        NPR=0
0052        TYPE *,' PRINT DATA??? 1=YES, 0=NO'
0053        ACCEPT *,NPR
0054        IF (NPR.EQ.0) GO TO 97
0055        PRINT 2012,(PTITLE(I),I=1,40)
0056        PRINT 2013,((APX(I,J),J=1,8),I=1,4)
0057        DO 96 I=1,LJ
0058        PRINT 2009,(C(J,I),J=1,4),I,(TITLE(J,I),J=1,8)
0059 2009   FORMAT(' ',4F8.3,I8,5X,8A4)
0060   96   CONTINUE
```

```
0063  97    CONTINUE
0064        CLOSE(UNIT=NFILE)
0065   1    CONTINUE
0066        TYPE *,' TYPE 0 TO EXIT, 1 FOR TETRAHEDRAL PLOT,'
0067        TYPE *,' 2 FOR TRIANGULAR PLOT, 3 TO INPUT NEW DATA FILE'
0068        ACCEPT *,ISW
0069        IF (ISW.EQ.0)GO TO 999
0071        GO TO (3100,5000,22)ISW
0072        GO TO 1
      C
      C     THIS SECTION IS FOR TETRAHEDRAL PLOTS
      C
0073 3100   CONTINUE
0074        TYPE *,' TETRAHEDRON PLOTTING ROUTINE'
0075        TYPE *,' YOUR OPTIONS ARE: 0=EXIT, 1=INPUT ANGLES,'
            *,' 2=PLOT TETRAHEDRON'
0076        TYPE *,' 3=PLOT ASSEMBLAGES, 4=PLOT POINTS'
0077        TYPE *,' INPUT OPTION...'
0078        ACCEPT *,IOPT
0079        IF (IOPT.EQ.0) GO TO 1
0081        GO TO (3300,3400,3500,3600)IOPT
0082        GO TO 3100
      C
      C     ROUTINE TO INPUT ANGLES
      C
0083 3300   CONTINUE
0084        TYPE *,' INPUT ALPHA,THETA,GAMMA,EYE DISTANCE AND SCALE '
0085        TYPE *,' (IF IN DOUBT, TYPE 0,0.5,4000)'
0086        ACCEPT *,ALPHA,THETA,GAMMA,EYE,SCALE
0087        ALPHA=ALPHA/57.29578
0088        THETA=THETA/57.29578
0089        GAMMA=GAMMA/57.29578
0090        GO TO 3100
      C
      C     ROUTINE TO PLOT TETRAHEDRON
      C
0091 3400   CONTINUE
0092        DO 60 I=1,4
0093        DO 62 K=1,4
0094   62   BARY(K)=C1(I,K)
0095        CALL PROJEC(BARY,ALPHA,THETA,GAMMA,EYE,XL,YL,XR,YR)
0096        XLA(I)=XL
0097        YLA(I)=YL
0098        XRA(I)=XR
0099        YRA(I)=YR
0100   60   CONTINUE
0101        NOPEN=1
0102        GO TO 6000
0104 3410   CONTINUE
0105        CALL PON
0106        CALL LABSIZ(100.,200.)
0107        DO 3412 I=1,4
0108        X=XLA(I)*SCALE+3750+ISHIFT

0109        Y=YLA(I)*SCALE+5000
0110        CALL MOVE(X,Y,1)
0111        WRITE(NPOUT,333)126,39,(APX(I,J),J=1,8),3
0112  333   FORMAT(' ',60A1,$)
0113 3412   CONTINUE
0114        DO 3415 I=1,4
0115        X=XRA(I)*SCALE+11250+ISHIFT
0116        Y=YRA(I)*SCALE+5000
0117        CALL MOVE(X,Y,1)
0118        WRITE(NPOUT,333)126,39,(APX(I,J),J=1,8),3
0119 3415   CONTINUE
0120        CALL MOVE(2000.,9000.,1)
0121        WRITE(NPOUT,333)126,39,(PTITLE(J),J=1,40),3
0122        RD=57.29578
0123        WRITE(NPOUT,334)126,39,32,32,32,32,32,32,32,32,
            1  ALPHA*RD,THETA*RD,GAMMA*RD,EYE,SCALE,3
0124  334   FORMAT(' ',11A1,5F8.1,A1)
0125        CALL POFF
0126        GO TO 3100
      C
      C     ROUTINE TO PLOT ASSEMBLAGES CONNECTED BY TIE-LINES
      C
0127 3500   CONTINUE
0128        TYPE *,' INPUT MINERAL NUMBERS TO PLOT'
0129        TYPE *,' FOLLOW EACH WITH A RETURN. END WITH 0'
0130        NM=1
0131   11   ACCEPT *,MINNO(NM)
0132        IF (MINNO(NM).EQ.0)GO TO 200
0134        NM=NM+1
0135        GO TO 11
0136  200   CONTINUE
0137        NM=NM-1
0138        IF (NM.EQ.0)GO TO 3100
0140        J=MINNO(I)
0141        DO 24 I=1,NM
0142        J=MINNO(I)
0143        DO 25 K=1,4
0144   25   BARY(K)=C(K,J)
0145        CALL PROJEC(BARY,ALPHA,THETA,GAMMA,EYE,XL,YL,XR,YR)
0146        XLA(I)=XL
0147        YLA(I)=YL
0148        XRA(I)=XR
0149        YRA(I)=YR
0150   24   CONTINUE
0151        TYPE *,' INPUT PEN #'
0152        ACCEPT *,NOPEN
0153        GO TO 6000
      C
      C     ROUTINE TO PLOT INDIVIDUAL POINTS WITH A LABEL
      C
0154 3600   CONTINUE
0155        DO 3610 I=1,8
0156        LAB(I)=32
0157 3610   TYPE *,' INPUT MINERAL # TO PLOT, 0 TO EXIT'
            ACCEPT *,K
```

```
0158        IF(K.EQ.0)GO TO 3100
0160        TYPE *, ' INPUT LABEL ',
0161        ACCEPT 1012,(LAB(J),J=1,8)
0162        DO 3612 I=1,4
0163   3612 BARY(I)=C(I,K)
0164        CALL PROJEC(BARY,ALPHA,THETA,GAMMA-EYE,XL,YL,XR,YR)
0165        X=XL*SCALE+3750
0166        Y=YL*SCALE+5000.
0167        CALL PON
0168        CALL MOVE(X,Y,1)
0169        CALL SYMB(2,35.,35.)
0170        X=X+ISHIFT
0171        CALL MOVE(X,Y,1)
0172        CALL LABSIZ(75.,150.)
     C THIS WRITE STATEMENT WRITES A LABEL ONTO THE PLOT
0173        WRITE(NPOUT,333)126,39,(LAB(I),I=1,8),3
0174        X=XR*SCALE+11250
0175        Y=YR*SCALE+5000
0176        CALL MOVE(X,Y,1)
0177        CALL SYMB(2,35.,35.)
0178        X=X+ISHIFT
0179        CALL MOVE(X,Y,1)
0180        CALL LABSIZ(75.,150.)
0181        WRITE(NPOUT,333)126,39,(LAB(I),I=1,8),3
0182        CALL POFF
0183        GO TO 3600
     C
     C THIS SECTION PLOTS LEFT AND RIGHT STEREO VIEWS FROM COORDINATES
     C CONTAINED IN THE ARRAYS XLA AND XRA.  THERE ARE NM POINTS AND ALL
     C POINTS ARE CONNECTED BY TIE-LINES
     C
0184   6000 CONTINUE
0185        CALL PON
0186        CALL NEWPEN(NOPEN)
     C PLOT LEFT STEREO VIEW
0187        NL=NM-1
0188        DO 30 I=1,NL
0189        L=I+1
0190        DO 30 K=L,NM
0191        X=XLA(I)*SCALE+3750
0192        Y=YLA(I)*SCALE+5000
0193        CALL MOVE(X,Y,1)
0194        X=XLA(K)*SCALE+3750
0195        Y=YLA(K)*SCALE+5000
0196     30 CALL MOVE(X,Y,0)
     C PLOT RIGHT VIEW
0197        DO 40 I=1,NL
0198        L=I+1
0199        DO 40 K=L,NM
0200        X=XRA(I)*SCALE+11250
0201        Y=YRA(I)*SCALE+5000
0202        CALL MOVE(X,Y,1)
0203        X=XRA(K)*SCALE+11250
0204        Y=YRA(K)*SCALE+5000

0205     40 CALL MOVE(X,Y,0)
0206        CALL PENUP
0207        CALL POFF
0208        GO TO (3100,3410,3500,3100)IOPT
0209        GO TO 3100
     C
     C SECTION FOR TRIANGLE PLOTS
     C
0210   5000 CONTINUE
0211        TYPE *,' TRIANGLE PLOT ROUTINE'
0212        TYPE *,'YOUR CHOICES ARE: 0=EXIT, 1=PLOT TRIANGLE; 2=PLOT DATA'
0213        ACCEPT *,ITRI
0214        IF (ITRI.EQ.0) GO TO 1
0216        GO TO (5100,5200)ITRI
0217        GO TO 5000
     C
     C ROUTINE TO PLOT A TRIANGLE
     C
0218   5100 CONTINUE
0219        TYPE *,'INPUT XOR,YOR AND SCALE FACTOR (e.g. 4000,4000,5000)'
0220        ACCEPT *,XOR,YOR,SCALE
0221        XLA(1)=1.0
0222        YLA(1)=0
0223        XLA(2)=-.5
0224        YLA(2)=.866025
0225        XLA(3)=0
0226        YLA(3)=0
0227        CALL PON
0228        CALL NEWPEN(1)
0229        X=1.0*SCALE+XOR
0230        Y=0.0*SCALE+YOR
0231        CALL MOVE(X,Y,1)
0232        X=0.5*SCALE+XOR
0233        Y=0.866025*SCALE+YOR
0234        CALL MOVE(X,Y,0)
0235        CALL MOVE(XOR,YOR,0)
0236        X=1.0*SCALE+XOR
0237        Y=0.0*SCALE+YOR
0238        CALL MOVE(X,Y,0)
0239        CALL LABSIZ(100.,200.)
0240        CALL MOVE(XOR,(YOR-300.),1)
0241        WRITE(NPOUT,333)126,39,(APX(3,J),J=1,8),3
0242        X=1.0*SCALE+XOR
0243        CALL MOVE(X,(YOR-300.),1)
0244        WRITE(NPOUT,333)126,39,(APX(1,J),J=1,8),3
0245        X=0.5*SCALE+XOR
0246        Y=0.866025*SCALE+YOR+50.
0247        CALL MOVE(X,Y,1)
     C THESE WRITE STATEMENTS WRITE A LABEL ONTO THE PLOT
0248        WRITE(NPOUT,333)126,39,(APX(2,J),J=1,8),3
0249        CALL MOVE(2000.,9000.,1)
0250        WRITE(NPOUT,333)126,39,(PTITLE(J),J=1,40),3
0251        WRITE(NPOUT,5134)126,39,32,32,32,32,32,32,32,32,
```

```
0252  5134  FORMAT(' ',11A1,'SCALE= ',F8.1,A1)
0253        CALL POFF
0254        GO TO 5000
      C
      C     ROUTINE TO PLOT DATA
      C
0255  5200  CONTINUE
0256        TYPE *,'CHOOSE: 0=EXIT; 1=PLOT POINTS; 2=PLOT ASSEMBLAGES'
0257        ACCEPT *,IPL
0258        IF (IPL.EQ.0)GO TO 5000
0260        GO TO (5300,5400)IPL
0261        GO TO 5200
      C
      C     ROUTINE TO PLOT INDIVIDUAL POINTS WITH A LABEL
      C
0262  5300  CONTINUE
0263        TYPE *,'INPUT LABEL WIDTH AND HEIGHT (e.g. 100,100)'
0264        ACCEPT *,W,H
0265        H=H*3./2.
0266        DO 5310 I=1,8
0267  5310  LAB(I)=32
0268  5323  TYPE *,' INPUT MINERAL # TO PLOT, 0 TO EXIT'
0269        ACCEPT *,K
0270        IF(K.EQ.0)GO TO 5200
0272        TYPE *,' INPUT LABEL ,
0273        ACCEPT 1012,(LAB(J),J=1,8)
0274        Y=.866025*C(2,K)*SCALE+YOR
0275        X=(C(1,K)+.5*C(2,K))*SCALE+XOR
0276        CALL PON
0277        CALL MOVE(X,Y,1)
0278        CALL SYMB(2,35.,35.,)
0279        X=X+50.
0280        CALL MOVE(X,Y,1)
0281        CALL LABSIZ(W,H)
0282        WRITE(NPOUT,333)126.39,(LAB(I),I=1,8).3
0283        CALL POFF
0284        GO TO 5323
      C
      C     ROUTINE TO PLOT ASSEMBLAGES CONNECTED BY TIE-LINES
      C
0285  5400  CONTINUE
0286        TYPE *,' INPUT MINERAL NUMBERS TO PLOT FOLLOWED BY A RETURN'
0287        TYPE *,' WHEN DONE TYPE 0'
0288        NM=1
0289  5411  ACCEPT *,MINNO(NM)
0290        IF (MINNO(NM).EQ.0)GO TO 5420
0292        NM=NM+1
0293        GO TO 5411
0294  5420  CONTINUE
0295        NM=NM-1
0296        IF (NM.EQ.0)GO TO 5200
0298        DO 5424 I=1,NM
0299        J=MINNO(I)

0300        YLA(I)=.866025*C(2,J)
0301        XLA(I)=C(1,J)+.5*C(2,J)
0302  5424  CONTINUE
0303        TYPE *,' INPUT PEN #'
0304        ACCEPT *,NOPEN
0305        CALL PON
0306        CALL NEWPEN(NOPEN)
0307        NL=NM-1
0308        DO 5430 I=1,NL
0309        L=I+1
0310        DO 5430 K=L,NM
0311        X=XLA(I)*SCALE+XOR
0312        Y=YLA(I)*SCALE+YOR
0313        CALL MOVE(X,Y,1)
0314        X=XLA(K)*SCALE+XOR
0315        Y=YLA(K)*SCALE+YOR
0316  5430  CALL MOVE(X,Y,0)
0317        CALL PENUP
0318        CALL POFF
0319        GO TO 5400
0320  999   CALL POFF
0321        END


0001        SUBROUTINE PROJEC(BARY,A,T,G,E,XL,YL,XR,YR)
      C
      C     THIS SUBROUTINE CONVERTS THE 4 BARYCENTRIC COORDINATES CONTAINED
      C     IN THE ARRAY BARY INTO CARTESIAN COORDINATES FOR LEFT (XL,YL)
      C     AND RIGHT (XR,YR) STEREO VIEWS.
      C     TETRAHEDRON IS ROTATED AND PROJECTED FROM A,T,G,AND E
      C     WHERE  A=ALPHA, T=THETA, G=GAMMA, E=EYE DISTANCE.
      C     ALGORITHM IS AFTER SPEAR (1980).AM. MINERAL., V. 65, 1291-1293.
      C
0002        DIMENSION BARY(4)
0003        X=((BARY(1)-.25)+.5*(BARY(2)-.25)+.5*(BARY(4)-.25))/0.8165
0004        Y=((BARY(2)-.25)+.333333*(BARY(4)-.25))/0.9428
0005        Z=BARY(4)-.25
0006        X2=X*COS(A)-Y*SIN(A)
0007        Y2=(X*SIN(A)+Y*COS(A))*COS(T)+Z*SIN(T)
0008        Z2=Z*COS(T)-SIN(T)*(X*SIN(A)+Y*COS(A))
0009        ZL=Z2*COS(G/2)-X2*SIN(G/2)
0010        XL=(X2*COS(G/2)+Z2*SIN(G/2))*E/(E-ZL)
0011        YL=Y2*E/(E-ZL)
0012        ZR=X2*SIN(G/2)+Z2*COS(G/2)
0013        XR=(X2*COS(G/2)-Z2*SIN(G/2))*E/(E-ZR)
0014        YR=Y2*E/(E-ZR)
0015        RETURN
0016        END
```

APPENDIX B: PROGRAM LINEQU: A FORTRAN Program to Solve a System of Linear Equations

This program solves a system of linear equations by matrix inversion techniques. The program has two options: invert matrix or solve a system of linear equations of the form $A \cdot X = Y$. The program can be used in component transformation problems or to balance chemical reactions.

Input in straightforward and self-explanatory. The program requests the dimensions of the matrix and then asks for the coefficient matrix A to be input by columns (i.e., mineral by mineral). Input is free format. If a system of equations is to be solved, the program also requests the data vector Y. The program then prints out the input matrix A, the inverse of this matrix A^{-1}, the determinant of A, and the data vector Y and solution vector X if necessary.

```
C    PROGRAM CALCULATES EITHER THE INVERSE OF A SQUARE MATRIX
C    OR SOLVES A SYSTEM OF LINEAR EQUATIONS BY POST MULTIPLICATION
C    OF THE INVERSE OF THE COEFFICIENTS MATRIX BY THE DATA VECTOR Y
       REAL*8 A,DETA,Y,SUM,X
       DIMENSION A(20,20),IR(20),IC(20),Y(20),X(20)
       DIMENSION TITLE(20)
1      CONTINUE
       TYPE *,'INPUT TITLE'
       READ (5,1026,END=999) TITLE
1026   FORMAT (20A4)
       WRITE (6,1234) TITLE
1234   FORMAT (' ',////,20A4)
       TYPE *,'INPUT MATRIX DIMENSION (N FOR NXN);TYPE 0 TO EXIT'
       READ(5,*,END=999) MA
       IF(MA.EQ.0)GO TO 999
       TYPE *,'INPUT MATRIX BY COLUMNS, A RETURN AFTER EACH COLUMN'
       DO 10 I=1,MA
10     READ(5,*) (A(J,I),J=1,MA)
       TYPE *,'TYPE 0 FOR INVERSE ONLY, 1 FOR SOLUTION TO EQUATIONS'
       ACCEPT *,ITYPE
       IF (ITYPE.EQ.0) GO TO 100
       TYPE *,'INPUT DATA VECTOR, Y'
       ACCEPT *,(Y(I),I=1,MA)
100    CONTINUE
       WRITE (6,2005)
2005   FORMAT(' ',//,' INPUT COEFFICIENTS MATRIX ',//)
       DO 15 I=1,MA
15     WRITE(6,2010)(A(I,J),J=1,MA)
2010   FORMAT(' ',14F9.5)
       DETA=0.0
       IER=0.0
       CALL MINVRS(A,20,MA,DETA,IER,IR,IC)
       IF (IER.EQ.1) GO TO 50
       WRITE(6,2043)
2043   FORMAT(' ',//,' INVERSE OF COEFFICIENTS MATRIX',//)
       DO 20 I=1,MA
20     WRITE(6,2010)(A(I,J),J=1,MA)
       WRITE (6,2020) DETA
2020   FORMAT (' ',' DETERMINANT= ',F10.4)
       IF (ITYPE.EQ.0) GO TO 1
       WRITE (6,2314)(Y(I),I=1,MA)
2314   FORMAT(' ',////,'.......DATA VECTOR..Y.....',/,14F9.5)
       DO 150 I=1,MA
       SUM=0.0
       DO 155 J=1,MA
155    SUM=SUM+A(I,J)*Y(J)
150    X(I)=SUM
       WRITE (6,2313)(X(I),I=1,MA)
2313   FORMAT(' ',////,'..... SOLUTION VECTOR .....',/,14F9.5)
       GO TO 1
50     WRITE(6,2000)
2000   FORMAT (' ',////,'MATRIX IS SINGULAR....ABORT')
       GO TO 1
999    CONTINUE
       END
```

```
       SUBROUTINE MINVRS (A,IA,MA,DETA,IER,IR,IC)
       REAL*8 A,PIV,DETA,PIV1,TEMP
       DIMENSION A(IA,IA),IR(MA),IC(MA)
       IER=0
       DO 1 I=1,MA
       IR(I)=0
1      IC(I)=0
       DETA=1.0
       DO 123 IJKL=1,MA
       CALL SUBMXS (A,IA,IA,MA,MA,IR,IC,I,J)
       PIV=A(I,J)
       DETA=PIV*DETA
       IF (PIV.EQ.0.0) GO TO 17
       IR(I)=J
       IC(J)=I
       PIV1=1.0/PIV
       DO 5 K=1,MA
5      A(I,K)=A(I,K)*PIV
       A(I,J)=PIV
       DO 9 K=1,MA
       IF (K.EQ.I) GO TO 9
       PIV1=A(K,J)
       A(K,J)=A(I,J)
6      DO 8 L=1,MA
8      A(K,L)=A(K,L)-PIV1*A(I,L)
       A(K,J)=PIV1
9      CONTINUE
       PIV1=A(I,J)
       DO 11 K=1,MA
11     A(K,J)=-PIV1*A(K,J)
       A(I,J)=PIV1
123    CONTINUE
12     DO 16 I=1,MA
       K=IC(I)
       M=IR(I)
       IF (K.EQ.I) GO TO 16
       DETA=-DETA
       DO 14 L=1,MA
       TEMP=A(K,L)
       A(K,L)=A(I,L)
14     A(I,L)=TEMP
       DO 15 L=1,MA
       TEMP=A(L,M)
       A(L,M)=A(L,I)
15     A(L,I)=TEMP
       IC(M)=K
       IR(K)=M
16     CONTINUE
       RETURN
17     IER=1
       RETURN
       END
```

```
       SUBROUTINE SUBMXS (A,IA,JA,MA,NA,IR,IC,I,J)
       REAL*8 A,X,TEST
       DIMENSION A(IA,JA),IR(MA),IC(NA)
       I=0
       J=0
       TEST=0.0
       DO 5 K=1,MA
       IF (IR(K).NE.0) GO TO 5
       DO 4 L=1,NA
       IF(IC(L).NE.0) GO TO 4
       X=DABS(A(K,L))
       IF(X.LT.TEST) GO TO 4
       J=L
       I=K
       TEST=X
4      CONTINUE
5      CONTINUE
       RETURN
       END
```

```
( PROGRAM LINEQU
  CODED BY F. SPEAR
  UPDATE APRIL 20, 1982 )
```

Chapter 4

ANALYTICAL FORMULATION of PHASE EQUILIBRIA: The GIBBS' METHOD

F.S. Spear, J.M. Ferry, and D. Rumble III

INTRODUCTION

A basic goal in the study of metamorphic rocks is to characterize intensive variables during metamorphic events. The fundamental assumption that
allows extraction of estimates of intensive variables from mineral assemblages
is that the mineral phases attained a state of chemical equilibrium at some
condition of elevated temperature and pressure during metamorphism. Furthermore, it is assumed that the compositions of minerals in rocks have not significantly changed from the time that they attained this high temperature-
pressure equilibrium state. There are obviously numerous objections to these
two assumptions: (a) Metamorphism is a dynamic process and there is little
to assure that all or even most mineral compositions freeze in at the same
point on the pressure-temperature-time path of a particular metamorphic rock;
(b) rocks may have a polymetamorphic history with some mineral compositions
the product of one event and other compositions the product of a second event;
(c) reaction rates, diffusion rates, and/or rates of heat transfer may be too
slow for minerals in a rock to ever attain chemical equilibrium at any time
during a metamorphic event; etc. While many metamorphic rocks undoubtedly
suffer from these problems, it is remarkable the number of studies that have
found metamorphic rocks to be consistent with attainment of chemical equilibrium at some condition of elevated temperature and pressure.

Classical chemical thermodynamics constitutes a comprehensive mathematical framework in which intensive variables -- temperature, pressure,
chemical potential, phase composition -- are related in a system at chemical
equilibrium. Much of our modern understanding of metamorphism derives from
a two-step exercise: (1) assumption that minerals in metamorphic rocks constitute a chemical system in a fossilized state of equilibrium; and (2) application of chemical thermodynamics to that rock system. The compositions
of minerals in metamorphic rocks constitute certain intensive variables of
the high temperature-pressure metamorphic state of equilibrium. Thermodynamics links these composition variables to other intensive variables of interest, which no longer may be directly determined (e.g., temperature and
pressure). Through these links, it is thus possible to use mineral assemblages
and compositions of minerals to reconstruct the value of those intensive

NOTATION

f   degrees of freedom

c   number of components in a system

p   number of phases in a system

n   number of linearly independent stoichiometric relations that can be written among phase components in a system

r   number of phase components in a system

P   pressure (bars)

$P_{(i)}$ partial pressure of component i (bars)

P_{total} total or lithostatic pressure (bars)

P_{fluid} fluid pressure (bars)

T   temperature (°K)

R   universal gas constant (cal/degree-mole)

$\bar{G}$ molar Gibbs energy (cal)

$\bar{G}^j$ molar Gibbs energy of phase j (cal)

$\bar{G}_i^j$ partial molar Gibbs energy of component i in phase j (cal)

μ chemical potential (cal)

μ_i chemical potential of component i in a system (cal)

μ_i^o chemical potential of component i in a phase with composition pure i (cal)

μ_i^j chemical potential of component i in phase j (cal)

X   mole fraction

X_i mole fraction of component i

X_i^j mole fraction of component i in phase j

$\bar{S}$ molar entropy (cal/degree-mole)

$\bar{S}^j$ molar entropy of phase j (cal/degree-mole)

$\bar{S}_i^j$ partial molar entropy of component i in phase j (cal/degree-mole)

$\bar{V}$ molar volume (cal/bar)

$\bar{V}^j$ molar volume of phase j (cal/bar)

$\bar{V}_i^j$ partial molar volume of component i in phase j (cal/bar)

K_D^{j-k} distribution coefficient for cation exchange between phases j and k

f(i)   fugacity of component i

Subscript notation for components

Fa Fe_2SiO_4

Fo Mg_2SiO_4

Fs $FeSiO_3$

En $MgSiO_3$

Fe   iron component of mineral solid solution

Al   aluminum component of mineral solid solution

Ti   titanium component of mineral solid solution

Mg   magnesium component of mineral solid solution

Superscript notation for phases

Qt   quartz   Im   ilmenite

Px   pyroxene   Sta   staurolite

Ol   olivine   Chl   chlorite

Ky   kyanite   Gar   garnet

Si   sillimanite   Bio   biotite

o   denotes phase with composition pure i where i is the associated component variables that cannot be measured. The purpose of this chapter is to develop a family of equations, grounded in chemical thermodynamics, which are useful in extracting information about intensive variables during metamorphism from minerals in metamorphic rocks and their compositions. We present numerous examples of how these equations can be and have been applied to the study of metamorphic rocks. From the beginning we emphasize that the material is not original to us; we act principally as compilers and expositors. Chemical thermodynamics of heterogeneous systems may be largely considered the contribution of a single man: J. Willard Gibbs. In his honor we refer to the methods discussed in this chapter as the Gibbs method. Further, we owe a great debt to J.B. Thompson, Jr. whose class lectures and publications constitute in large part the groundwork for the following treatment of thermodynamics and its application to geologic systems.

An important place to begin the discussion of chemical thermodynamics is the Gibbs phase rule. The Gibbs phase rule ($f = c - p + 2$) specifies, under most circumstances, the number of degrees of freedom (f), or variance, of a chemical system of c components and p phases in heterogeneous equilibrium. The phase rule should be thought of as a tally of the number of independent variables and the number of independent constraints in such a system. Specifically, the variables are temperature, pressure and the chemical potentials of the various chemical components in the different phases of the system, and the constraints are equations that specify the mathematical relations among these different variables. Put very simply, the variance as calculated from the phase rule is a statement of the number of variables minus the number of equations and is exactly analogous to the variance in mathematical parlance of a system of simultaneous equations.

The phase rule, although specifying the *number* of equations and unknowns, does not specify the explicit mathematical relationship among these variables. Explicit relationships among T, P and chemical potential do exist, however. These equations are the various different "equilibrium conditions" that must hold true for a system in chemical equilibrium.

In this chapter we describe how mathematically to formulate the relationships among the various intensive parameters such as T, P, μ_i and X_i. These relationships will be formulated as a system of simultaneous linear equations that completely describe the variance of the system. The number of equations and unknowns will be such that the variance of the system of equations will be the same as the variance calculated from the phase rule. This system of

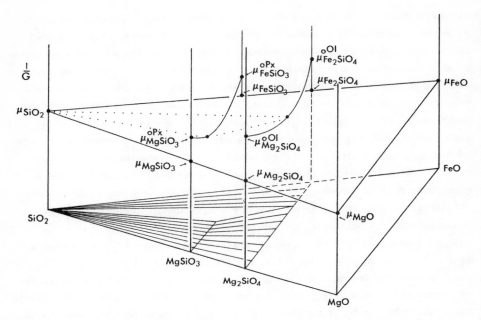

Figure 1. Ḡ-X diagram at constant P and T for the system SiO₂-FeO-MgO. Orthopyroxene (Px) and olivine (Ol) are shown as binary solid solutions and the Ḡ-surfaces for these phases are drawn as "knife edges". The tangent plane is tangent to the Ḡ-surfaces for Px + Ol + quartz (Qt). The chemical potentials of various components, as determined from the intersection of the tangent plane with the Ḡ-axis of a particular composition, are also shown.

equations can then be solved to derive specific equations that relate changes in T, P, μ_i or X_i in an assemblage of phases.

The method discussed in this chapter is an outgrowth of Gibbs' derivation of the phase rule and has enormous potential for extracting information on variations in P, T and μ_i during crystallization of rocks. As an example of the type of information that can be obtained from application of the Gibbs method, consider the Ḡ-X diagram for the ternary system SiO₂-FeO-MgO shown in Figure 1. This system will be discussed in great detail later in the chapter. The phases orthopyroxene and olivine are represented as binary solid solutions and the Gibbs (Ḡ-X) surfaces for these phases are shown as "knife edges"; quartz is shown as a phase of fixed composition with a fixed Gibbs free energy at constant T and P. The base of the prism is the three-component isothermal, isobaric phase diagram for the system SiO₂-FeO-MgO. Tie lines in the phase diagram may be viewed as projections of lines connecting points on the various Ḡ-X surfaces that are simultaneously tangent to a common plane. For example, the three-phase field orthopyroxene ÷ olivine + quartz results when the plane is simultaneously tangent to the Ḡ-surfaces for all three phases; in each two-phase field, orthopyroxene + quartz, orthopyroxene + olivine, and quartz + olivine, the plane is tangent to only two phases at a time. The intersection of the tangent plane with the $\bar{G}$-axis at each apex of the triangular prism gives the value of the chemical potential (or partial molar Gibbs free energy) of SiO_2, FeO or MgO in each phase that lies on that particular tangent plane. Moreover, the intersection of the tangent plane with the $\bar{G}$ axis at the composition $FeSiO_3$, $MgSiO_3$, Fe_2SiO_4, Mg_2SiO_4 (or any other composition) defines the value of the chemical potential of that component in the phases in equilibrium. The fact that all phases in an equilibrium assemblage lie on the same tangent plane ensures that the chemical potential of each component will be equal in all three phases (e.g., $\mu_{SiO_2}^{Qt} = \mu_{SiO_2}^{Px} = \mu_{SiO_2}^{Ol}$). This relation is one of the fundamental conditions of chemical equilibrium.

The intensive variables of interest in this system are T, P, and the chemical potentials (or compositions) of the three components in the various phases. The Gibbs method enables us to formulate analytically, and thereby state explicitly, the interdependency of these variables. It also allows us to calculate in a quantitative manner how these different variables change with respect to one another. For example, in the three-phase region of Figure 1 (Px + Ol + Qt) the chemical potentials of SiO_2, FeO and MgO in these phases (as well as the chemical potentials of components such as $MgSiO_3$ and Fe_2SiO_4) are fixed at constant P and T. But as T and P are changed, the $\bar{G}$-surfaces for these phases will go up or down as dictated by the equation $d\bar{G} = -\bar{S}dT + \bar{V}dP$. Thus the tangent plane, the compositions of orthopyroxene and olivine, and the chemical potentials of SiO_2, FeO and MgO in the three phases will change. The Gibbs method permits explicit determination of how the variables composition (X_i), chemical potential (μ_i), temperature (T), and pressure (P) are interrelated and how they change with respect to one another. In the three-phase (divariant) region this is done through derivation of expressions such as $(\partial T/\partial X)_P$, $(\partial P/\partial X)_T$ and $(\partial P/\partial T)_X$. It can also be seen in Figure 1 that in any of the two-phase (trivariant) regions (e.g., orthopyroxene + olivine) the tangent plane is free to roll across the two Gibbs surfaces as the Fe/Mg ratio is changed. As the bulk Fe/Mg is changed, the plane is tangent to phases of different Fe/Mg and the intercepts of the tangent plane to the three $\bar{G}$-axes (i.e., the chemical potentials) change. In other words, there is a direct correlation between the chemical potentials of the various components and the Fe/Mg of the phases in equilibrium; the change in chemical potential with respect to composition can be calculated through derivation of expressions such as $(\partial \mu/\partial X)_{P,T}$.

Differentials with respect to *composition* such as $(\partial\mu/\partial X)$ or $(\partial T/\partial X)$ are especially useful because mineral composition is a *measurable* quantity. This is probably the most powerful application of the Gibbs method -- it provides the theoretical framework whereby a measurable quantity (i.e., mineral composition) can be used to calculate differences in quantities that are not directly measurable (i.e., temperature, pressure, or chemical potential at the time of equilibration). In application to natural assemblages, therefore, *the measured mineral composition can be used to calculate differences in P, T or μ_i of crystallization between similar assemblages of different composition.*

It is useful to point out that the equations of thermodynamics used in the Gibbs method describe relations among surfaces, slopes, tangent planes, and intercepts; as such, these relations must obey all of the rules of differential calculus. It turns out that the different variables, functions, slopes, and intercepts are all quantities with thermodynamic significance such as P, T, μ_i, X_i, $\bar{G}$, $\bar{S}$ and $\bar{V}$. The Gibbs method, therefore, can be thought of as a mathematical description of the relations among surfaces, planes, slopes, and intercepts. Of course, this is not to imply that thermodynamics is not fundamentally rooted in experiment and observation, but only that the mathematical relations among variables, which will be described in this chapter, also obey the rules of calculus.

In the next four sections we derive the necessary equations for the analytical formulation of phase equilibria. Some techniques for solution are given in the appendices. This will be followed by some examples of applications. The last section is a review of the literature pertaining to this type of analysis and the information that can be gained from rocks through its application. [A discussion of the Gibbs method and some applications can also be found in Rumble (1974, 1976a).]

CONDITIONS OF HOMOGENEOUS EQUILIBRIUM: THE GIBBS-DUHEM EQUATION

The Gibbs-Duhem equation

$$0 = \bar{S}dT - \bar{V}dP + \Sigma X_i d\mu_i \tag{1}$$

is a relationship among the intensive variables of a single phase in internal, homogeneous equilibrum. "Homogeneous" equilibrium refers to equilibrium within a specific phase, and the conditions of homogeneous equilibrium are the equations that specify that each phase is in internal equilibrium with respect to processes such as internal diffusion, cation ordering, and chemical speciation. If a phase is in homogeneous equilibrium, the Gibbs-Duhem equations must hold true. A rigorous derivation of the equation is given by Tunell (1979).

The Gibbs-Duhem equation has a very specific geometric significance as well. Consider the geometric construction of the $\bar{G} - X - T$ diagram shown in Figure 2A, which is drawn for a fixed pressure. The diagram shows a two-component system with one phase (A) with a fixed composition of X_2^A (for example, component 1 = SiO_2, component 2 = Al_2O_3, and phase A = mullite with $X_2^A = 0.6$). The *molar Gibbs free energy* of A is μ_A; this quantity changes as a function of changing T along the line indicated. The tangent line drawn through phase A defines the quantities μ_1^A and μ_2^A where it intercepts the $\bar{G}$ axis at $X_2 = 0$ and $X_2 = 1$. These two quantities are the *partial molar Gibbs free energies* of components 1 and 2, respectively, in phase A, and these also change as a function of T. Thus a slice through this diagram at $X_2 = X_2^A$ is a $\bar{G} - T$ diagram for phase A and would appear as shown in Figure 2B.

As the temperature is changed, the molar Gibbs free energy of phase A (μ_A) will "track" down the line labeled μ_A in Figure 2B and as it does, the intercepts of the tangent line (μ_1^A and μ_2^A) will change. In fact, all three values (μ_A, μ_1^A, and μ_2^A) must change sympathetically. This relationship can be expressed by a simple algebraic construction: The tangent line that passes through A at some arbitrary T can be described by the equation of a straight line: $y = mx + b$. That is,

$$\mu_A = \left(\frac{\mu_2^A - \mu_1^A}{1}\right) X_2^A + \mu_1^A \tag{2}$$

where the slope is $(\mu_2^A - \mu_1^A)$, the intercept is μ_1^A, and $y = \mu_A$ when $x = X_2^A$. Differentiation of this equation with respect to T yields

$$\left(\frac{d\mu_A}{dT}\right) = X_2^A \left(\frac{d\mu_2^A}{dT}\right) + (1 - X_2^A)\left(\frac{d\mu_1^A}{dT}\right). \tag{3}$$

Thus far, this derivation involves only analytical geometry. Now it is simply necessary to note that the change in the chemical potential of A with respect to T is a quantity of thermodynamic significance known as entropy (that is, the slope of the curve defined by a phase on a $\mu_A - T$ diagram is equal to $-\bar{S}^A$; or $\frac{d\mu_A}{dT} \equiv -\bar{S}^A$. Substituting, we have

$$0 = \bar{S}^A + X_2^A \left(\frac{d\mu_2^A}{dT}\right) + (1 - X_2^A)\left(\frac{d\mu_1^A}{dT}\right) = \bar{S}^A dT + X_2^A d\mu_2^A + X_1^A d\mu_1^A \tag{4}$$

which is the Gibbs-Duhem equation at constant P. A similar operation substituting $\frac{d\mu_A}{dP} = \bar{V}^A$ will yield the full Gibbs-Duhem equation.

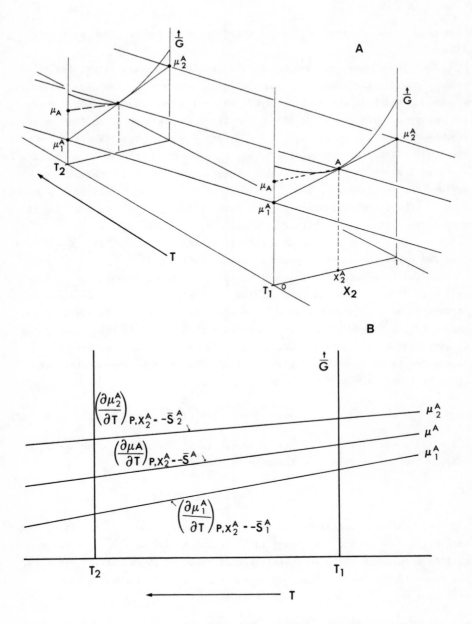

Figure 2. (A) $\bar{G}$-X-T construction at constant P for a two-component system containing one phase (A) of variable composition intermediate between components 1 and 2. A line is drawn through phase A in a direction along the T-axis showing the variation of the Gibbs free energy of A (μ_A) with respect to T. A tangent line is drawn through μ_A at two arbitrary temperatures showing the intercepts with the $\bar{G}$-axis at $X_2 = 0$ (μ_1^A) and $X_2 = 1^A$ (μ_2^A). (B) A section through $\bar{G}$-T-X space at constant X_2^A. The slopes ($\partial\mu/\partial T$) are either the molar entropy of phase A ($\bar{S}^A$) or the partial molar entropies of components 1 and 2 in phase A ($\bar{S}_1^A$ or $\bar{S}_2^A$).

112

Thus the Gibbs-Duhem equation, in a geometric sense, specifies the relationship between the molar Gibbs free energy of a phase (μ_A) and the intercepts of the tangent line with the axes at $X_2 = 0(\mu_2^A)$ and $X_2 = 1(\mu_2^A)$ as the phase changes composition, temperature, or pressure. In this regard the Gibbs-Duhem equation is a "lever rule" of $\bar{G}$-X-T-P space and simply says that not all of the parameters change independently. *The Gibbs-Duhem equation ensures that the tangent line remains "attached" to the $\bar{G}$-surface for a phase as T, P and X are changed.*

CONDITIONS OF HETEROGENEOUS EQUILIBRIUM

For a system in heterogeneous equilibrium there is a Gibbs-Duhem equation that must be valid for each phase present in the equilibrium assemblage.

The equations that ensure that all phases lie on the same tangent plane are the conditions of heterogeneous equilibrium. This can easily be seen from the construction in Figure 3 where it is clear that for phase A and B to lie on the same tangent plane, the equations

$$\mu_1^A = \mu_1^B \quad \text{and} \quad \mu_2^A = \mu_2^B \tag{5}$$

must hold true. If these equations are valid, then their derivatives are also valid:

$$0 = d\mu_1^A - d\mu_1^B \tag{6}$$

$$0 = d\mu_2^A - d\mu_2^B . \tag{7}$$

For the two-phase assemblage shown in Figure 3, we now have four linear equations that can be written in six unknowns $(T, P, \mu_1^A, \mu_2^A, \mu_1^B, \mu_2^B)$

$$0 = \bar{S}^A dT - \bar{V}^A dP + X_1^A d\mu_1^A + X_2^A d\mu_2^A + \quad 0 \quad + \quad 0 \tag{8}$$

$$0 = \bar{S}^B dT - \bar{V}^B dP + \quad 0 \quad + \quad 0 \quad + X_1^B d\mu_1^B + X_2^B d\mu_2^B \tag{9}$$

$$0 = \quad 0 \quad + \quad 0 \quad + d\mu_1^A + \quad 0 \quad - \quad d\mu_1^B + \quad 0 \tag{10}$$

$$0 = \quad 0 \quad + \quad 0 \quad + \quad 0 \quad + d\mu_2^A + \quad 0 \quad - \quad d\mu_2^B \tag{11}$$

[Note: Rumble (1974, 1976a) writes the Gibbs-Duhem equations and the conditions of heterogeneous equilibrium in a way that removes the dependent compositional variable, X_1 $(X_1 = 1 - X_2)$. His formulation of these equations is

$$0 = \bar{S}^A dT - \bar{V}^A dP + d\mu_1^A + X_2^A d(\mu_2^A - \mu_1^A) + \quad 0 \quad + \quad 0 \tag{12}$$

$$0 = \bar{S}^B dT - \bar{V}^B dP + \quad 0 \quad + \quad 0 \quad + d\mu_1^B + X_2^B d(\mu_2^B - \mu_1^B) \tag{13}$$

113

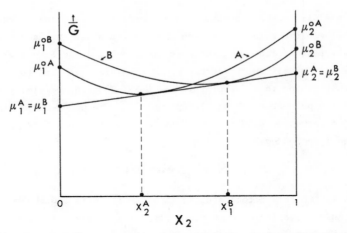

Figure 3. $\bar{G}$-X construction at constant P and T showing two phases A and B in equilibrium. For the two phase assemblage A + B, the composition of the coexisting phases are x_2^A and x_2^B and the conditions $\mu_1^A = \mu_1^B$ and $\mu_2^A = \mu_2^B$ must be valid.

$$0 = 0 + 0 + d\mu_1^A + 0 - d\mu_1^B + 0 \qquad (14)$$

$$0 = 0 + 0 + 0 + d(\mu_2^A - \mu_1^A) + 0 - d(\mu_2^B - \mu_1^B) \qquad (15)$$

There is no particular advantage to removing the dependent compositional term at this point, and either procedure may be followed. It is an easy enough matter to remove this dependent parameter later during solution of the equations, if desired.]

THE ADDITION OF A NEW VARIABLE — dX

The above system of equations is the formal mathematical equivalent of the phase rule variance of the assemblage A + B. Note that the phase rule tells us that there are two degrees of freedom in a two-component system containing two phases, and the above system of equations also has two degrees of freedom (6 unknowns - 4 equations). In essence, *these four equations are all that is necessary to ensure that the $\bar{G}$-surfaces for the phases A and B remain attached to the same tangent plane at all times.*

At this point the chemical system is defined and further manipulation of the set of equations depends entirely on what result is desired. Any new unknown may be introduced, and subsequently solved for, as long as the same number of equations as unknowns is introduced and the variance of the system does not change. For example, it might be desirable to monitor the chemical potential of a component in a phase that has a composition intermediate between

114

components 1 and 2 (e.g., phase α with composition $X_1^\alpha = \nu_1$ and $X_2^\alpha = \nu_2$ where ν_1 and ν_2 are the stoichiometric coefficients of components 1 and 2 in phase α) and the equation

$$d\mu_\alpha = \nu_1 d\mu_1 + \nu_2 d\mu_2$$

could thus be added to the list. The importance of adding these types of equations will be seen in the examples which follow.

Another desirable variable to add to the system of equations is a compositional term such as dX_2^A because we can directly measure the composition of minerals in metamorphic rocks. Only by the addition of composition variables can changes in intensive parameters such as $d\mu$ or dT be monitored by changes in composition (e.g., $d\mu/dX$). The choice of which compositional parameters are to be incorporated is dictated solely by which phases are to be used as monitors and thus is completely at the discretion of the user.

An equation that incorporates derivatives of compositional variables into the system of equations without adding any other new variables is one that defines the total differential of the slope of the tangent to the $\bar{G}$-surface of the phase in question. For the example of phase A in Figure 3, this slope is $\mu_2^A - \mu_1^A$. Because the Gibbs free energy surface for a binary phase such as A is a function of only P, T and X_2^A (only one of the two compositional parameters is independent), μ_1^A and μ_2^A are also functions of only P, T and X_2^A as is the difference, $\mu_2^A - \mu_1^A$. Thus $(\mu_2^A - \mu_1^A) = f(P, T, X_2^A)$. The total differential of this quantity is an equation depicting how this slope changes as P, T and X_2^A are changed:

$$d(\mu_2^A - \mu_1^A) = \left(\frac{\partial(\mu_2^A - \mu_1^A)}{\partial T}\right)_{P,X_2^A} dT + \left(\frac{\partial(\mu_2^A - \mu_1^A)}{\partial P}\right)_{T,X_2^A} dP + \left(\frac{d(\mu_2^A - \mu_1^A)}{dX_2^A}\right)_{P,T} dX_2^A \quad (17)$$

Each of the three parts of this differential can be written in a different form. μ_1^A is the partial molar Gibbs free energy of component 1 in phase A and thus $\left(\frac{\partial \mu_1^A}{\partial T}\right)_{P,X_2^A} = -\bar{s}_1^A$, the partial molar entropy of component 1 in phase A. Similarly,

$$\left(\frac{\partial \mu_2^A}{\partial T}\right)_{P,X_2^A} = -\bar{s}_2^A \quad \text{and} \quad \left(\frac{\partial(\mu_2^A - \mu_1^A)}{\partial T}\right)_{P,X_2^A} = -(\bar{s}_2^A - \bar{s}_1^A) \ . \quad (18)$$

The term $-(\bar{s}_2^A - \bar{s}_1^A)$ has a particular geometric significance as well, as can be seen by reference to Figure 2. $-\bar{s}_1^A$ and $-\bar{s}_2^A$ are the slopes that the traces of the intercepts of the tangent line (μ_1^A and μ_2^A) make on a $\bar{G}$-T projection at constant P and X_2^A (see Figure 2B). These slopes are, in general, relatively constant. If $S_2^A < S_1^A$, as is the case in Figure 2B, then the difference is negative and the difference $\mu_2^A - \mu_1^A$ increases with increasing T, as shown in Figure 2B. In other words, the first term of the total differential of $\mu_2^A - \mu_1^A$,

$$\left(\frac{\partial(\mu_2^A - \mu_1^A)}{\partial T}\right)_{P, X_2^A} dT = -(\bar{S}_2^A - \bar{S}_1^A) dT$$

is positive.

A similar consideration of the second term yields

$$\left(\frac{\partial(\mu_2^A - \mu_1^A)}{\partial P}\right)_{T, X_2^A} dP = (\bar{V}_2^A - \bar{V}_1^A) dP \text{ by noting that } \left(\frac{\partial \mu_2^A}{\partial P}\right)_{T, X_2^A} = \bar{V}_2^A, \text{ etc.}$$

The third term can easily be evaluated by noting that $\mu_2^A - \mu_1^A$ at constant P and T is simply the slope of the tangent line at constant P and T. This slope is the first derivative of $\bar{G}^A$ with respect to composition. Thus $(\mu_2^A - \mu_1^A)_{P,T} = (\partial \bar{G}^A / \partial X_2^A)_{P,T}$ and

$$\left(\frac{\partial(\mu_2^A - \mu_1^A)}{\partial X_2^A}\right)_{P,T} dX_2^A = \left(\frac{\partial^2(\bar{G}^A)}{\partial(X_2^A)^2}\right)_{P,T} dX_2^A \tag{19}$$

(Haase, 1948). Equation (19) states that the rate of change of the slope $d(\mu_2 - \mu_1)$ at constant P and T depends on the curvature of the Gibbs function. In final form the total differential is reduced to

$$d(\mu_2^A - \mu_1^A) = -(\bar{S}_2^A - \bar{S}_1^A) dT + (\bar{V}_2^A - \bar{V}_1^A) dP + \left(\frac{\partial^2(\bar{G}^A)}{\partial(X_2^A)^2}\right)_{P,T} dX_2^A . \tag{20}$$

The equation introduces the variable dX_2^A, but also provides an additional constraint. Hence, the variance of the system of equations has not changed despite the addition of equation (20).

Equation (20) is for a two-component solid solution phase (A), which has only one independent compositional parameter and only one independent tangent slope. For solid solutions of higher order, the above equation is modified slightly and additional equations are required. For example, in a three-component solid solution, there are two independent composition terms, X_2^A and X_3^A. There are also two independent slopes to the tangent plane, $(\mu_2^A - \mu_1^A)$ and $(\mu_3^A - \mu_1^A)$, each of which is a function of P, T, X_2^A, and X_3^A. Therefore, we need to write the total differential for both $(\mu_2^A - \mu_1^A)$ and $(\mu_3^A - \mu_1^A)$ so that we introduce two new equations with the two additional compositional terms dX_2^A and dX_3^A. As before,

$$d(\mu_2^A - \mu_1^A) = \left(\frac{\partial(\mu_2^A - \mu_1^A)}{\partial T}\right)_{P,X_2^A,X_3^A} dT + \left(\frac{\partial(\mu_2^A - \mu_1^A)}{\partial P}\right)_{T,X_2^A,X_3^A} dP$$

$$+ \left(\frac{\partial(\mu_2^A - \mu_1^A)}{\partial X_2^A}\right)_{P,T,X_3^A} dX_2^A + \left(\frac{\partial(\mu_2^A - \mu_1^A)}{\partial X_3^A}\right)_{P,T,X_2^A} dX_3^A \qquad (21)$$

and

$$d(\mu_3^A - \mu_1^A) = \left(\frac{\partial(\mu_3^A - \mu_1^A)}{\partial T}\right)_{P,X_2^A,X_3^A} dT + \left(\frac{\partial(\mu_3^A - \mu_1^A)}{\partial P}\right)_{T,X_2^A,X_3^A} dP$$

$$+ \left(\frac{\partial(\mu_3^A - \mu_1^A)}{\partial X_2^A}\right)_{P,T,X_3^A} dX_2^A + \left(\frac{\partial(\mu_3^A - \mu_1^A)}{\partial X_3^A}\right)_{P,T,X_2^A} dX_3^A \; . \qquad (22)$$

By the same reasoning as used above, these two equations can be rewritten as

$$d(\mu_2^A - \mu_1^A) = -(\bar{S}_2^A - \bar{S}_1^A)dT + (\bar{V}_2^A - \bar{V}_1^A)dP + \left(\frac{\partial^2(\bar{G}^A)}{\partial(X_2^A)^2}\right)_{P,T,X_3^A} dX_2^A + \left(\frac{\partial^2(\bar{G}^A)}{\partial X_2^A \partial X_3^A}\right)_{P,T,X_2^A} dX_3^A$$

$$\qquad (23)$$

and

$$d(\mu_3^A - \mu_1^A) = -(\bar{S}_3^A - \bar{S}_1^A)dT + (\bar{V}_3^A - \bar{V}_1^A)dP + \left(\frac{\partial^2(\bar{G}^A)}{\partial X_2^A \partial X_3^A}\right)_{P,T,X_3^A} dX_2^A + \left(\frac{\partial^2(\bar{G}^A)}{\partial(X_3^A)^2}\right)_{P,T,X_2^A} dX_3^A \; .$$

$$\qquad (24)$$

(Note that with ternary and higher order solid solutions, there are cross derivatives of the Gibbs function.) For this three-component solid solution we have therefore introduced two new variables dX_2^A and dX_3^A, but also two new constraints as well. Solid solutions of higher order require additional compositional derivatives, additional curvature and cross curvature terms, and additional equations.

PHASE COMPONENTS AND THE GIBBS' METHOD

The above discussion makes no reference to how components should be chosen in phases, and system components (e.g., MgO, FeO) were tacitly used in all derivations. However, there are very good reasons why it is *not* always desirable to use system components to describe the chemical variation of phases, such as when the system components cannot be varied independently of

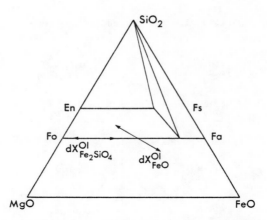

Figure 4. The triangle SiO_2-MgO-FeO showing the difference between changing FeO in olivine (dX_{FeO}^{Ol} – a system component) and changing Fe_2SiO_4 (dX_{Fa}^{Ol} – a phase component).

other components in a phase (also, see Chapter 1 by J.B. Thompson, this volume). For example, consider a situation where it is desired to monitor the change in an intensive parameter with respect to change in the composition of an olivine. With reference to Figure 4, the differential dX_{FeO}^{Ol} represents changes in the composition of olivine along a line that is radial to the FeO apex. However, it is clear from the figure that olivine cannot change its composition in this manner because stoichiometric olivines are constrained to lie on the join Mg_2SiO_4-Fe_2SiO_4. The only way an olivine can change composition off this join is to incorporate vacancies and destroy the olivine stoichiometry. In contrast, components Fe_2SiO_4 and Mg_2SiO_4 can be added to and subtracted from an olivine without destroying the stoichiometry of the phase. Changes in the mole fraction of these components (e.g., dX_{Fa}^{Ol}) simply represent changes along the Fo-Fa join. Components such as Fe_2SiO_4 and Mg_2SiO_4 are called independently variable components of a phase or, more simply, *phase components*.

If we wish to use phase components, both the Gibbs-Duhem equations and the conditions of heterogeneous equilibrium must be modified so that the variables are the chemical potentials of the phase components in the phases under consideration. Before this can be done, however, it is necessary to choose the independently variable components of each phase. To do this, one must have a knowledge of the crystal chemistry of the phase and the allowable chemical variations. As an example, garnet may be modeled as a quaternary solid solution among the components $Fe_3Al_2Si_3O_{12}$, $Mg_3Al_2Si_3O_{12}$, $Ca_3Al_2Si_3O_{12}$ and $Mn_3Al_2Si_3O_{12}$. The only criteria that must be satisfied for the phase components is that they describe the chemical variability of the phase and that they be linearly independent. The Gibbs-Duhem equations can then be written with the chemical potentials and mole fractions of the phase components with no other modification.

The conditions of heterogeneous equilibrium now become a linearly independent set of stoichiometric relations among the chemical potentials of the phase components. It is an outgrowth of Gibbs' derivation of the phase rule that *for every stoichiometric relation that can be written among the phase components of a system in heterogeneous equilibrium, an equivalent relation can be written among the chemical potentials of these phase components.* It is these relations that form the conditions of heterogeneous equilibrium.

This relationship can be seen graphically by referring to Figure 1. For the assemblage quartz + orthopyroxene + olivine to be in heterogeneous equilibrium, the Gibbs surfaces for these three phases must lie on the same tangent plane. This ensures that the chemical potentials of the three system components, SiO_2-FeO-MgO, are equal in all three phases. The *phase components* in this example are SiO_2 (quartz), $FeSiO_3$ and $MgSiO_3$ (pyroxene), and Fe_2SiO_4 and Mg_2SiO_4 (olivine), and there are three stoichiometric relations that can be written among these phase components:

$$SiO_2 + Fe_2SiO_4 = 2FeSiO_3 \tag{25}$$

$$SiO_2 + Mg_2SiO_4 = 2MgSiO_3 \tag{26}$$

$$Fe_2SiO_4 + 2MgSiO_3 = Mg_2SiO_4 + 2FeSiO_3 . \tag{27}$$

The chemical potential of each of these phase components is defined by the intersection of the tangent plane with the composition axis for that phase component. It can be seen from the figure that because the chemical potentials of each phase component all lie on the same tangent plane, the same stoichiometric relations must exist among the chemical potentials of the phase components. Specifically,

$$\mu_{SiO_2} + \mu_{Fe_2SiO_4} = 2\mu_{FeSiO_3} \tag{28}$$

$$\mu_{SiO_2} + \mu_{Mg_2SiO_4} = 2\mu_{MgSiO_3} \tag{29}$$

$$\mu_{Fe_2SiO_4} + 2\mu_{MgSiO_3} = \mu_{Mg_2SiO_4} + 2\mu_{FeSiO_3} . \tag{30}$$

Only two of these three equilibrium conditions are linearly independent (28 - 29 = 30) and only two are needed to specify the conditions of heterogeneous equilibrium. The choice of which two of the three equations are to be

used is completely arbitrary. The number of linearly independent stoichiometric relations that can be written among any set of phases containing r
phase components in a system of c system components is given as

$$n = r - c .\qquad(31)$$

Equation (31) can be derived by noting that the phase rule variance must equal
the variance of the system of equations ($f = c + 2 - p = r + 2 - p - n$) (see Chapter 1 by J.B. Thompson, this volume and Thompson et al., 1982). In the above
example, $c = 3$, $r = 5$ and n, the number of independent relations, $= 2$.
The *total* number of stoichiometric equations that can be written among r
phase components may be much larger than n. Finger and Burt (1972) have
written a computer program to determine all of the stoichiometric relations
that may be written among any set of phase components r. Choosing a set that
is linearly independent from this total set is largely a matter of trial and
error, but can usually be done by inspection. Testing for linear independence
can be done in the manner described in the chapter on Linear Algebraic Manipulation of Composition Space. The choice of which equations to use as an
independent set is completely arbitrary.

The stoichiometric relations that form the basis of the conditions of
heterogeneous equilibrium are equations such as exchange reactions (e.g.,
Fe-Mg exchange) or net-transfer reactions (using the terminology of Thompson
and Thompson, 1976). It is an interesting aside that these very same stoichiometric relations for net-transfer reactions form the axes of reaction
space (see Chapter 2 by J.B. Thompson, this volume). When written in the
form $\Delta\mu° = -RT\ln K$ the relations form the basis for calculating the composition of equilibrium metamorphic fluids (see Chapter 6 by Ferry and Burt, this
volume), for geothermometry and geobarometry (see Chapter 5 by Essene, this
volume) and for evaluating buffering phenomena (see Chapter 7 by Rice and
Ferry, this volume). The important point to recognize is that out of all
possible equilibria, only a subset is linearly independent. Thus, if the
thermodynamic properties of any one independent set in a system is known, the
thermodynamic properties of all other equilibria in the system can be calculated [as has been done, for example, by Skippen (1971) and Hoschek (1980)].

EXAMPLES

We now have a system of equations that describes analytically the variance of a chemical system. In summary, for a system of n-system components

120

Table 1. System of equations for the univariant equilibrium
kyanite + sillimanite

<u>Full Equation Set</u>

$$0 = \overline{S}^{Ky} dT - \overline{V}^{Ky} dP + d\mu_{Al}^{Ky} + 0$$

$$0 = \overline{S}^{Si} dT - \overline{V}^{Si} dP + 0 + d\mu_{Al}^{Si}$$

$$0 = 0 + 0 + d\mu_{Al}^{Ky} - d\mu_{Al}^{Si}$$

<u>Modified Equations for Solution of dP/dT</u>

$$-\overline{S}^{Ky} = -_V{}^{Ky}\frac{dP}{dT} + \frac{d\mu_{Al}^{Ky}}{dT} + 0$$

$$-\overline{S}^{Si} = -\overline{V}^{Si}\frac{dP}{dT} + 0 + \frac{d\mu_{Al}^{Si}}{dT}$$

$$0 = 0 + \frac{d\mu_{Al}^{Ky}}{dT} - \frac{d\mu_{Al}^{Si}}{dT}$$

<u>Modified Equations in Matrix Form</u>

$$\begin{bmatrix} -\overline{V}^{Ky} & 1 & 0 \\ -\overline{V}^{Si} & 0 & 1 \\ 0 & 1 & -1 \end{bmatrix} \cdot \begin{bmatrix} dP/dT \\ d\mu_{Al}^{Ky}/dT \\ d\mu_{Al}^{Si}/dT \end{bmatrix} = \begin{bmatrix} -\overline{S}^{Ky} \\ -\overline{S}^{Si} \\ 0 \end{bmatrix}$$

with p-phases that contain a total of r-phase components, the unknowns are:
dP, dT, and dμ of each r-phase component along with dX_i for one solid solution, which is chosen to relate to the variables dT, dP, and dμ. The equations are: one Gibbs-Duhem equation for each of the p-phases, (r – c) conditions of heterogeneous equilibria, and one equation for each dX_i of interest.
Several examples will now be worked out that detail specific calculations and
applications.

The system Al_2SiO_5

As a first example, consider the analytical formulation of the univariant equilibrium kyanite + sillimanite. The number of system components is
one (Al_2SiO_5) and the total number of phase components is two (Al_2SiO_5 in
kyanite and Al_2SiO_5 in sillimanite). Thus, in addition to the two Gibbs-
Duhem equations that can be written for this assemblage (one for each phase),
we can also write one independent condition of heterogeneous equilibrium

$$\mu_{Al_2SiO_5}^{Ky} = \mu_{Al_2SiO_5}^{Si}$$

resulting in a total of three equations in four unknowns (see top panel of

Table 1). Note that the variance of this system of equations is one and is the same obtained from application of the phase rule.

A solution to this system of equations may be obtained by taking the ratio of one variable to the others. For instance, the slope of the univariant kyanite + sillimanite equilibrium curve, dP/dT, can be calculated by dividing each equation by dT and rearranging as shown in the middle and bottom panels of Table 1. The solution dP/dT can be obtained by simultaneous solution of these three equations:

$$\frac{dP}{dT} = \frac{\bar{S}^{Si} - \bar{S}^{Ky}}{\bar{V}^{Si} - \bar{V}^{Ky}} = \frac{\Delta \bar{S}}{\Delta \bar{V}}$$

which is, of course, the Clapeyron equation. [Note: A discussion of techniques for solving these systems of equations is given in Appendix A.]

The system Al_2SiO_5-SiO_2-Fe_2O_3-$FeTiO_3$

As a slightly more complicated example, consider the analytical formulation of equilibrium among aluminosilicate, Fe-Ti oxide, and quartz. Grew (1980) has discussed the solubility of Fe_2SiO_5 in sillimanite and has used the Gibbs method to derive equations relating the mole fraction of Fe_2SiO_5 in sillimanite (x_{Fe}^{Si}) to T, P, and the mole fraction of Fe_2O_3 in coexisting Fe-Ti oxide. This example is particularly illustrative because it demonstrates how P-T regions may be contoured with isopleths of constant mineral composition.

Consider first the coexistence of sillimanite + kyanite + ilmenite + quartz and the question of how the Fe_2SiO_5 content of sillimanite changes with temperature in this assemblage. Following Grew (1980), the four-phase assemblage sillimanite + kyanite + ilmenite + quartz can be treated in terms of the four-component system: SiO_2-Al_2O_3-Fe_2O_3-$FeTiO_3$. The phase rule variance of this assemblage is two; therefore, the assemblage is stable in an area in P-T space.

Assuming that both the aluminosilicates and ilmenite are binary solid solutions, there are seven phase components in this system: SiO_2 (quartz), Al_2SiO_5 and Fe_2SiO_5 (sillimanite); Al_2SiO_5 and Fe_2SiO_5 (kyanite); and Fe_2O_3 and $FeTiO_3$ (ilmenite). There are three conditions of heterogeneous equilibrium (n = c - r = 3):

$$0 = d\mu_{SiO_2}^{Qt} + d\mu_{Fe_2O_3}^{Im} - d\mu_{Fe}^{Si} \tag{33}$$

$$0 = d\mu_{Al}^{Ky} - d\mu_{Al}^{Si} \tag{34}$$

$$0 = d\mu_{Fe}^{Ky} - d\mu_{Fe}^{Si} . \tag{35}$$

If we wish to monitor the mole fraction of Fe_2SiO_5 in sillimanite in this assemblage, it is necessary to introduce an equation that introduces the unknown dX_{Fe}^{Si}; namely,

$$d(\mu_{Fe}^{Si} - \mu_{Al}^{Si}) = -(\bar{S}_{Fe}^{Si} - \bar{S}_{Al}^{Si})dT + (\bar{V}_{Fe}^{Si} - \bar{V}_{Al}^{Si})dP + \left(\frac{\partial^2(\bar{G}^{Si})}{\partial(X_{Fe}^{Si})^2}\right)_{P,T} dX_{Fe}^{Si} . \tag{36}$$

The subscripts Fe and Al refer to the components Fe_2SiO_5 and Al_2SiO_5 in sillimanite, respectively. The full set of equations is listed in Table 2. These equations can be solved for the equilibrium dP/dT slope for the assemblage at constant X_{Fe}^{Si}, by setting $dX_{Fe}^{Si} = 0$ (the column containing this variable therefore is canceled), dividing through by dT, and moving the column containing the entropies to the other side of the equal sign. Following these manipulations, there are eight non-homogeneous equations in eight unknowns. Solution of these equations yields

$$\left(\frac{dP}{dT}\right)_{X_{Fe}^{Si}} = \frac{(\bar{S}^{Si} - \bar{S}^{Ky}) - (\bar{S}_{Fe}^{Si} - \bar{S}_{Al}^{Si})(X_{Fe}^{Si} - X_{Fe}^{Ky})}{(\bar{V}^{Si} - \bar{V}^{Ky}) - (\bar{V}_{Fe}^{Si} - \bar{V}_{Al}^{Si})(X_{Fe}^{Si} - X_{Fe}^{Ky})} . \tag{37}$$

Similar procedures holding P constant (dP = 0) yield the results

$$\left(\frac{\partial T}{\partial X_{Fe}^{Si}}\right)_P = \frac{-\left(\dfrac{\partial^2(\bar{G}^{Si})}{\partial(X_{Fe}^{Si})^2}\right)_{P,T}(X_{Fe}^{Si} - X_{Fe}^{Ky})}{(\bar{S}^{Si} - \bar{S}^{Ky}) - (\bar{S}_{Fe}^{Si} - \bar{S}_{Al}^{Si})(X_{Fe}^{Si} - X_{Fe}^{Ky})} \tag{38}$$

Table 2. System of equations for divariant equilibrium sillimanite + kyanite + ilmenite + quartz

$$
\begin{bmatrix}
\bar{S}^{Si} & -\bar{V}^{Si} & X_{Al}^{Si} & X_{Fe}^{Si} & 0 & 0 & 0 & 0 & 0 \\
\bar{S}^{Ky} & -\bar{V}^{Ky} & 0 & 0 & X_{Al}^{Ky} & X_{Fe}^{Ky} & 0 & 0 & 0 \\
\bar{S}^{Qt} & -\bar{V}^{Qt} & 0 & 0 & 0 & 0 & 1 & 0 & 0 \\
\bar{S}^{Im} & -\bar{V}^{Im} & 0 & 0 & 0 & 0 & 0 & X_{Fe}^{Im} & X_{Ti}^{Im} & 0 \\
0 & 0 & -1 & 0 & 1 & 0 & 0 & 0 & 0 \\
0 & 0 & 0 & -1 & 0 & 1 & 0 & 0 & 0 \\
0 & 0 & 0 & -1 & 0 & 0 & 1 & 1 & 0 \\
-(\bar{S}_{Fe}^{Si}-\bar{S}_{Al}^{Si}) & (\bar{V}_{Fe}^{Si}-\bar{V}_{Al}^{Si}) & 1 & -1 & 0 & 0 & 0 & 0 & \left(\dfrac{\partial^2(\bar{G}^{Si})}{\partial(X_{Fe}^{Si})^2}\right)_{P,T}
\end{bmatrix}
\cdot
\begin{bmatrix}
dT \\ dP \\ d\mu_{Al}^{Si} \\ d\mu_{Fe}^{Si} \\ d\mu_{Al}^{Ky} \\ d\mu_{Fe}^{Ky} \\ d\mu_{SiO_2}^{Qt} \\ d\mu_{Fe}^{Im} \\ d\mu_{Ti}^{Im} \\ dX_{Fe}^{Si}
\end{bmatrix}
=
\begin{bmatrix}
0 \\ 0 \\ 0 \\ 0 \\ 0 \\ 0 \\ 0 \\ 0
\end{bmatrix}
$$

---

and with constant T ($dT = 0$),

$$\left(\frac{\partial P}{\partial X_{Fe}^{Si}}\right)_T = \frac{\left(\frac{\partial^2 (\bar{G}^{Si})}{\partial (X_{Fe}^{Si})^2}\right)_{P,T} (X_{Fe}^{Si} - X_{Fe}^{Ky})}{(\bar{V}^{Si} - \bar{V}^{Ky}) - (\bar{V}_{Fe}^{Si} - \bar{V}_{Al}^{Si})(X_{Fe}^{Si} - X_{Fe}^{Ky})} \; . \tag{39}$$

These are equations that define isopleths of X_{Fe}^{Si} for the four-phase assemblage in P-T space. The first equation gives the slopes of the isopleths, and the second and third equations give the spacing between isopleths at either constant P or T. Note that the slope of the isopleths depends on the entropies, volumes, partial molar entropies and volumes, and composition; whereas, the spacing between isopleths depends also on the curvature of the Gibbs function for sillimanite.

The above example is for the divariant assemblage sillimanite + kyanite + ilmenite + quartz, but Grew (1980) also treated the trivariant assemblage sillimanite (or kyanite) + ilmenite (or hematite) + quartz. In order to obtain a solution for a trivariant assemblage, it is necessary to hold two independent variables constant. A logical choice would be to hold both X_{Fe}^{Si} and X_{Fe}^{Im} constant; this requires generating an additional equation that introduces the variable dX_{Fe}^{Im} so that dX_{Fe}^{Im} can be set equal to zero. The full set of equations is shown in Table 3. Setting $dX_{Fe}^{Si} = dX_{Fe}^{Im} = 0$, one can solve for the following slopes:

$$\left(\frac{\partial P}{\partial T}\right)_{X_{Fe}^{Si}, X_{Fe}^{Im}} = \frac{(\bar{S}^{Si} - \bar{S}^{Im} - \bar{S}^{Qt}) + X_{Al}^{Si}(\bar{S}_{Fe}^{Si} - \bar{S}_{Al}^{Si}) - X_{Ti}^{Im}(\bar{S}_{Fe}^{Im} - \bar{S}_{Ti}^{Im})}{(\bar{V}^{Si} - \bar{V}^{Im} - \bar{V}^{Qt}) + X_{Al}^{Si}(\bar{V}_{Fe}^{Si} - \bar{V}_{Al}^{Si}) - X_{Ti}^{Im}(\bar{V}_{Fe}^{Im} - \bar{V}_{Ti}^{Im})} \tag{40}$$

Table 3. System of equations for trivariant equilibrium sillimanite + ilmenite + quartz

$$
\begin{bmatrix}
\bar{S}^{Si} & -\bar{V}^{Si} & X_{Al}^{Si} & X_{Fe}^{Si} & 0 & 0 & 0 & 0 & 0 \\
\bar{S}^{Qt} & -\bar{V}^{Qt} & 0 & 0 & 1 & 0 & 0 & 0 & 0 \\
\bar{S}^{Im} & -\bar{V}^{Im} & 0 & 0 & 0 & X_{Fe}^{Im} & X_{Ti}^{Im} & 0 & 0 \\
0 & 0 & 0 & -1 & 1 & 1 & 0 & 0 & 0 \\
-(\bar{S}_{Fe}^{Si} - \bar{S}_{Al}^{Si}) & (\bar{V}_{Fe}^{Si} - \bar{V}_{Al}^{Si}) & 1 & -1 & 0 & 0 & 0 & \left(\frac{\partial^2(\bar{G}^{Si})}{\partial(X_{Fe}^{Si})^2}\right)_{P,T} & 0 \\
-(\bar{S}_{Fe}^{Im} - \bar{S}_{Ti}^{Im}) & (\bar{V}_{Fe}^{Im} - \bar{V}_{Ti}^{Im}) & 0 & 0 & 0 & -1 & 1 & 0 & \left(\frac{\partial^2(\bar{G}^{Im})}{\partial(X_{Fe}^{Im})^2}\right)_{P,T}
\end{bmatrix}
\begin{bmatrix}
dT \\
dP \\
d\mu_{Al}^{Si} \\
d\mu_{Fe}^{Si} \\
d\mu_{SiO_2}^{Qt} \\
d\mu_{Fe}^{Im} \\
d\mu_{Ti}^{Im} \\
dX_{Fe}^{Si} \\
dX_{Fe}^{Im}
\end{bmatrix}
=
\begin{bmatrix}
0 \\
0 \\
0 \\
0 \\
0 \\
0 \\
0 \\
0 \\
0
\end{bmatrix}
$$

124

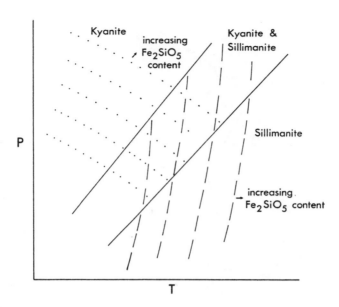

Figure 5. Schematic P-T diagram showing isopleths of constant X^{Si}_{Fe} in both sillimanite (dashed lines) and kyanite (dotted lines). Solid lines mark the range of P-T conditions for the co-existence of sillimanite + kyanite + ilmenite + quartz. Slopes are schematic only.

$$\left(\frac{\partial T}{\partial X^{Si}_{Fe}}\right)_{P,X^{Im}_{Fe}} = \frac{X^{Si}_{Al}\left(\frac{\partial^2(\bar{G}^{Si})}{\partial(X^{Si}_{Fe})^2}\right)_{P,T}}{(\bar{S}^{Si}-\bar{S}^{Im}-\bar{S}^{Qt}) + X^{Si}_{Al}(\bar{S}^{Si}_{Fe}-\bar{S}^{Si}_{Al}) - X^{Im}_{Ti}(\bar{S}^{Im}_{Fe}-\bar{S}^{Im}_{Ti})} \tag{41}$$

$$\left(\frac{\partial P}{\partial X^{Si}_{Fe}}\right)_{T,X^{Im}_{Fe}} = \frac{-X^{Si}_{Al}\left(\frac{\partial^2(\bar{G}^{Si})}{\partial(X^{Si}_{Fe})^2}\right)_{P,T}}{(\bar{V}^{Si}-\bar{V}^{Im}-\bar{V}^{Qt}) + X^{Si}_{Al}(\bar{V}^{Si}_{Fe}-\bar{V}^{Si}_{Al}) - X^{Im}_{Ti}(\bar{V}^{Im}_{Fe}-\bar{V}^{Im}_{Ti})} \cdot \tag{42}$$

These equations, taken with the previous set, permit contouring of both the divariant and trivariant regions of P-T space for the system SiO_2-Al_2O_3-Fe_2O_3-$FeTiO_3$ containing the phases sillimanite (and/or kyanite) + ilmenite (or hema-tite) + quartz. Figure 5 is such a schematic P-T grid with isopleths taken in part from the grid of Grew (1980, Fig. 7). Because there are insufficient entropy and volume data to permit calculation of the grid, the slopes shown are schematic only. However, Grew (1980) was able to conclude with the avail-able data that the Fe_2O_3 content of sillimanite coexisting with ilmenite and quartz decreases with pressure and increases with temperature and that the

125

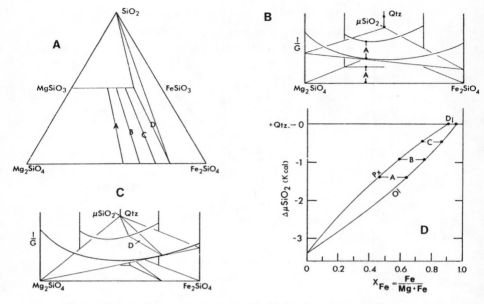

Figure 6. (A) Hypothetical phase diagram for the system SiO_2 - Mg_2SiO_4 - Fe_2SiO_4 showing two-phase orthopyroxene + olivine assemblages (A,B,C) and a three-phase orthopyroxene + olivine + quartz assemblage (D). (B) Perspective view of $\bar{G}$-X construction showing plane tangent to assemblage (A) and its intercept with the SiO_2 axis (= μ_{SiO_2}). (C) Similar $\bar{G}$-X construction for assemblage (D). Note that μ_{SiO_2} is higher here than for assemblage (A). (D) μ_{SiO_2} - X_{Mg} diagram for coexisting orthopyroxene + olivine. Numbers calculated as discussed in text.

rate of increase with temperature is greater in hematite-bearing rocks than in ilmenite-bearing rocks.

The system SiO_2-FeO-MgO

This example is taken from Rumble (1976) and is especially illustrative because it is possible to verify the results of the analytical formulation graphically. Consider again the three-component system SiO_2-MgO-FeO that has been discussed with reference to Figures 1 and 4 and a portion of which (SiO_2-Mg_2SiO_4-Fe_2SiO_4) is shown in Figure 6A. The assemblages A-D are distinct assemblages consisting of orthopyroxene + olivine ± quartz with different Fe/Mg ratios in the pyroxene and olivine. Each assemblage, for example, may have accumulated in a different layer of a stratiform ultramafic body or may represent different compositional layers in a high-grade iron formation. The first question to be answered is: How does the chemical potential of SiO_2 -- μ_{SiO_2} -- vary from layer to layer, assuming that all assemblages crystallized at the same P and T?

The graphical solution to this question can be seen in Figure 6B,C and D. In Figure 6B a plane is drawn tangent to the $\bar{G}$-surfaces for orthopyroxene and olivine at the Fe/Mg of assemblage (A) and the tangent plane intersects the SiO_2 axis at μ_{SiO_2}. In Figure 6C, the tangent plane is drawn for assemblage D, which also contains quartz, and μ_{SiO_2} for this assemblage is $\mu_{SiO_2}^D = \mu_{SiO_2}^{Qt}$. It is clear that $\mu_{SiO_2}^D$ is higher than $\mu_{SiO_2}^A$. Also, because assemblage D contains free quartz, this is the highest value of μ_{SiO_2} than can be attained by any assemblage at this P and T. The conclusion to be drawn is that samples with more Fe-rich orthopyroxene + olivine crystallized at higher values of μ_{SiO_2} as depicted in Figure 6D. Note that *the critical factor* in determining whether μ_{SiO_2} will increase or decrease with a given change in Fe/Mg *is the partitioning of Fe and Mg* between olivine and orthopyroxene. In Figure 6A, $X_{Fe}^{Px} < X_{Fe}^{Ol}$, which is how natural samples behave. However, had the partitioning been reversed (say, for a different mineral pair), the change in μ_{SiO_2} would have been in the opposite sense.

The analytical formulation of this problem proceeds as follows. First, it is necessary to choose the phase components for olivine and orthopyroxene, for example, $MgSiO_3$, $FeSiO_3$, Mg_2SiO_4 and Fe_2SiO_4. Note that it would have made no difference if we had chosen $Mg_2Si_2O_6$ and $Fe_2Si_2O_6$ for pyroxene components -- we would simply have had to alter the stoichiometric coefficients in the conditions of heterogeneous equilibria. For the two-phase region olivine + orthopyroxene, there are two Gibbs-Duhem equations that can be written, and there is one independent condition of heterogeneous equilibrium ($n = r - c = 4 - 3 = 1$), based on the stoichiometric relation

$$\underset{\text{olivine}}{\underset{[Fa]}{Fe_2SiO_4}} + \underset{\text{pyroxene}}{\underset{[En]}{2MgSiO_3}} = \underset{\text{olivine}}{\underset{[Fo]}{Mg_2SiO_4}} + \underset{\text{pyroxene}}{\underset{[Fs]}{2FeSiO_3}} , \qquad (43)$$

which can be written in terms of chemical potentials as

$$0 = d\mu_{Fa}^{Ol} - d\mu_{Fo}^{Ol} - 2d\mu_{Fs}^{Px} + 2d\mu_{En}^{Px} \qquad (44)$$

These three equations are sufficient to describe the variance of the system.

The goal of this exercise is to derive an expression that describes the change in μ_{SiO_2} with respect to changing Fe/Mg in pyroxene (or olivine) [e.g., $(\partial\mu_{SiO_2}/\partial X_{Fs}^{Px})_{P,T}$]. It is therefore necessary to introduce the variables $d\mu_{SiO_2}$ and dX_{Fs}^{Px} with suitable equations. $d\mu_{SiO_2}$ can be introduced with the relation

$$SiO_2 + Mg_2SiO_4 = 2MgSiO_3 \qquad (45)$$

which requires, at equilbrium, that

Table 4. System of homogeneous equations in matrix form for the equilibrium orthopyroxene + olivine

$$
\begin{bmatrix}
\overline{S}^{Px} & -\overline{V}^{Px} & X_{Fs}^{Px} & X_{En}^{Px} & 0 & 0 & 0 & 0 \\
\overline{S}^{01} & -\overline{V}^{01} & 0 & 0 & X_{Fa}^{01} & X_{Fo}^{01} & 0 & 0 \\
0 & 0 & 2 & -2 & -1 & 1 & 0 & 0 \\
0 & 0 & 0 & -2 & 0 & 1 & 1 & 0 \\
-(\overline{S}_{Fs}^{Px}-\overline{S}_{En}^{Px}) & (\overline{V}_{Fs}^{Px}-\overline{V}_{En}^{Px}) & -1 & 1 & 0 & 0 & 0 & \left(\dfrac{\partial^2(\overline{G}^{Px})}{\partial(X_{Fs}^{Px})^2}\right)_{P,T}
\end{bmatrix}
\begin{bmatrix}
dT \\ dP \\ d\mu_{Fs}^{Px} \\ d\mu_{En}^{Px} \\ d\mu_{Fa}^{01} \\ d\mu_{Fo}^{01} \\ dX_{Fs}^{Px}
\end{bmatrix}
=
\begin{bmatrix}
0 \\ 0 \\ 0 \\ 0 \\ 0
\end{bmatrix}
$$

Table 5. System of non-homogeneous equations for the equilibrium orthopyroxene + olivine

$$
\begin{bmatrix}
X_{Fs}^{Px} & X_{En}^{Px} & 0 & 0 & 0 \\
0 & 0 & X_{Fa}^{01} & X_{Fo}^{01} & 0 \\
2 & -2 & -1 & 1 & 0 \\
0 & -2 & 0 & 1 & 1 \\
-1 & 1 & 0 & 0 & 0
\end{bmatrix}
\begin{bmatrix}
d\mu_{Fs}^{Px}/dX_{Fs}^{Px} \\
d\mu_{En}^{Px}/dX_{Fs}^{Px} \\
d\mu_{Fa}^{01}/dX_{Fs}^{Px} \\
d\mu_{Fo}^{01}/dX_{Fs}^{Px} \\
d\mu_{SiO_2}/dX_{Fs}^{Px}
\end{bmatrix}
=
\begin{bmatrix}
0 \\ 0 \\ 0 \\ 0 \\ -\left(\dfrac{\partial^2(\overline{G}^{Px})}{\partial(X_{Fs}^{Px})^2}\right)_{P,T}
\end{bmatrix}
$$

$$0 = d\mu_{SiO_2} + d\mu_{Fo}^{01} - 2d\mu_{En}^{Px} \tag{46}$$

(note that the equivalent expression involving the Fe-end member could also have been used). The equation needed to introduce dX_{Fs}^{Px} is

$$d(\mu_{Fs}^{Px}-\mu_{En}^{Px}) = -(\overline{S}_{Fs}^{Px}-\overline{S}_{En}^{Px})dT + (\overline{V}_{Fs}^{Px}-\overline{V}_{En}^{Px})dP + \left(\frac{\partial^2(\overline{G}^{Px})}{\partial(X_{Fs}^{Px})^2}\right)_{P,T} dX_{Fs}^{Px} . \tag{47}$$

The resulting five homogeneous equations in eight unknowns are shown in matrix form in Table 4. Note that the phase rule tells us that the two-phase assemblage Px + 01 has three degrees of freedom, as does the system of equations.

A solution to these equations can be obtained by first fixing P and T ($dT = dP = 0$) and then dividing each equation by dX_{Fs}^{Px}. The result is five non-homogeneous equations in five unknowns as shown in Table 5. There are many ways to obtain a solution to the equations in Table 5, and some techniques are given in Appendix A. The analytical solution to these equations is

$$\left(\frac{\partial \mu SiO_2}{\partial X_{Fs}^{Px}}\right)_{P,T} = 2 \cdot \left(\frac{\partial^2 (\bar{G}^{Px})}{\partial (X_{Fs}^{Px})^2}\right)_{P,T} (X_{Fa}^{Ol} - X_{Fs}^{Px}) \ . \tag{48}$$

The sign of this expression may be deduced because the curvature of the Gibbs function must be positive for a stable binary two-component solid solution (e.g., sign of $[\partial^2 (\bar{G}^{Px})/\partial (X_{Fs}^{Px})^2]_{P,T}$ is positive). Moreover, because $X_{Fa}^{Ol} >$ X_{Fs}^{Px}, $(X_{Fa}^{Ol} - X_{Fs}^{Px})$ is positive. Thus $(\partial \mu_{SiO_2}/\partial X_{Fs}^{Px})_{P,T}$ must be positive, which is the same result obtained graphically. Note that the algebraic solution also reveals the importance of the Fe-Mg partitioning between the coexisting phases. Had X_{Fa}^{Ol} been less than X_{Fs}^{Px}, the result would have had the opposite sign.

The above equation gives the slope of the olivine-pyroxene equilibrium curve on an isothermal, isobaric $\mu_{SiO_2} - X_{Fs}^{Px}$ diagram. If a numerical value for this slope is desired, it is necessary to know the composition of co-existing orthopyroxene and olivine at the T and P of interest and to have an analytical expression for the Gibbs function for orthopyroxene (e.g., $\bar{G}^{Px} = X_{Fs}^{Px}\mu_{Fs}^{Px} + X_{En}^{Px}\mu_{En}^{Px})$ that can be differentiated twice to get the curvature. If orthopyroxene is modeled as an ideal one-site Fe-Mg solid solution (which is clearly not correct), then

$$\bar{G}^{Px} = X_{Fs}^{Px}\mu_{Fs}^{\circ} + X_{En}^{Px}\mu_{En}^{\circ} + RT(X_{Fs}^{Px} ln X_{Fs}^{Px} + X_{En}^{Px} ln X_{En}^{Px}) \tag{49}$$

and

$$\left(\frac{\partial^2 (\bar{G}^{Px})}{\partial (X_{Fs}^{Px})^2}\right)_{P,T} = \frac{RT}{X_{Fs}^{Px} X_{En}^{Px}} \ . \tag{50}$$

More complicated expressions can also be differentiated to yield curvatures.

The solution can be taken one step farther and numerical values on a μ - X diagram (e.g., Fig. 6D) can be computed by integration of this differential equation:

$$\int d\mu_{SiO_2} = \int \left(\frac{2RT}{X_{Fs}^{Px} X_{En}^{Px}} (X_{Fa}^{Ol} - X_{Fs}^{Px})\right) dX_{Fs}^{Px} \ . \tag{51}$$

A discussion of how to integrate equations such as these is given in Appendix B.

Results of this integration are presented in Table 6 and Figure 6D. Also shown in Table 6 is the slope, $(\partial \mu_{SiO_2}/\partial X_{Fa}^{Ol})_{P,T}$, calculated in the same manner as that for orthopyroxene. Note that the $(\partial \mu/\partial X)_{P,T}$ slope for orthopyroxene *steepens* with increasing X_{Fs}^{Px} whereas the slope for olivine becomes more gentle. This feature gives rise to the "closed loop" appearance of Figure 6D that is

Table 6. Values computed from numerical integration of olivine-orthopyroxene example

X_{Fs}^{Px}	X_{Fa}^{Ol}	$\left(\dfrac{\partial\mu_{SiO_2}}{\partial X_{Fs}^{Px}}\right)_{P,T}$	$\Delta\mu_{SiO_2}$	$\Sigma\Delta\mu_{SiO_2}$	$\left(\dfrac{\partial\mu_{SiO_2}}{\partial X_{Fa}^{Ol}}\right)_{P,T}$
0.9	0.95	2689			5289
			277	277	
0.8	0.90	2854			4984
			295	572	
0.7	0.84	3040			4679
			315	887	
0.6	0.77	3252			4374
			337	1224	
0.5	0.69	3497			4068
			364	1588	
0.4	0.59	3780			3764
			395	1983	
0.3	0.48	4113			3458
			431	2414	
0.2	0.35	4512			3153
			475	2889	
0.1	0.20	4995			2848
			526	3415	
0.01	0.02	5528			2573

Table 7. Slopes of P-T-X loops for the equilibrium orthopyroxene + olivine + quartz

$$\left(\frac{\partial P}{\partial X_{Fs}^{Px}}\right)_T = \frac{2\ [\partial^2(\overline{G}^{Px})/\partial(X_{Fs}^{Px})^2]_{P,T}(X_{Fa}^{Ol} - X_{Fs}^{Px})}{\overline{V}^Q + \overline{V}^{Ol} - 2\overline{V}^{Px} - 2\cdot(\overline{V}_{Fs}^{Px} - \overline{V}_{En}^{Px})(X_{Fa}^{Ol} - X_{Fs}^{Px})}$$

$$\left(\frac{\partial T}{\partial X_{Fs}^{Px}}\right)_P = \frac{-2\ [\partial^2(\overline{G}^{Px})/\partial(X_{Fs}^{Px})^2]_{P,T}(X_{Fa}^{Ol} - X_{Fs}^{Px})}{\overline{S}^Q + \overline{S}^{Ol} - 2\overline{S}^{Px} - 2\cdot(\overline{S}_{Fs}^{Px} - \overline{S}_{En}^{Px})(X_{Fa}^{Ol} - X_{Fs}^{Px})}$$

$$\left(\frac{\partial T}{\partial P}\right)_{X_{Fs}^{Px}} = \frac{\overline{S}^Q + \overline{S}^{Ol} - 2\overline{S}^{Px} - 2\cdot(\overline{S}_{Fs}^{Px} - \overline{S}_{En}^{Px})(X_{Fa}^{Ol} - X_{Fs}^{Px})}{\overline{V}^Q + \overline{V}^{Ol} - 2\overline{V}^{Px} - 2\cdot(\overline{V}_{Fs}^{Px} - \overline{V}_{En}^{Px})(X_{Fa}^{Ol} - X_{Fs}^{Px})}$$

characteristic of diagrams of this sort where intensive parameters are plotted against composition (e.g., T-X, P-X, and μ-X diagrams).

P-T-X derivatives. There are many other derivatives that can be derived in the system SiO_2-MgO-FeO shown in Figure 6. In particular, derivatives involving P, T and X can be quite informative and may have application in geo-thermometry and geobarometry. Indeed, Bohlen et al. (1980) and Bohlen and Boettcher (1981) have recently discussed the experimental calibration and application of the assemblage orthopyroxene + olivine + quartz as a geobarom-eter.

Consider the divariant three-phase assemblage orthopyroxene + olivine + quartz shown in Figure 6A. The three-phase triangle is fixed at constant P and T, but will shift in composition in response to changes in these vari-ables. To monitor these changes in composition, the three derivatives,

$$(\partial P / \, X_{Fs}^{Px})_T \, , \ (\partial T / \partial X_{Fs}^{Px})_P \, , \ (\partial P / \partial T)_{X_{Fs}^{Px}}$$

can be derived as shown in Table 7 (note that only two of these three deriva-tives are independent). Similar derivates involving X_{Fa}^{Ol} can also be computed.

Examination of the equations listed in Table 7 reveals that derivatives involving pressure, temperature, and composition always involve terms resem-bling $\Delta \bar{V}$, $\Delta \bar{S}$, and the curvature of the Gibbs function. The slope, therefore, $(\partial P / \partial T)$ looks very similar to the Clapeyron equation, but in this case the slope is for phases of variable composition. It is informative to recast the equations to examine the symmetry of the numerators and denominators of the derivative. For example, the derivative

$$(\frac{\partial P}{\partial X_{Fs}^{Px}})_T =$$

$$\frac{2\left[(\frac{\partial \mu_{Fs}^{Px}}{\partial X_{Fs}^{Px}})_{T,P} - (\frac{\partial \mu_{En}^{Px}}{\partial X_{Fs}^{Px}})_{T,P}\right](X_{Fa}^{Ol}-X_{Fs}^{Px})}{(\frac{\partial \mu^{Qt}}{\partial P})_{T,X_{Fs}^{Px}} + (\frac{\partial \mu^{Ol}}{\partial P})_{T,X_{Fs}^{Px}} - 2(\frac{\partial \mu^{Px}}{\partial P})_{T,X_{Fs}^{Px}} - 2\left[(\frac{\partial \mu_{Fs}^{Px}}{\partial P})_{T,X_{Fs}^{Px}} - (\frac{\partial \mu_{En}^{Px}}{\partial P})_{T,X_{Fs}^{Px}}\right](X_{Fa}^{Ol}-X_{Fs}^{Px})} \quad (52)$$

by substituting back the appropriate derivatives for the partial molar quan-tities. Thus it can be seen that the terms in these derivatives are all slopes on μ-P, μ-X or μ-T diagrams, appropriately weighted by either the stoichiometry of the heterogeneous equilibria or by variables of mineral composition.

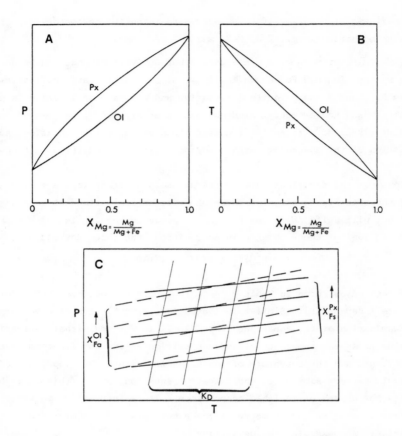

Figure 7. Schematic P-X, T-X and P-T diagram for the equilibrium orthopyroxene + olivine + quartz. Lines on Figure 7C are contours at constant X_{Fs}^{Px}, X_{Fa}^{Ol} and K_D^{Px-Ol}.

The equations listed in Table 7 can be integrated as before to yield P-X, T-X and P-T diagrams as shown schematically in Figure 7. A discussion of how to integrate these equations is given in Appendix B. Note the similarity between the P-X and T-X diagram and the μ-X diagram of Figure 6D. On the P-T diagram are shown contours of constant X_{Fs}^{Px} and X_{Fa}^{Ol}. In general, the slopes of these contours will be different and thus the intersection of the two contours gives a unique P-T point, which has considerable application in geothermometry and geobarometry. Note that the P-T region can also be contoured with lines of constant K_D^{Px-Ol}, which will intersect the contours of constant mineral composition at the appropriate values as shown in Figure 7C. Thus the Gibbs method and the alternative method of equilibrium constants both can be used to contour P-T space with isopleths of constant mineral composition or K_D. Although the results appear in somewhat different form, they are internally consistent.

The AFM diagram

A third example to consider is the behavior of three- and four-phase assemblages on the A-F-M diagram for pelitic schists (J.B. Thompson, 1957). An analysis of P-X and T-X equilibria in this system has been given by A.B. Thompson (1976), but this analysis was not cast in the analytical formulation presented here. A few examples will illustrate how the Gibbs method can be used to decipher the behavior of mineral assemblages in pelitic schists with respect to changes in P, T, and μ_{H_2O}.

Consider the four-phase A-F-M assemblage garnet + biotite + chlorite + staurolite (+ muscovite + quartz). In the model pelitic system, SiO_2-Al_2O_3-MgO-FeO-K_2O-H_2O, this assemblage is divariant with respect to P, T, and μ_{H_2O}, and it buffers μ_{H_2O} at a specific value at constant P and T. The four different assemblages containing each combination of three of the mafic minerals -- garnet + biotite + staurolite, garnet + biotite + chlorite, garnet + staurolite + chlorite, and biotite + staurolite + chlorite -- are trivariant. The Fe/Mg of coexisting minerals in the trivariant assemblages shift with changing P, T, and μ_{H_2O} forming isothermal P-X and isobaric T-X loops at constant μ_{H_2O} or isobaric, isothermal μ_{H_2O}-X loops (Thompson, 1976, Fig. 1B). Using the Gibbs method we can derive analytical expressions for how the trivariant three-phase triangles shift in Fe/Mg with changing P, T, and μ_{H_2O}. The phase compositions and the stoichiometry of the continuous reactions, as given by Thompson (1976), are presented in Table 8. All phases are treated as binary Fe-Mg solid solutions with the compositions shown.

Derivatives with μ_{H_2O} at constant P and T. The system of equations necessary to describe changes in the composition of garnet in the assemblage garnet + staurolite + chlorite with changing P, T, or μ_{H_2O} is shown in Table 9. The first five equations are Gibbs-Duhem equations for quartz, muscovite, garnet, staurolite, and chlorite. The next two equations are conditions of heterogeneous equilibrium. There are six system components and eight phase components for this assemblage and, therefore, two linearly independent conditions of heterogeneous equilibrium. The two that were chosen are Fe-Mg exchange between garnet and chlorite and between garnet and staurolite. This yields seven equations in 10 unknowns, which correctly describes the variance of the system.

133

Table 8.

Phase components and stoichiometric relations for garnet-staurolite-biotite-chlorite-muscovite-quartz assemblages in the system $SiO_2-Al_2O_3-MgO-FeO-K_2O-H_2O$ (after Thompson, 1976)

Mineral Name	Abbreviation	Formula	Activity-composition Model
Garnet	Gar	$(Fe,Mg)_3Al_2Si_3O_{12}$	X_{Fe}^3
Staurolite	Sta	$(Fe,Mg)_2Al_{26/3}Si_4O_{22}(OH)_2$	X_{Fe}^2
Biotite	Bio	$K_3(Fe,Mg)_8AlAl_4Si_8O_{30}(OH)_6$	X_{Fe}^8
Chlorite	Chl	$(Fe,Mg)_7Al_2Al_2Si_4O_{15}(OH)_{12}$	X_{Fe}^7
Muscovite	Mus	$KAl_3Si_3O_{10}(OH)_2$	1

$$14\ Chl + 3\ Mus + 33\ Qt = 30\ Gar + 1\ Bio + 84\ H_2O$$
$$42\ Sta + 33\ Bio + 213\ Qt = 116\ Gar + 99\ Mus + 42\ H_2O$$
$$3\ Sta + 33\ Chl + 93\ Qt = 79\ Gar + 201\ H_2O$$
$$116\ Chl + 237\ Mus = 90\ Sta + 79\ Bio + 183\ Qt + 606\ H_2O$$

Table 9.

System of homogeneous equations for the equilibrium quartz (Qt) + muscovite (Mus) + garnet (Gar) + Chlorite (Chl) + staurolite (Sta)

$$
\begin{bmatrix}
\bar{S}^{Qt} & -\bar{V}^{Qt} & 1 & 0 & 0 & 0 & 0 & 0 & 0 & 0 & 0 & 0 \\
\bar{S}^{Mus} & -\bar{V}^{Mus} & 0 & 1 & 0 & 0 & 0 & 0 & 0 & 0 & 0 & 0 \\
\bar{S}^{Gar} & -\bar{V}^{Gar} & 0 & 0 & X_{Fe}^{Gar} & X_{Mg}^{Gar} & 0 & 0 & 0 & 0 & 0 & 0 \\
\bar{S}^{Chl} & -\bar{V}^{Chl} & 0 & 0 & 0 & 0 & X_{Fe}^{Chl} & X_{Mg}^{Chl} & 0 & 0 & 0 & 0 \\
\bar{S}^{Sta} & -\bar{V}^{Sta} & 0 & 0 & 0 & 0 & 0 & 0 & X_{Fe}^{Sta} & X_{Mg}^{Sta} & 0 & 0 \\
0 & 0 & 0 & 0 & 7 & -7 & -3 & 3 & 0 & 0 & 0 & 0 \\
0 & 0 & 0 & 0 & 2 & -2 & 0 & 0 & -3 & 3 & 0 & 0 \\
0 & 0 & 93 & 0 & -79 & 0 & 33 & 0 & 3 & 0 & -201 & 0 \\
-(\bar{S}_{Mg}^{Gar}-\bar{S}_{Fe}^{Gar}) & (\bar{V}_{Mg}^{Gar}-\bar{V}_{Fe}^{Gar}) & 0 & 0 & 1 & -1 & 0 & 0 & 0 & 0 & 0 & \left(\dfrac{\partial^2(\bar{G}^{Gar})}{\partial(X_{Mg}^{Gar})^2}\right)_{P,T}
\end{bmatrix}
\begin{bmatrix}
dT \\
dP \\
d\mu^{Qt} \\
d\mu^{Mus} \\
d\mu_{Fe}^{Gar} \\
d\mu_{Mg}^{Gar} \\
d\mu_{Fe}^{Chl} \\
d\mu_{Mg}^{Chl} \\
d\mu_{Fe}^{Sta} \\
d\mu_{Mg}^{Sta} \\
d\mu_{H_2O} \\
dX_{Mg}^{Gar}
\end{bmatrix}
=
\begin{bmatrix}
0 \\
0 \\
0 \\
0 \\
0 \\
0 \\
0 \\
0 \\
0
\end{bmatrix}
$$

Note that we have not as yet incorporated μ_{H_2O} as a variable of the system. To do so we must specify the behavior of μ_{H_2O} in rocks during metamorphism. The least restrictive strategy to adopt is to make no assumption about H_2O at all and to simply examine how μ_{H_2O} changes with mineral composition in much the same way that we examined how μ_{SiO_2} changed in the orthopyroxene-olivine example. In other words, we examine how the tangent plane for this assemblage intersects the μ_{H_2O} axis, and how this intersection changes with changing mineral composition, without regard for what the ultimate significance of these changes might be. The appropriate equation in this case (written in terms of the Fe end-members) is

$$0 = 3d\mu_{Fe}^{Sta} + 33d\mu_{Fe}^{Chl} + 93d\mu^{Qt} - 79d\mu_{Fe}^{Gar} - 201d\mu_{H_2O} . \qquad (53)$$

This approach requires no assumption as to whether or not a fluid phase is present. In a graphical sense, we are assuming *only* that the intercept of the tangent plane with the μ_{H_2O} axis can vary independently, subject only to the constraints imposed by the particular equilibrium assemblage. This is not the only way to treat H_2O, however. If we wished to treat H_2O as a pure phase present at the P and T of interest, we would write a Gibbs-Duhem equation for pure H_2O. If we were to assume a binary H_2O-CO_2 phase present at P and T of interest, we would write a Gibbs-Duhem equation for this binary fluid. If we were to assume that μ_{H_2O} is externally controlled, then we would define μ_{H_2O} with the same equation used above, and then hold μ_{H_2O} constant by setting $d\mu_{H_2O} = 0$. In this fashion, it is possible to explicitly state any type of assumptions about the chemical variability and behavior of the "fluid phase" in metamorphism by incorporating the appropriate equations and variables.

The final equation is the total differential $d(\mu_{Mg}^{Gar}-\mu_{Fe}^{Gar})$ and is needed to incorporate the variable dX_{Mg}^{Gar} into the system of equations. We have chosen the composition of garnet to monitor changes in μ_{H_2O}, but we could just as easily have chosen staurolite or chlorite.

Solutions involving $d\mu_{H_2O}$ at constant P and T will be examined first. The differential $(\partial\mu_{H_2O}/\partial X_{Mg}^{Gar})_{P,T}$ can easily be obtained by setting $dP = dT = 0$, dividing each equation by dX_{Mg}^{Gar} and solving by Cramer's rule. The solution to this, and differentials for the other three, three-phase triangles are given in Table 10. Since the curvature of each Gibbs durface must be positive in each case, the disposition of each three-phase triangle with respect to μ_{H_2O} can be ascertained through only a knowledge of the Fe-Mg partitioning between coexisting mafic minerals. It turns out that each of these

Table 10. μ_{H_2O} - X slopes for 4 different AFM assemblages

Garnet + staurolite + biotite + muscovite + quartz

$$\left(\frac{\partial \mu_{H_2O}}{\partial X_{Mg}^{Gar}}\right)_{P,T} = \left(\frac{\partial^2(\overline{G}^{Gar})}{\partial(X_{Mg}^{Gar})^2}\right)_{P,T} \frac{(58X_{Mg}^{Gar} - 44X_{Mg}^{Bio} - 14X_{Mg}^{Sta})}{21}$$

Garnet + biotite + chlorite + muscovite + quartz

$$\left(\frac{\partial \mu_{H_2O}}{\partial X_{Mg}^{Gar}}\right)_{P,T} = \left(\frac{\partial^2(\overline{G}^{Gar})}{\partial(X_{Mg}^{Gar})^2}\right)_{P,T} \frac{(45X_{Mg}^{Gar} + 4X_{Mg}^{Bio} - 49X_{Mg}^{Chl})}{126}$$

Garnet + staurolite + chlorite + muscovite + quartz

$$\left(\frac{\partial \mu_{H_2O}}{\partial X_{Mg}^{Gar}}\right)_{P,T} = \left(\frac{\partial^2(\overline{G}^{Gar})}{\partial(X_{Mg}^{Gar})^2}\right)_{P,T} \frac{(79X_{Mg}^{Gar} - 2X_{Mg}^{Sta} - 77X_{Mg}^{Chl})}{201}$$

Biotite + chlorite + staurolite + muscovite + quartz

$$\left(\frac{\partial \mu_{H_2O}}{\partial X_{Mg}^{Bio}}\right)_{P,T} = \left(\frac{\partial^2(\overline{G}^{Bio})}{\partial(X_{Mg}^{Bio})^2}\right)_{P,T} \frac{(158X_{Mg}^{Bio} + 45X_{Mg}^{Sta} - 203X_{Mg}^{Chl})}{1212}$$

three-phase triangles becomes more Mg-rich with *decreasing* μ_{H_2O}.

A more quantitative solution may be obtained if the compositions of co-existing phases are known at a specific P and T. Guidotti (1974) presents data on coexisting garnet + staurolite + biotite + chlorite + muscovite + quartz from rocks in Maine where crystallization temperatures are believed to have been ≈600°C ($X_{Mg}^{Gar} \approx 0.1$, $X_{Mg}^{Sta} \approx 0.15$, $X_{Mg}^{Bio} \approx 0.4$, $X_{Mg}^{Chl} \approx 0.5$). Using these values, and assuming curvature of the Gibbs function consistent with ideal garnet and biotite solid solutions, the equations in Table 10 can be integrated numerically to give the μ_{H_2O}-X loops shown in Figure 8. A reference value of $\Delta\mu_{H_2O} = 0$ was chosen for the coexistence of the four-phase assemblage.

The μ_{H_2O}-X diagram of Figure 8A can be compared with the isothermal, isobaric projection from quartz, muscovite and staurolite of the phase relations as shown in Figure 8B. By considering a plane tangent to the different assemblages in Figure 8B and the intersection of this plane with the H_2O axis, it can be ascertained that the assemblage garnet + chlorite + staurolite is stable at high values of μ_{H_2O} (as is the assemblage garnet + chlorite + biotite, which does not show up in a projection from staurolite). In fact, in sufficiently Fe-rich compositions, the assemblage garnet +

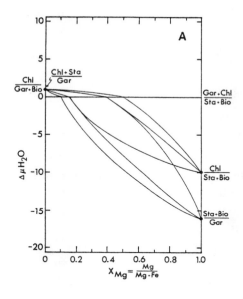

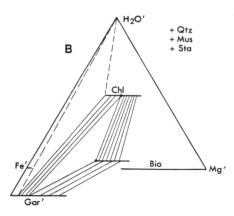

Figure 8. (A) Isothermal, isobaric μ_{H_2O} - X_{Mg} projection showing μ-X loops for the four, three-phase AFM assemblages: garnet + chlorite + staurolite, garnet + chlorite + biotite, garnet + biotite + staurolite, and chlorite + staurolite + biotite. Note: Because the μ - X loops for the assemblages garnet + biotite + chlorite and garnet + chlorite + staurolite are practically coincident, only one loop has been drawn. (B) Isothermal, isobaric projection of phase relations from quartz, muscovite, and staurolite onto the plane FeO–MgO–H_2O.

chlorite + staurolite will coexist with a pure H_2O fluid phase, as shown schematically by the dashed triangle in Figure 8B. The value of μ_{H_2O} for this assemblage will be $\mu_{H_2O}^o$ at the specified P and T and forms an upper limit to the μ-X loop for this assemblage.

The four-phase assemblage garnet + chlorite + biotite + staurolite occurs at an intermediate value of μ_{H_2O}, and the two assemblages garnet + biotite + staurolite and biotite + chlorite + staurolite occur at lower values of μ_{H_2O}. It should be noted that there is only one specific value of μ_{H_2O} at this P and T where the four-phase assemblage can exist. Therefore, it would be very unlikely to see this four-phase assemblage if μ_{H_2O} were externally controlled during metamorphism, but it would be quite reasonable to see it if μ_{H_2O} were internally controlled by the mineral assemblage (see Rice and Ferry, this volume). It would also be impossible to see different samples of the same three-phase assemblage with different Fe-Mg compositions if μ_{H_2O} were externally controlled.

Derivatives with P and T. Derivatives with respect to X_{Mg}^{Gar} involving T and P can be obtained in a similar manner, but it is again necessary to hold two variables constant at a time. For example, in the assemblage garnet + staurolite + chlorite + quartz, the

137

change in X_{Mg}^{Gar} with respect to changing T at constant P and μ_{H_2O} is given as

$$\left(\frac{\partial T}{\partial X_{Mg}^{Gar}}\right)_{P,\mu_{H_2O}} =$$

$$\frac{\left(\frac{\partial^2(\bar{G}^{Gar})}{\partial(X_{Mg}^{Gar})^2}\right)_{P,T}(79X_{Mg}^{Gar}-77X_{Mg}^{Chl}-2X_{Mg}^{Sta})}{(79\bar{S}^{Gar}-33\bar{S}^{Chl}-3\bar{S}^{Sta}-9\bar{S}^{Qt}) - (\bar{S}_{Mg}^{Gar}-\bar{S}_{Fe}^{Gar})(79X_{Mg}^{Gar}-77X_{Mg}^{Chl}-2X_{Mg}^{Sta})} \tag{54}$$

which represents the slope of a curve defined by the assemblage on a T-X diagram at constant P and μ_{H_2O}. Equation (54), the analogous one for $\left(\partial P/ \partial X_{Mg}^{Gar}\right)_{T,\mu_{H_2O}}$, and others like it have obvious applications in characterizing temperature and pressure during metamorphism. The magnitude of this slope is impossible to evaluate without data on the entropies of the various phases, but it is possible to determine the sign of the slope by making some reasonable assumptions: (1) The curvature term for the Gibbs energy of garnet must be positive inasmuch as garnet is a stable solid solution; (2) the composition term in the numerator and denominator is negative, as determined from the measured mineral composition data of Guidotti (1974); (3) the first entropy term in the denominator is negative because both chlorite and staurolite contain H_2O and should therefore have a larger entropy than garnet, which is anhydrous; (4) the difference in the partial molar entropies of the magnesian and iron end members of garnet can be evaluated using the equation for the slope $(\partial\bar{S}/\partial X)$ in Appendix B and noting that the entropy of Fe-Mg exchange is approximately 6.42 eu/atom (see Brady and Stout, 1980; Helgeson et al., 1978). The value calculated for $(\bar{S}_{Mg}^{Gar}-\bar{S}_{Fe}^{Gar})$ is -6.916 eu.

Using these assumptions, the isobaric $\partial T/\partial X$ slope for the equilibrium is negative, as shown in Figure 9A. This T-X loop reveals a somewhat paradoxical result: The rock hydrates as the temperature is raised. This can be seen in Figure 9A where, as T is raised, the hydrous phases chlorite + staurolite are produced at the expense of garnet. This occurs because if μ_{H_2O} is held constant, as T is increased, the activity of H_2O relative to a standard state of P and T of interest (i.e., the relative humidity) increases, as pointed out by J.B. Thompson, Jr. (1957, p. 844). This is not a situation likely to occur in metamorphic processes and thus gives an apparently contradictory result.

A more reasonable variable to hold constant on a T-X diagram would be $P(H_2O)$, which is independent of T. This can be done algebraically by assuming that H_2O is present as a phase at some arbitrary value of $P(H_2O)$ ($\neq P_{total}$) and writing a Gibbs-Duhem equation for H_2O where the $\bar{V}dP$ term is $\bar{V}^{H_2O}dP(H_2O)$.

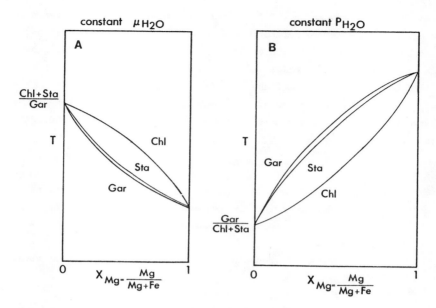

Figure 9. Schematic $T-X_{Mg}$ loops for the assemblage garnet + chlorite + staurolite + quartz + muscovite. (A) T-X loop at constant P and μH_2O. (B) T-X loop at constant P and P_{H_2O}. Note that at constant μ_{H_2O} the rock hydrates with increasing T, but at constant $P(H_2O)$ the rock dehydrates with increasing T (see text).

This introduces an additional equation as well as an additional variable $[P(H_2O)]$, which can then be held constant $[dP(H_2O) = 0]$. The T-X derivative at constant P and $P(H_2O)$ is given as

$$\left(\frac{\partial T}{\partial X_{Mg}^{Gar}}\right)_{P,P(H_2O)} =$$

$$\frac{\left(\frac{\partial^2(\bar{G}^{Gar})}{\partial(X_{Mg}^{Gar})^2}\right)_{P,T}(79X_{Mg}^{Gar}-77X_{Mg}^{Chl}-2X_{Mg}^{Sta})}{(201\bar{S}^{H_2O}+79\bar{S}^{Gar}-33\bar{S}^{Chl}-3\bar{S}^{Sta}-93\bar{S}^{Qt}) - (\bar{S}_{Mg}^{Gar}-\bar{S}_{Fe}^{Gar})(79X_{Mg}^{Gar}-77X_{Mg}^{Chl}-2X_{Mg}^{Sta})} \cdot$$

$$\tag{55}$$

In this case, the numerator has the same sign as in equation (54) but the denominator, which now contains a term for the entropy of H_2O, has changed sign because the entropy of H_2O is considerably larger than the entropy change in the solids. Therefore, the T-X loop at constant $P(H_2O)$ has a positive slope and increasing T will result in dehydration (Fig. 9B).

One last derivative that is of considerable interest to metamorphic petrologists is $(dP/dT)_{X_{Mg}^{Gar}}$. This is the slope of isopleths of constant

139

pyrope content in garnet on a P–T diagram at constant μ_{H_2O}. Consider a garnet growing during a prograde metamorphic event. If the metamorphic P–T path of the rock is the same as the slope $(dP/dT)_{x^{Gar}_{Mg}}$, then the garnet will grow at a fixed composition. The implication of this result is that there are specific P–T paths, as indicated by the above differential, that will result in the growth of unzoned porphyroblasts. This may in part explain why some garnets are strongly zoned whereas others from similar metamorphic grades but different areas are unzoned.

SUMMARY OF PROCEDURE

To summarize the procedure involved in application of the Gibbs method, the following steps must be taken for any system under consideration.

(1) Choose the system components that describe the chemical variability of the system.

(2) Choose the phase components that describe the chemical variablility of the phases.

(3) Write a Gibbs–Duhem equation for each phase.

(4) Write a linearly independent set of heterogeneous equilibrium conditions.

This is the basic set of equations necessary to describe the metamorphic assemblage, and the variance of this set of equations should be the same as the phase rule variance of the assemblage.

(5) Decide what derivatives would be useful to examine and add appropriate equations as necessary.

(6) Solve for the appropriate ratio of differentials by holding variables constant as desired.

(7) Integrate if desired. The integrated forms of the solutions permit one to directly relate compositions of minerals in an equilibrium assemblage to the intensive variables, T, P, and μ_i.

LITERATURE REVIEW

The above examples were presented mainly for illustrative purposes. There are several studies in the literature that have employed the Gibbs method with the goal of characterizing physical conditions during metamorphism, and these are briefly reviewed here.

Most of the applications of the Gibbs method have been directed towards using phase equilibria and mineral chemistry to decipher differences in the chemical potentials of H_2O, CO_2 and O_2 in metamorphic rocks. Moreover, most of this work has been done in geographically restricted areas such as single outcrops, so that the assumption of constant P and T could be employed. Because in these cases the only thermodynamic information necessary to obtain a numerical solution is the curvature of the Gibbs function, significant results were obtained with very few assumptions and little or no thermochemical data.

The first published study that employed the Gibbs method was Rumble's (1971) thermodynamic analysis of phase equilibria in the Fe-Ti oxide system. Using published experimental data on Fe-Ti oxides, Rumble was able to conclude that the composition of coexisting ilmenite and magnetite, and the $f(O_2)$ calculated from their coexistence, were substantially unaffected by pressure in the range of 1-10 kbar. In a later paper, Rumble (1973) applied a similar analysis to Fe-Ti oxides from the regionally metamorphosed Clough quartzite, New Hampshire. In this study, he was able to analyze the effect of solid solution of Mn-components on the μ_{O_2} values recorded by coexisting Fe-Ti oxides and to document the existence of gradients in the chemical potential of μ_{O_2} between adjacent beds in the Clough quartzite.

Rumble (1974, 1976, 1978) extended his algebraic analysis to include silicate mineral assemblages common to pelitic rocks. The study was conducted on a small outcrop of the Clough quartzite on Black Mountain in New Hampshire where P and T could be assumed to be constant over the study area. The graphical analysis of the pelitic assemblages indicated that at least one assemblage acted as a water buffer, and that the chemical compositions of the coexisting mafic silicates was inconsistent with the entire outcrop having crystallized at the same value of μ_{H_2O} (see discussion in Chapter 3 by Spear et al., this volume). The algebraic analysis demonstrated that variations in the chemical potentials of H_2O, H_2 and O_2 from bed to bed are indeed recorded by the silicate mineral assemblages. Chemical potential differences as large as 0.65 Kcal for μ_{H_2O} and 42 Kcal for μ_{O_2} were measured between beds.

Rumble also integrated the equations relating the chemical potentials of H_2O and O_2 to mineral composition and constructed the $\mu_{H_2O}-\mu_{O_2}$ diagram shown in Figure 10. This diagram is a Schreinemakers net that is contoured for mineral composition. With this diagram, the compositions of the silicate and oxide phases in two assemblages can be used to quantitatively estimate differences in the chemical potentials of H_2O and O_2 (assuming a temperature of 495°C and that the silicate or oxide phase of interest obeys ideal solution behavior). The beauty of this diagram, and others like it, is that it

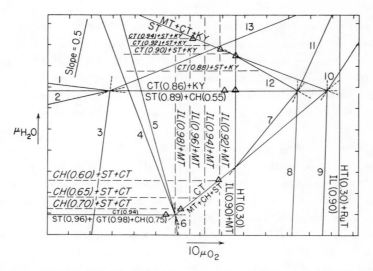

Figure 10. Isobaric, isothermal μ_{H_2O} versus μ_{O_2} diagram for silicate and oxide minerals (plus quartz and muscovite) in the system $SiO_2-TiO_2-Al_2O_3-Fe_2O_3-FeO-MgO-K_2O-H_2O$ from Rumble (1978). The relative positions of samples collected from Black Mountain, New Hampshire are denoted by Δ. **Number** in parentheses following IL (ilmenite) or HT (hematite) gives $FeTiO_3/(FeTiO_3 + Fe_2O_3)$ ratio. **Number** in parentheses following CH (chlorite), CT (chloritoid), GT (garnet), or ST (staurolite) gives $Fe/(Fe + Mg)$ ratio. Other symbols are KY (kyanite), MT (magnetite), and RUT (rutile).

enables estimation of differences in intensive parameters during metamorphism, with only the measured mineral compositions as input. Diagrams such as these can be constructed in a quantitative fashion with little or no thermochemical data.

Ferry (1979) used the Gibbs method to map chemical potential differences in H_2O and CO_2 in an outcrop of metamorphosed impure carbonate rock from the Waterville Formation, Maine. The results of this study are particularly interesting because evidence was found for substantial μ_{H_2O} and μ_{CO_2} differences between sedimentary beds, but little or no difference was found along strike within a single bed. The implication of this result is that the sedimentary layering acted as channelways for fluid migration during devolatilization reactions. Another significant result of this study is that the metamorphism may have taken place under conditions of $P_{fluid} < P_{total}$.

Three other studies have also been directed towards determination of the behavior of volatile species in different metamorphic rock types. Grambling (1980) found evidence for differences in μ_{H_2O} of 0.9 Kcal in pelitic schists of the Truchas Peaks area, north-central New Mexico. Spear (1977, 1981, 1982)

142

reported differences in μ_{H_2O} of up to 2 Kcal in amphibolites of the Post Pond Volcanics, Vermont. Grew (1981) used the Gibbs method to document differences in μ_{H_2O} and μ_{O_2} between mineral assemblages in granulite facies rocks from East Antarctica and used these differences to explain the observed variations in mineral assemblages and mineral chemistry.

Three other applications are of note. Rumble (1976b) used the Gibbs method to derive a rigorous formulation of the adiabatic gradient in multi-component, multiphase systems. Sack (1982) applied the methodology to a study of the development of metasomatic zones in orthopyroxene-bearing reaction skarns from the Adirondack highlands, New York. Grew (1980) derived slopes of isopleths of the Fe_2SiO_5 content in sillimanite on a P-T diagram for the assemblage sillimanite + ilmenite + quartz (see earlier discussion).

From the studies cited above it is clear that differences in the chemical potentials of volatile components are common during metamorphism over distances measured in meters in a wide variety of rock types. These differences in turn require the existence of gradients in chemical potentials. We have attempted to quantitatively compute gradients from the various studies cited above by dividing reported differences in μ_{H_2O} by the distance separating the samples that record the difference. In Figure 11 we illustrate maximum values of ($\Delta\mu_{H_2O}$/distance) consistent with data for amphibolites, quartzites, carbonates, and pelitic schist reported by Spear (1977), Rumble (1978), Ferry (1979), and Grambling (1981), respectively. Although values in Figure 10 are maximum ($\Delta\mu_{H_2O}$/distance), they are *minimum* estimates of gradients in μ_{H_2O} because there is no way to evaluate from the studies how μ_{H_2O} changes with distance between samples that record the differences in μ_{H_2O} (for example, between two samples that record a difference in μ_{H_2O}, μ_{H_2O} may change continuously and linearly with distance or μ_{H_2O} may change discontinuously creating a step function with distance).

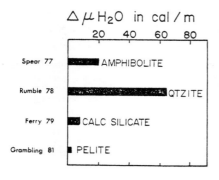

Figure 11. Plot of μ_{H_2O}/meter for four different rock types described in the literature. Gradients are averages and thus represent minimum values.

In spite of the uncertain magnitude of the gradients in Figure 11, it is interesting to note that gradients differ by approximately an order of magnitude. The size and preservation of the gradients is controlled by the interplay of numerous factors: (a) rock permeability; (b) diffusion rates; (c) composition and amount of any through-flowing metamorphic fluid; and (d) the capacity of mineral assemblages in individual rock samples to buffer the chemical potential of volatile components. The gradients therefore serve as a record (albeit a complicated one) of metamorphic processes, and the differences between gradients in Figure 11 record how these processes differ between one rock type and another and between locations in a metamorphic terrain. One important application in a comparative study of gradients in μ_i is qualitative comparison of rock permeability during metamorphism. If the buffer capacity of individual rocks can be characterized (by modal analysis, see Rice and Ferry in this volume) and if samples are collected from the same outcrop (so that it can be safely assumed that they experienced the same sort of history of infiltration, if any), then the gradients record relative differences in rock permeability during metamorphism. Characterization of permeability is important because it will lead to a better understanding of metamorphism as a dynamic process and a better understanding of the interrelation between the evolution of metamorphic mineral assemblages and deformational history. Both are goals that we will likely see metamorphic petrologists increasingly concerned with in future years.

CONCLUSION

The Gibbs method is a powerful technique for the analysis of phase equilibria in metamorphic or igneous rocks. The major advantage of this approach over more conventional approaches of determining properties of the fluid phase during metamorphism (e.g., Ferry and Burt, this volume) is that in some cases, little or no thermodynamic data is necessary. For example, absolutely no thermochemical data are necessary to determine the *sign* of a slope such as $(\partial \mu_{H_2O}/\partial X)_{P,T}$. The existence of chemical potential gradients can thus be deduced with nothing more than data on compositions of coexisting minerals.

ACKNOWLEDGMENTS

The authors wish to acknowledge the helpful reviews of J. Rice, H. Lang, J. Selverstone, K. Kimball and K. Hodges. Special thanks go to D.L. Frank for patiently typing and retyping several drafts and to D. Hall for drafting. This work was supported in part by NSF grant EAR-8108617 (Spear) and a Joseph H. Defrees grant of the Research Corporation (Spear).

145

APPENDIX A

A NOTE ON SOLVING SYSTEMS OF EQUATIONS

There are two general approaches that can be taken to obtain solutions to the systems of equations derived here. The first approach is to obtain a numerical solution by substituting values for the various coefficients in the system of equations and then solving the equations by standard numerical techniques such as matrix inversion. This approach is quick and amenable to computer applications and is sometimes the only recourse when large systems are involved.

The second approach is to obtain an analytical solution, such as has been done in the examples in this chapter. The advantage of an analytical solution is that it is possible to see the exact functional dependency of the derived slope on the various coefficients, which can be a considerable aid to one's intuition. Obtaining an analytical solution is straightforward, but it is not always easy because of the amount of algebra involved. The purpose of this appendix is to point out some tricks that can make solutions considerably easier.

All of the analytical solutions presented in this chapter were found by using Cramer's Rule. A discussion of Cramer's Rule is given in Chapter 3 by Spear et al., this volume. Basically, this involves solving two determinants, one for the numerator and one for the denominator. The determinants are all solved by *expanding by signed minors* (a minor of a matrix is the matrix with a row and column crossed out). Expansion by minors is greatly facilitated if there are rows or columns that contain only one coefficient with the rest of the coefficients zero. It is a property of matrices that any multiple of a row (or column) can be added to or subtracted from any other row (or column) without changing the determinant. Using this rule, it is often possible to alter the matrix so that several rows and columns have only one element that is non-zero. Moreover, it is always possible using this rule to remove the *dependent* compositional variable in each solid solution by adding all the columns containing the mole fractions for that phase.

As an example, consider solution of the system of equations shown in Table 5 for the term $(d\mu_{SiO_2}/dX_{Fs}^{Px})$. The *numerator* of this equation will be the coefficient matrix with the "Y" vector substituted for the last column of the coefficient matrix; that is,

$$\begin{vmatrix} X_{Fs}^{Px} & X_{En}^{Px} & 0 & 0 & 0 \\ 0 & 0 & X_{Fa}^{Ol} & X_{Fo}^{Ol} & 0 \\ 2 & -2 & -1 & 1 & 0 \\ 0 & -2 & 0 & 1 & 0 \\ -1 & 1 & 0 & 0 & -(\partial^2\bar{G}) \end{vmatrix}$$

where $(\partial^2\bar{G})$ represents the second derivative of the Gibbs function with respect to X_{Fs}^{Px}. Note that columns 1 and 2 can be added together as can columns 3 and 4 to yield.

$$\begin{vmatrix} X_{Fs}^{Px} & 1 & 0 & 0 & 0 \\ 0 & 0 & X_{Fa}^{Ol} & 1 & 0 \\ 2 & 0 & -1 & 0 & 0 \\ 0 & -2 & 0 & 1 & 0 \\ -1 & 0 & 0 & 0 & -(\partial^2\bar{G}) \end{vmatrix}$$

Two times row 1 can be added to row 4 and row 2 can be subtracted from row 4 to give

$$\begin{vmatrix} X_{Fs}^{Px} & 1 & 0 & 0 & 0 \\ 0 & 0 & X_{Fa}^{Ol} & 1 & 0 \\ 2 & 0 & -1 & 0 & 0 \\ 2X_{Fs}^{Px} & 0 & -X_{Fa}^{Ol} & 0 & 0 \\ -1 & 0 & 0 & 0 & -(\partial^2\bar{G}) \end{vmatrix}$$

This matrix can now be expanded by elements $(1,2)$, $(2,4)$ and $(5,5)$ to give

$$(-1)(1)(-\partial^2\bar{G}) \begin{vmatrix} 2 & -1 \\ 2X_{Fs}^{Px} & -X_{Fa}^{Ol} \end{vmatrix} = -2(\partial^2\bar{G})(X_{Fa}^{Ol}-X_{Fs}^{Px}) \ .$$

A similar solution will reveal that the determinant of the *denominator* $= -1$, thus yielding the solution shown earlier in the text.

APPENDIX B

INTEGRATING THE SLOPE EQUATIONS

The slope equations [e.g., (dP/dT)] give the instantaneous slope at a particular P, T, and composition of phases. In order to calculate numerical *differences* in intensive parameters (e.g., chemical potential differences, T-X or μ-X loops, etc.), it is necessary to integrate the slope equations.

The simplest procedure is to calculate the slope and assume it is constant (i.e., a straight line). This is probably sufficiently accurate over short extrapolations, but is clearly a poor assumption over a wide interval of integration (e.g., T-X and μ-X "loops" would never close).

A rigorous integration can be obtained numerically by the following method. Consider the differential equation derived for the pyroxene-olivine equilibrium in the system SiO_2-FeO-MgO:

$$\int d\mu_{SiO_2} = \int \frac{2RT}{X_{Fs}^{Px} X_{En}^{Px}} (X_{Fa}^{Ol} - X_{Fs}^{Px}) \ dX_{Fs}^{Px} \ . \tag{B1}$$

In all examples of this type the mineral compositions X_{Fa}^{Ol} and X_{Fs}^{Px} are *not* independent but are related by partition coefficients. In this example, the pyroxene and olivine are related by the expression (assuming ideal mixing)

$$K_D^{Px-Ol} = \left(\frac{X_{Fs}^{Px}}{X_{En}^{Px}}\right) \cdot \left(\frac{X_{Fo}^{Ol}}{X_{Fa}^{Ol}}\right) \ . \tag{B2}$$

The simplest approach to integrating these equations is to assume that K_D is a constant over the range of integration; then the entire differential can be cast in terms of X_{Fs}^{Px} and integrated. Also, since the constant of integration is not known, a suitable reference point must be chosen. A convenient point might be to choose $\mu_{SiO_2} = 0$ at quartz saturation; an alternative choice might be to choose the value of $\mu_{SiO_2}^o$ at P and T of quartz saturation, if this value were known from thermochemical tables.

Integration of the above differential equation was done in the following manner. First, the following assumptions were made: (1) The composition of coexisting orthopyroxene and olivine at quartz saturation are $X_{Fs}^{Px} = 0.9$ and $X_{Fa}^{Ol} = 0.951$; (2) therefore, $K_D^{Px-Ol} = 2.21$ (e.g., Bohlen and Boettcher, 1981) and is constant over the entire Fe-Mg range; (3) the value of μ_{SiO_2} at quartz saturation is 0; and (4) T = 900°C. The integration was done for increments of $X_{Fs}^{Px} = 0.01$ between $X_{Fs}^{Px} = 0.9$ and 0.0. For each increment the following were calculated: (a) the composition of coexisting olivine (from the K_D expression); (b) the $(\partial\mu_{SiO_2}/\partial X_{Fs}^{Px})_{P,T}$; (c) the incremental $\Delta\mu_{SiO_2}$ for each ΔX_{Fs} using the *average* slope over the interval of ΔX_{Fs}. The values of this integration have been tabulated in Table 4 and graphed in Figure 6D.

The above integration, which was done at constant T and P, involved only a knowledge of the curvature of the Gibbs function, the Fe-Mg partition coefficient, and T. Integration of T-X, P-X, or P-T slopes are more difficult, because the entropy and volume terms must also be computed. Consider the equations in Table 5 for the orthopyroxene-olivine-quartz equilibrium. Calculation of numerical values of these slopes requires knowledge of (1) the entropies, volumes, partial molar entropies and volumes, and Gibbs-function curvature of the phases, and (2) how these parameters change with T, P, and X. The problem is not as intractable as it might at first seem.

Consider first the volume terms. The molar volume of a pure phase such as quartz can be found in tables of thermodynamic data (e.g., Helgeson et al., 1978) and the molar volume of solid solutions can be calculated as a mechanical mixture of the end-members. That is, for a two-component solid solution,

$$\bar{V}^\alpha = X_1^\alpha \bar{V}_1^\circ + X_2^\alpha \bar{V}_2^\circ . \tag{B3}$$

If there is a positive or negative volume of mixing, this equation will have to be changed accordingly. If the above equation holds, then the *partial molar volumes*, as calculated by the intercepts of the tangent plane (e.g., see Fig. 2) are equal to the molar volumes of the pure end-members. That is, for α

$$\bar{V}_1^\alpha = \bar{V}_1^\circ \quad \text{and} \quad \bar{V}_2^\alpha = \bar{V}_2^\circ . \tag{B4,B5}$$

The entropies are not quite as straightforward because, in addition to the entropy of *mechanical* mixture, there is also the entropy of mixing (i.e., configurational entropy) to consider. Thus the entropy of a two-component solid solution becomes

$$\bar{S}^\alpha = X_1^\alpha S_1^\circ + X_2^\alpha \bar{S}_2^\circ - n(X_1^\alpha R \ln X_1^\alpha + X_2^\alpha R \ln X_2^\alpha) \tag{B6}$$

where n is the number of energetically equivalent sites per mineral formula over which the species mix. The relationship of these parameters can be seen in Figure B1 and is similar to the $\bar{G}$-X construction of Figure 3. The entropy of the mixture ($\bar{S}^\alpha$) can easily be calculated from the above equation if the phase compositions and the standard state entropies are known. The terms involving partial molar entropies, which are the intercepts of the tangent line to the respective axes, are also easy to evaluate because the

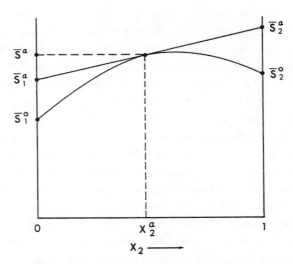

Figure B1. $\bar{S}$ (entropy) versus X diagram showing relations among the entropy of a solid solution ($\bar{S}^{\alpha}$), the partial molar entropies in the solution ($\bar{S}_1^{\alpha}$ and $\bar{S}_2^{\alpha}$), and the standard state entropies ($\bar{S}_1^{\circ}$ and $\bar{S}_2^{\circ}$).

equations in Table 5 only contain terms for *differences* in partial molar quantities. These differences are simply the *slope* of the tangent line. For example,

$$\bar{S}_2^{\alpha} - \bar{S}_1^{\alpha} = \left(\frac{\partial \bar{S}^{\alpha}}{\partial X_2^{\alpha}}\right)_{T,P} = \bar{S}_2^{\circ} - \bar{S}_1^{\circ} - nR(ln X_2^{\alpha} - ln X_1^{\alpha}) \ . \qquad (B7)$$

The only other point to consider is the integration of these values from the standard state reference values to T and P of interest. This can be done if the heat capacity (Cp) is known. However, because all equations involve only differences in entropies, the usual assumption to make is that $\Delta Cp = 0$, and thus all minerals extrapolate by the same amount from the standard state to T and P. The *differences* in the values at P and T will therefore be the same as at the standard state values (e.g., see Wood and Fraser, 1977, Chapter 1).

An an example, consider the equations in Table 5 for the orthopyroxene + olivine + quartz equilibrium. Assuming the values $X_{Fs}^{Px} = 0.9$ and $X_{Fa}^{Ol} = 0.951$ at 900°C, numerical values of the slopes can be calculated from data in Helgeson et al. (1978) and the above equations. The results are:

$$\left(\frac{dP}{dX_{Fs}^{Px}}\right)_T = 34,305 \text{ bar/mole fraction} \qquad (B8)$$

$$\left(\frac{dT}{dX_{Fs}^{Px}}\right)_P = -40,022 \text{ deg/mole fraction} \tag{B9}$$

$$\left(\frac{dP}{dT}\right)_{X_{Fs}^{Px}} = 0.857 \text{ bar/deg .} \tag{B10}$$

Note that these are instantaneous slopes at a particular P and T. To integrate the entire P-X or T-X loop it is necessary to increment P or T and X,
and calculate a new slope. If the assumption that $\Delta Cp = 0$ is valid, then the
entropy and volume terms do not change. However, the Fe-Mg partitioning between pyroxene and olivine does change (it must, to make a closed loop) and
thus a function that relates K_D^{Px-Ol} to P and T is needed; namely,

$$\Delta\mu° = \Delta\bar{H}° - T\Delta\bar{S}° + (P-1)\Delta\bar{V}° = -RT\ell nK_D^{Px-Ol} . \tag{B11}$$

With these values, $\Delta\bar{H}°$, $\Delta\bar{S}°$, and $\Delta\bar{V}°$ of exchange, the entire P-X or T-X loop
can be calculated.

Even without these values it is possible to draw the P-X, T-X and P-T
diagrams qualitatively with only a single, instantaneous slope, as shown in
Figure 7. Note that these calculations imply that the composition of pyroxene
in equilibrium with olivine in this assemblage is almost independent of temperature, but very sensitive to pressure.

Comparison of the three slopes calculated from the data of Helgeson et
al. (1978) with the experimental data of Bohlen and Boettcher (1981) is revealing. The experimentally derived slopes of Bohlen and Boettcher, dP/dX,
dT/dX, and dP/dT, are approximately 33,750 bar/mole fraction, -3375 deg/mole
fraction and 10 bar/deg, respectively. The $(\partial T/\partial X)$ and $(\partial P/\partial T)$ slopes derived
for theory and experiment do not agree well (although the signs are correct)
primarily becuase small errors in entropy estimates greatly affect the result.
Again, the need for carefully reversed experimental brackets, such as those
of Bohlen and Boettcher (1981), is emphasized. Of particular note is the
similarity of the two dP/dX slopes (34,305 and 33,750 bar/mole fraction),
both of which are quite large, as is desirable for a useful geobarometer.
The power of the Gibbs method in this example is that it enables prediction
of this slope with only a temperature estimate, molar volume data, and the
composition of coexisting orthopyroxene and olivine at the temperature of
interest.

CHAPTER 4 REFERENCES

Bohlen, S.R. and A.L. Boettcher (1981) Experimental investigations and geological applications of orthopyroxene geobarometry. Am. Mineral. 66, 951-964.

_____, E.J. Essene, and A.L. Boettcher (1980) Reinvestigation and application of olivine-quartz-orthopyroxene barometry. Earth Planet. Sci. Lett. 47, 1-10.

Brady, J.B. and J.H. Stout (1980) Normalizations of thermodynamic properties and some implications for graphical and analytical problems in petrology. Am. J. Sci. 280, 173-189.

Ferry, J.M. (1979) A map of chemical potential differences within an outcrop. Am. Mineral. 64, 966-985.

Finger, L.W. and D.M. Burt (1972) REACTION, a Fortran IV computer program to balance chemical reactions. Carnegie Inst. Wash. Year Book 71, 616-620.

Grambling, J.A. (1981) Kyanite, andalusite, sillimanite, and related mineral assemblages in the Truchas Peaks region, New Mexico. Am. Mineral. 66, 702-722.

Grew, E.S. (1980) Sillimanite and ilmenite from high-grade metamorphic rocks of Antarctica and other areas. J. Petrol. 21, 39-68.

_____ (1981) Granulite-facies metamorphism at Molodezhnaya station, East Antarctica. J. Petrol. 22, 297-336.

Guidotti, C.V. (1974) Transition from staurolite to sillimanite zone, Rangely Quadrangle, Maine. Geol. Soc. America Bull. 85, 475-490.

Haase, V.R. (1948) Einige allgemeine Beziehungen fur Zustandsfunktionen bei Vielkomponenten systemem, Z. Naturforschg. 3a, 285-290.

Helgeson, H.C., J.M. Delany, H.W. Nesbitt, and D.K. Bird (1978) Summary and critique of the thermodynamic properties of rock-forming minerals. Am. J. Sci. 278A, 229 pp.

Hoschek, G. (1980) Phase relations of a simplified marly rock system with application to the western Hohe Tauern (Austria). Contrib. Mineral. Petrol. 73, 53-68.

Rumble, D., III (1971) Thermodynamic analysis of phase equilibria in the system Fe_2TiO_4-Fe_3O_4-TiO_2. Carnegie Inst. Wash. Year Book 69, 198-207.

_____ (1973) Fe-Ti oxide minerals from regionally metamorphosed quartzites of western New Hampshire. Contrib. Mineral. Petrol. 42, 181-195.

_____ (1974) Gibbs phase rule and its application in geochemistry. J. Wash. Acad. Sci. 64, 199-208.

_____ (1976a) The use of mineral solid solutions to measure chemical potential gradients in rocks. Am. Mineral. 61, 1167-1174.

_____ (1976b) The adiabatic gradient and adiabatic compressibility. Carnegie Inst. Wash. Year Book 75, 651-655.

_____ (1978) Mineralogy, petrology and oxygen isotopic geochemistry of the Clough Formation, Black Mountain, western New Hampshire, U.S.A. J. Petrol. 19, 317-340.

Sack, R.O. (1982) Reaction skarns between quartz-bearing and olivine-bearing rocks. Am. J. Sci., in press.

Skippen, G.B. (1971) Experimental data for reactions in siliceous marbles. J. Geol. 79, 451-481.

Spear, F.S. (1977) Phase equilibria of amphibolites from the Post Pond Volcanics, Vermont. Carnegie Inst. Wash. Year Book 76, 613-619.

_____ (1981) $\mu(H_2O)$-$\mu(CO_2)$-X(Fe-Mg) relations in amphibolite assemblages. Geol. Soc. America Abstr. Prog. 13, 559.

_____ (1982) Phase equilibria of amphibolites from the Post Pond Volcanics, Mt. Cube Quadrange, Vermont. J. Petrol., in press.

Thompson, A.B. (1976) Mineral reactions in pelitic rocks: I. Prediction of P-T-X(Fe-Mg) phase relations. Am. J. Sci. 276, 401-424.

Thompson, J.B., Jr. (1957) The graphical analysis of mineral assemblages in pelitic schists. Am. Mineral. 42, 842-858.

_____, J. Laird, and A.B. Thompson (1982) Reactions in amphibolite, greenschist and blueschist. J. Petrol. 23, 1-27.

_____ and A.B. Thompson (1976) A model system for mineral facies in pelitic schists. Contrib. Mineral. Petrol. 58, 243-277.

Tunell, G. (1979) On the mathematical derivation of Gibbs' equation (98) from experimentally determinable thermodynamic relations. Carnegie Inst. Wash. Supp. Publ. No. 49, 46 pp.

Chapter 5
GEOLOGIC THERMOMETRY and BAROMETRY
E.J. Essene

INTRODUCTION

Geologic barometry and thermometry, the estimates of metamorphic pressures and temperatures, have had a history nearly as old as metamorphic petrology itself. Arguments about the importance of burial versus thermal metamorphism concerned the relative roles of pressure and temperature. Early studies using index minerals, and metamorphic facies were involved with qualitative estimates of conditions of metamorphism. Index minerals, which appear and/or disappear systematically in common rock types, were seen repeated in similar sequences in wholly unrelated terranes.[1] Without necessarily understanding the reactions which generated index minerals, the implicit assumption of similar metamorphic grade for a given metamorphic zone wherever it is found is a simple, yet powerful, concept. The facies concept went still further in emphasizing repeatable *assemblages* in similar rock types, so that the ingredients for tagging metamorphic reactions -- the phases in equilibrium -- became recognized. A three-pronged attack on the problems of metamorphism by laboratory experiments, by thermodynamic modelling and calculations, and by careful studies of metamorphic assemblages with the advent of the electron microprobe, has led to a much better understanding of the intensive chemical parameters prevailing during metamorphism. The experienced petrologist can often estimate metamorphic conditions to within a few kbar and 100°C by identifying key assemblages without recourse to elaborate analytical data, but more accurate estimates require careful observation and analysis.

The accuracy with which comparative metamorphic thermobarometry can be applied in the absence of sophisticated experimental or thermodynamic data is well illustrated by two examples: Bowen's (1940) estimation of phase equilibria in the system $CaO-MgO-SiO_2-CO_2-(H_2O)$, and Schuiling's (1957) calibration of reactions in the Al_2SiO_5 system. Both of these writers turned geothermometry and geobarometry backwards, calibrating simple univariant equilibira from largely unstated and certainly prejedicial views of metamorphic conditions. Bowen (1940) published the first systematic exposition of

---

[1]The archaic spelling of the word terrane is here used to indicate bedrock of a region while the spelling of terrain is reserved for the geographic location.

univariant equilibria by comparison with analogous mapped isograds (his Table 1). Only the reaction

$$\text{calcite} + \text{quartz} = \text{wollastonite} + CO_2 \qquad (1)$$
$$CaCO_3 + SiO_2 = CaSiO_3 + CO_2$$

(Goldschmidt, 1912) and the reaction

$$\text{dolomite} = \text{periclase} + \text{calcite} + CO_2 \qquad (2)$$
$$CaMg(CO_3)_2 = MgO + CaCO_3 + CO_2$$

(Eitel, 1925, in Bowen, 1940) had been calculated at that time and the thermodynamic data base certainly was inaccurate, as conceded by Bowen. Nevertheless, he placed the first seven reactions within 50-60°C of their presently accepted locations (Harker and Tuttle, 1955a,b; Metz and Trommsdorff, 1968; Skippen, 1971; Kase and Metz, 1980; Jacobs and Kerrick, 1981), a remarkably prescient performance in the absence of experimental data. Schuiling (1957) located the reactions for

$$\text{kyanite} = \text{andalusite} = \text{sillimanite} \qquad (3)$$
$$Al_2SiO_5 = Al_2SiO_5 = Al_2SiO_5$$

on a P-T diagram by noting the metamorphic grade at which the isograds occurred and fixed the Al_2SiO_5 triple point at 3.5 kbar and 540°C. Experimental petrologists (Clark et al., 1957; Bell, 1963; Khitarov et al., 1963) later insisted that kyanite could not be stable below 10 kbar for reasonable geologic temperatures of 500-600°C. As it happened, they had erroneously relied on unreversed syntheses which were in part carried out in a non-hydrostatic pressure apparatus. Despite a strong cautionary note about relying on such procedures (Fyfe, 1960), many workers accepted these pressure estimates and expounded at length in error on possible tectonic mechanisms necessary to generate pressures of 10 kbar in ordinary kyanite-zone rocks (e.g., Pitcher and Flinn, 1965). In fact, Schuiling (1957) had placed the Al_2SiO_5 triple point within half a kilobar and 50°C of current best estimates (see section below on Al_2SiO_5 minerals). In retrospect, the accuracy in placement of phase equilibria by Bowen and by Schuiling suggests that experimentalists would be wise to check the sense of their results against field observations, and moreover, these two examples suggest that geologic barometry and thermometry can be quite accurate if properly calibrated mineral equilibria are used.

This paper cannot possibly cover all proposed systems or applications of such systems to the barometry and thermometry of metamorphites. Only those systems which have been quantitatively calibrated by experiment or thermodynamic calculation will be considered here, and systems generally applied

to mantle rocks, such as quartz-coesite-stishovite, $CaO-MgO-Al_2O_3-SiO_2$, etc.
have also been excluded. Although a large number of references are tabulated,
many others have been excluded through oversight, lack of space, and fatigue
of the typist. The interested reader is urged to travel backwards in time
with the aid of the bibliographies in the references provided and to scout
future references using Science Citation Indices. Various types of thermometers and barometers will be discussed taking as examples the most commonly
used systems. But first, generalities of using and testing thermobarometers
will be considered.

THE CONCEPT OF A PETROGENETIC GRID

"We may eventually be able to determine a large number of [uni-
variant] curves... The equilibria between solids ... will cut across
the curves involving a gas equilibrium. Of these latter it may eventually be possible to add many more, depicting the dissociation of
hydrates, sulfides, oxides, and other compounds. The intersecting
curves of the two classes will thus cut up the general P-T diagram
into a grid which we may call a *petrogenetic grid*. With the necessary data determined by experiment we might be able to locate very
closely on the grid both the temperature and the pressure of formation of those rocks and mineral deposits of any terrane...   The
determinations necessary for the production of such a grid constitute
a task of colossal magnitude, but the data will gradually be acquired.

N. L. Bowen (1940), p. 273-274.

With these prophetic words, Bowen laid out the path for the future work of
many experimental and field petrologists. While many petrologists continued
to map isograds and infer possible reactions among chemically equivalent metamorphic assemblages, experimental calibrations were begun, including intensive
efforts in the systems $MgO-SiO_2-H_2O$ initiated by experiments of Bowen and
Tuttle (1949), $CaO-MgO-CO_2$ first investigated by Harker and Tuttle (1955a,b)
and by Goldsmith and Heard (1961), and $CaO-SiO_2-CO_2$ begun by Harker and Tuttle
(1956) and by Harker (1959). In addition, simple solid-solid reactions, such
as those relating the polymorphs of Al_2SiO_5, $CaCO_3$, SiO_2, Ca_2SiO_4, $MgSiO_3$,
etc., were also investigated by various workers. The reactions in most of
these systems were not applicable to most mapped isograds and it became evident that it was difficult to locate equilibria without a great deal of care
(Fyfe, 1960). As late as 1965 it was still difficult to construct a calibrated petrogenetic grid for a given metamorphic suite. Most workers used
schematic grids constructed using various assumptions and simplifications and
superimposed upon their favorite Al_2SiO_5 diagram (e.g., Miyashiro, 1961; Albee,
1965, Fyfe and Turner, 1966). Various petrogenetic grids have been developed

for metapelites by many workers, including Thompson (1976), Albee (1965, 1972), Hess (1969, 1971) and Carmichael (1978). In the last 15 years, however, thermodynamic and experimental data have proliferated at an ever-increasing pace (e.g., Mueller and Saxena, 1977; Robie et al., 1978; Helgeson et al., 1978). Metamorphic temperatures and pressures in most terranes can now be characterized reasonably well if one is careful in selecting well calibrated and appropriate systems.

ASSUMPTIONS AND PRECAUTIONS IN THERMOBAROMETRY

In choosing among the current plethora of fashionable (and unfashionable) barometers and thermometers, the worker should keep in mind the various assumptions involved in constructing and applying them.

How well calibrated is a given system? The somewhat shady history of the Al_2SiO_5 polymorphs should be a cautionary example. The experiments that were used to calibrate a given system should be well reversed with careful characterization of run products and reaction directions.

Is the system insensitive to temperature (or pressure) changes at the expected range of conditions in the rocks? If so, it obviously is a poor choice. An example is diopside-enstatite solvus thermometry at metamorphic temperatures (T < 800°C) where the pyroxenes are located on the steep limbs of the solvus (Valley et al., 1982). In such cases even small analytical errors involved in the location of the solvus as well as in the pyroxene analyses will produce large errors in inferred temperatures.

Is the system too sensitive to chemical changes? If so, even small errors in the solution models adjusting for additional components will generate large errors in the inferred P or T. For instance, the reaction:

$$MnSiO_3 = MnSiO_3 \qquad (4)$$
$$\text{rhodonite = pyroxmangite}$$

calibrated by Maresch and Mottana (1976), has a very small $\Delta \bar{V}$ (~ 0.2 cc/mol). Even though pyroxmangite-rhodonite pairs are found in metamorphic rocks, correction for the other components in solid solution displaces the reaction erratically because of uncertainties in the activity models and because of the very small $\Delta \bar{V}$ (Winter et al., 1981). Similar problems will be encountered in potential geothermometers with a small $\Delta \bar{S}$; for instance, consider the high temperature reactions quartz = tridymite = cristobalite in the system SiO_2. The reaction quartz = tridymite has a $\Delta \bar{S} = -2.2$ joules/mol°K and tridymite = cristobalite has $\Delta \bar{S} = -1.0$ joules/mol°K at high temperatures (Robie

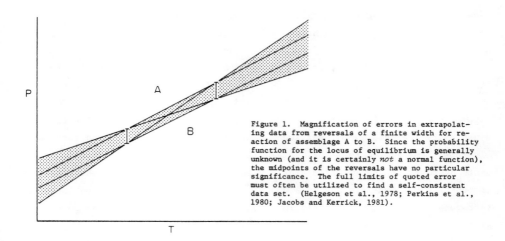

Figure 1. Magnification of errors in extrapolating data from reversals of a finite width for reaction of assemblage A to B. Since the probability function for the locus of equilibrium is generally unknown (and it is certainly *not* a normal function), the midpoints of the reversals have no particular significance. The full limits of quoted error must often be utilized to find a self-consistent data set. (Helgeson et al., 1978; Perkins et al., 1980; Jacobs and Kerrick, 1981).

et al., 1978). A differential reduction in SiO_2 activity of 0.01 by solid solution of the sort $(Na,K)AlO_2$ or $\square_{2/3}Al_{4/3}O_2$ will shift the first reaction by 45°C and the second by 140°C.[2] As a rule of thumb, reactions with a $\Delta\bar{S}$ < 4 joules/mol°K should not be used for thermometry, and those with a $\Delta\bar{V}$ < 2 cc/mol should be avoided in barometry.

Do the synthetic and/or natural minerals involved have variable structure states? The synthetic products should have the same structural state and chemistry as the metamorphic minerals had at the time of metamorphism; e.g., experiments on the stability of cordierite may not be applicable to cordierites in metamorphic rocks. Of course, the structural state of disordered metamorphic minerals may reset upon cooling, so one must know enough about the mineral system and the conditions of metamorphism to select the proper structural states; e.g., in using feldspar thermometry one must decide whether the metamorphic feldspars were ordered, partially disordered or completely disordered -- a difficult task in medium-grade (500-600°C) metamorphics.

Are long extrapolations in chemistry or in P-T required to apply the thermometer or barometer? Extrapolations should be avoided whenever possible because the "see-saw" effect works against accurate extrapolation (Fig. 1). This problem may be diminished if an accurate Clausius-Clapeyron slope calculation can constrain the range of possible slopes. Early calibrations of many barometers were made at higher P-T conditions than those of metamorphism,

----
[2]This may explain the widespread occurrence of cristobalite in lunar basalts which were never hotter than *ca.* 1200°C.

generating large errors when extrapolated down to crustal metamorphic P-T. Fortunately, many equilibria have now been determined near the P-T range of metamorphites minimizing this problem for many systems.

Extrapolation from simple mineral chemistries to complex natural solutions is a severe problem for many thermobarometers. Many systems are calibrated over a restricted chemical range and then applied to highly complex natural minerals. The effect of additional components must be modeled thermodynamically (Saxena, 1973), yet such relations are poorly understood for most minerals. One should test for systematic errors caused by additional components by plotting the inferred P or T versus the concentration of additional component. Multivariate analysis would be needed for adequate evaluation of n-component space, but obviously important chemical variables can be tested graphically.

Application of thermobarometers at large dilutions increases both systematic and random errors of extrapolation. For instance, the garnet-sillimanite-quartz-plagioclase thermobarometer[3] (Kretz, 1964; Ghent, 1976; Newton and Haselton, 1981) is commonly applied to garnets with only five percent grossular at pressures some 15 kbar away from the end-member reaction:

$$anorthite = grossular + sillimanite + quartz \qquad (4)$$
$$3CaAl_2Si_2O_8 = Ca_3Al_2Si_3O_{12} + 2Al_2SiO_5 + SiO_2$$

Quite apart from uncertainties generated by the long extrapolation in pressure and increased by analytical errors in the small amount of Ca in garnet, evaluation of grossular activity in garnet solutions at 95 percent dilution strains the credibility of our still imperfect thermodynamic knowledge. Similar comments apply to use of the diopside-enstatite thermometer on the steep limbs of the solvus at T < 800°C (Valley et al., 1982).

Are the thermobarometers retrograded or reset by a later metamorphism? This is a difficult question to answer in some cases, easy to confirm in others, and ultimately unknowable from an equilibrium standpoint. Using microscopic structural data (such as whether a phase or porphyroblast grew before, during, or after folding) can be misleading, because the chemistry can reset by diffusion upon prograde heating without necessarily reorienting the grain by recrystallization. Chemical resetting may be easy to identify if one obtains anomalously low P-T values, if an anomalous sample was collected from a late shear zone, or by finding late minerals with different chemistries

----

[3]This system is so labeled because it has too strong a temperature dependence to be called a barometer.

growing across the mesoscopic cleavage. It is much harder to identify retrogression if much of the area is reset, and one must find the occasionally preserved high-grade assemblage to prove the case. If every rock is reset and every chemical system reequilibrated, then certainly chemical evidence of retrogression is irretrievable and the earlier chemical history unknowable.

EXCHANGE THERMOMETRY

Introduction

Exchange reactions involve interchange of two similar atoms between different sites in one mineral (intracrystalline exchange) or between sites in two different minerals (intercrystalline exchange). While the elements (or isotopes) will redistribute between the sites or between the minerals with changing temperature, there is virtually no change in the amounts of the minerals from such reactions. The volume changes involved are very small and the entropy changes are relatively large, so that exchange reactions are largely independent of pressure and are ideal thermometers. Unfortunately, the small volume changes involved make these systems prone to resetting by diffusion alone during retrogression because back-reaction can proceed without recrystallization of the rock. Retrograde resetting of cations distributed between sites in a single phase requires diffusion to be operative over a distance some four orders of magnitude smaller than for cations distributed between metamorphic phases. Thus intercrystalline cation distributions are much less likely to be reset and potentially constitute much better metamorphic thermometers. Refractory phases such as garnet are best for exchange thermometry as they are less likely to be reset.

Intracrystalline exchange

There will generally be systematic partitioning of different elements between two available sites in a given structure. This partitioning is maximized at low temperatures and diminishes with increasing temperature as the structures expand and their entropy increases (Ghose, 1982). Minerals with at least two distinct sites of similar coordinations offer potential for intracrystalline exchange thermometry. Feldspars and pyroxenes have been the most carefully examined. Feldspars may partition Al systematically into the T_1 versus T_2 sites depending on temperature (Stewart and Ribbe, 1969), and simple powder X-ray measurements can yield estimates of composition and order (Wright and Stewart, 1968a,b). Workers should estimate the degree of order by determination of the unit cell dimension by least squares refinement, but

159

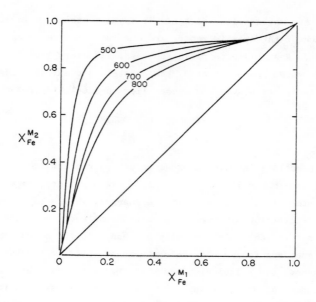

Figure 2. Distribution of Fe^{2+} betweeen the M(1) and M(2) sites in orthopyroxene as a function of temperature (°C) as determined by Mössbauer measurements (modified from Saxena and Ghose, 1971). Most metamorphic orthopyroxenes record temperatures of < 600°C but other geological thermometry indicates that they formed at T > 675°C (Bohlen and Essene, 1979; Hoffman and Essene, 1982; Perkins et al., 1982); this thermometer is therefore generally reset upon cooling.

even these measurements may be affected by possible submicroscopic coherent exsolution. Although many efforts have been made to compare feldspar dis-order with the grade of metamorphism, the measured parameters are not usually preserved in metamorphic systems at T > 500°C. The information obtained will provide limits on cooling rates, although it is probably also dependent on access of fluid during retrograde cooling as well. In any case, intracrys-talline exchange of Al and Si in feldspars is unsuitable for peak metamorphic thermometry.

Pyroxenes systematically partition cations between M(1) and M(2) sites (Ganguly, 1982). In orthopyroxene, magnesium and iron substitute on both sites with a temperature-dependent partition of iron into the M(2) site and magnesium into M(1) (Virgo and Hafner, 1969; Saxena and Ghose, 1971; see Fig. 2). Unfortunately, this cation exchange resets rather rapidly above 600°C so that all information about prior states (i.e., higher temperatures) is lost in metamorphic pyroxenes. This system and most other intracrystalline reac-tions are at best cooling rate speedometers rather than thermometers (Besancon, 1981, Ghose, 1982).

Intercrystalline exchange

Similar elements will systematically partition between sites of the same coordination in different minerals (Ramberg, 1952; Kretz, 1959). Intercrystalline exchange is generally considered in terms of isovalent elements which show some mutual solution, e.g., Na-K, Ca-Mg-Fe^{2+}-Mn, Al-Fe^{3+}-Cr, etc. Exchange thermometers are generally formulated in terms of the distribution coefficient, K_D, usually defined as $K_D = (a/b)^C/(a/b)^D$ where a and b are mole fractions of chemical components in the phases C and D. For exchange equilibria, the general relationship between K_D and temperature is:

$$\Delta \bar{H}^o - T\Delta \bar{S}^o + (P-1)\Delta \bar{V}^o + RT \ell n K_D + RT \ell n K_\gamma = 0 \; ,$$

where the superscript 'o' refers to the hypothetical end-member exchange reaction, and K_γ is the ratio of activity coefficients, in the same form as K_D. For most exchange reactions, $\Delta \bar{V}^o$ is small, and equilibrium composition is essentially only a function of temperature. To apply exchange reactions as thermometers, $\Delta \bar{H}^o$, $\Delta \bar{S}^o$ and K_γ must be known (from phase equilibrium experiments or calorimetry) or must be estimated. Some of the commonly used solution models used to obtain numerical values of K_γ are briefly described in the Appendix. K_D may be strongly affected by the presence of other substituents (e.g., Sr in feldspars or Mn in pyroxenes).

Only Mg-Fe^{2+} exchanges will be considered here because they encompass the most widely applied thermometers in metamorphic rocks, but the reader should be aware that dozens of other exchange thermometers have been proposed. Ferromagnesian minerals which have been used for Mg-Fe^{2+} exchange thermometry include olivine, garnet, clinopyroxene, orthopyroxene, spinel, ilmenite, cordierite, biotite, hornblende, and cummingtonite. Experimental calibrations of Mg-Fe^{2+} exchange are available for many systems, including olivine-orthopyroxene-spinel (Ramberg and Devore, 1951; Nafziger and Muan, 1967; Speidel and Osborn, 1967; Medaris, 1969; Nishizawa and Akimoto, 1973; Matsui and Nishizawa, 1974; Fujii, 1977; Engi, 1978), ilmenite-orthopyroxene and ilmenite-clinopyroxene (Bishop, 1980), garnet-olivine (Kawasaki and Matsui, 1977; O'Neill and Wood, 1979), orthopyroxene-clinopyroxene (Mori, 1977, 1978; Herzberg, 1978b), garnet-cordierite (Currie, 1971; Hensen and Green, 1973; Thompson, 1976; Hensen, 1977; Holdaway and Lee, 1977) as well as garnet-clinopyroxene and garnet-biotite (discussed below). Most of these experiments on the anhydrous pairs were obtained at T > 900°C and many were not compositionally reversed. Their best application is in high-temperature rocks such as mantle nodules and extrapolation of most of these data down to temperatures of crustal metamorphism is often unjustified. One potential problem with Mg/Fe^{2+} thermometry is in using microprobe analyses and assuming all iron to be ferrous iron. Adequate estimates of Fe^{3+} can be made in some phases if the analysis is normalized to cations and oxygen is raised to the expected stoichiometric levels by conversion of appropriate Fe^{2+} to Fe^{3+}. These phases include ilmenite, spinel (unless it shows $\gamma-R_2O_3$ solid solution), orthopyroxene, and clinopyroxene. Some phases, e.g., olivine, ordinarily carry no Fe^{3+} at peak metamorphic conditions. Chemical formulas for other phases, e.g., biotite and hornblende, cannot be adequately normalized with a probe analysis alone because there are too many possible substitutions for a unique solution, and Fe^{2+}/Fe^{3+} should be measured directly before attempting rigorous K_D thermometry.

Garnet-clinopyroxene. The garnet-clinopyroxene thermometer has been considered by many workers because of its potential importance for garnet-granulites, garnet-amphibolites, garnet-peridotites, and eclogites. Evans (1965b) and Banno and Matsui (1965) first pointed out that magnesium is partitioned preferentially into garnet over clinopyroxene with increasing temperature, and that pyrope solution in such garnets is more an indication of high temperatures than high pressures. Various workers correlated the garnet-clinopyroxene $K_D = (Fe/Mg)^{gn}/(Fe/Mg)^{cpx}$ with temperature for the reaction:

$$\text{pyrope} + \text{hedenbergite} = \text{almandine} + \text{diopside} \qquad (5)$$
$$\tfrac{1}{3}Mg_3Al_2Si_3O_{12} + CaFeSi_2O_6 = \tfrac{1}{3}Fe_3Al_2Si_3O_{12} + CaMgSi_2O_6$$

(Evans, 1965b; Essene and Fyfe, 1967; Saxena, 1968; Perchuk, 1968; Banno, 1970; Mysen and Heier, 1971; Irving, 1974; Oka and Matsumoto, 1974). Some workers have also inferred a dependence of this K_D upon additional components such as grossular in garnet or jadeite in pyroxene (Saxena, 1968; Banno, 1970; Oka and Matsumoto, 1974; Irving, 1974). Despite this, Raheim and Green (1974) calibrated the thermometer experimentally at high (mantle) P-T conditions using a Ca-poor garnet and Na-poor pyroxene and applied it to a variety of calcian and sodic eclogites formed at much lower (crustal) P-T (Raheim and Green, 1975; Raheim, 1975, 1976). Subsequently, Mori and Green (1978) recalibrated the thermometer based on additional experiments, and Ellis and Green (1979) experimentally evaluated the effect of grossular substitution on the garnet-clinopyroxene thermometer. Slavinskiy (1976), Ganguly (1979), Saxena (1979) and Dahl (1980) have reformulated the thermometer using various kinds

of thermodynamic, experimental, and field data. At least eight formulations
of this thermometer are thus available for testing and usage.

Comparative tests of the various formulations have been made (O'Hara
and Yarwood, 1978; Bohlen and Essene, 1980; Rollinson, 1980; Ernst, 1981).
It appears that the models of Raheim and Green (1974) and of Mori and Green
(1978) work best when applied to minerals of similar compositions equilibrated
at similar P-T as their experiments, but do not work well when applied to
granulites which equilibrated at much lower P-T. The formulation of Ellis
and Green (1979) appears to give much better (if somewhat scattered) temperatures in granulites (Johnson et al., 1979) and is given below:

$$T(°K) = (3104 \; X_{Ca}^{Gt} + 3030 + 10.86 \; P)/(lnK_D + 1.9039) \; .$$

The thermodynamic and calculated models of Slavinskiy (1976), Ganguly (1979),
Saxena (1979), and Dahl (1980) give widely scattered results when applied to
garnet-granulites of a wide composition range (Johnson et al., 1979). They
give more uniform though different temperatures when applied to rocks within
narrow compositions (Dahl, 1981). This suggests either that additional compositional parameters may be needed to make the thermometer generally applicable
or that retrogression has variably affected rocks of different compositions.

Application of garnet-clinopyroxene thermometry to crustal eclogites is
still speculative because of the larger extrapolations required both in temperature and in chemistry. In particular, eclogites associated with blueschists and amphibolites contain omphacitic pyroxenes and probably formed at
$T < 600°C$ (Essene and Fyfe, 1967; Black, 1973; Ernst, 1981). At these temperatures omphacites with >35 percent jadeite exhibit ordering of Mg and Fe^{2+}
versus Al in certain octahedral sites (Clark and Papike, 1968). This should
affect the activities of diopside and hedenbergite components and must change
the temperature dependence of the K_D. Thus garnet-clinopyroxene thermometry
(Raheim and Green, 1975; Raheim, 1975, 1976) should not be taken as quantitative for ordered omphacites, and as yet there are no other applicable thermometers available for these rocks. Experiments are unlikely to redress this
situation until laboratory equilibrium can be demonstrated with ordered
omphacites.

Garnet-biotite.   Garent-biotite thermometry has long been recognized
as having potential value based on observations in metamorphites (Kretz,
1959, 1964; Mueller, 1961; Perchuk, 1967; Saxena, 1969; Hietanen, 1969;
Thompson, 1976). It has been calibrated by field observations (Saxena, 1969;

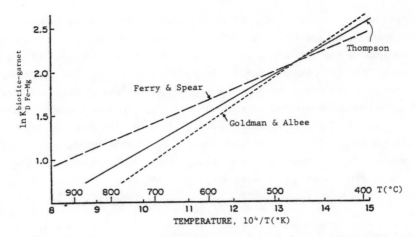

Figure 3. Various calibrations of the biotite-garnet K_D thermometer (modified from Bohlen and Essene, 1979). The Ferry and Spear (1979) version gives the highest temperatures; the Goldman and Albee (1977) calibration may be low because the isotope thermometers have reset from the higher temperatures (Ghent et al., 1979).

Thompson, 1976; Perchuk, 1977), by comparison with isotopic thermometry (Goldman and Albee, 1977), and by laboratory experimentation over a restricted compositional range (Ferry and Spear, 1978); not surprisingly, the calibrations are somewhat dissonant (Fig. 3). The exchange reaction is:

$$\text{phlogopite} \quad + \quad \text{almandine} \quad = \quad \text{annite} \quad + \quad \text{pyrope} \quad (6)$$
$$\text{KMg}_3\text{Si}_3\text{AlO}_{10}(\text{OH})_2 + \text{Fe}_3\text{Al}_2\text{Si}_3\text{O}_{12} = \text{KFe}_3\text{Si}_3\text{AlO}_{10}(\text{OH})_2 + \text{Mg}_3\text{Al}_2\text{Si}_3\text{O}_{12}$$

and the $K_D = (\text{Fe/Mg})^{\text{biot}}/(\text{Fe/Mg})^{\text{gn}}$. As with other exchange thermometers, it is likely that solid solution of additional components affects this thermometer, and Ferry and Spear (1978) suggest that it should be restricted for usage with garnets low in Ca and Mg and with biotites low in Ti. One serious problem, still unresolved, is the potential effect of unrecognized Fe^{3+} in biotite. The exchange equation strictly refers only to isovalent interchange, yet many workers have blithely used total iron for Fe^{2+} in biotite, which lowers the temperature estimate significantly if the biotite had in fact had 20-30 percent of the iron as Fe^{3+}. This problem needs to be evaluated with careful wet chemical or Mössbauer measurements. Fortunately, garnets and biotites in metapelites of the greenschist and amphibolite facies lie reasonably close to the respective binaries. In the uppermost amphibolite facies and granulite facies, however, biotites are increasingly rich in Ti, F, and Cl and garnet-biotite thermometry is increasingly erratic (Bohlen and Essene, 1980). In addition, garnet may show retrogressive exchange when in contact with adjacent biotites, and one must avoid these areas in analysis (Edwards

164

and Essene, 1981). In general, the garnet-biotite thermometer does not appear to work well in high-grade rocks. On the other hand, it yields reasonable results for lower-grade rocks (Ghent et al., 1979; Labotka, 1980; Ferry, 1980; Novak and Holdaway, 1981) and must be regarded as the thermometer of choice for medium-grade regional metamorphic rocks using Ferry and Spear's (1978) experimental calibration.

Isotopic thermometry. Isotopic thermometry is based on the temperature dependence of equilibrium partition of light stable isotopes between two phases (Urey, 1947, 1948). The very small volume changes attendent with exchange mean that the exchange process is almost completely independent of pressure. Stable isotopes are chosen so that their ratios will not vary with time due to radioactive decay. Heavy isotopes are not useful for thermometry because they do not fractionate as much as light isotopes. Light isotopes (typically of the elements C, O, S) may fractionate between one tenth and one percent between two phases at low temperatures; fractionation decreases to almost zero at high temperatures (800-1000°C). Fractionations tend to be largest between phases of widely different chemistries and structures and are most sensitive to temperature changes at low temperature (in contrast to solvus) thermometry. The $^{18}O/^{16}O$ fractionations for calcite-quartz, magnetite-quartz, magnetite-feldspar, and rutile-feldspar pairs show significant temperature dependencies. In addition, there are $n-1$ potential thermometers in a n-phase rock. With a few hundredths of a percent accuracy in isotopic ratios now routinely attained in many mass spectrometric laboratories, these differences are easily measured.

At first glance this technique would appear to be the ideal thermometer for metamorphic rocks, and one might wonder why there is any need for other geological thermometry. However, isotopic thermometry is plagued by several problems. For instance, there is a lack of calibration of some systems, and diagreement in the available calibrations for some minerals including quartz, pyroxene, and dolomite (Friedman and O'Neil, 1977). Experiments generally fail to demonstrate isotopic exchange equilibrium (exceptions are those of O'Neil and Taylor, 1968, on feldspar-H_2O). Retrograde isotopic exchange during cooling of high-temperature metamorphites often yields low inferred temperatures (Ghent et al., 1979). Nevertheless, isotopic thermometry still can generate data in agreement with that obtained from other thermometers (O'Neil and Clayton, 1964; Shepard and Schwarcz, 1970; Goldman and Albee, 1977; Taylor and O'Neil, 1977; Hoernes and Friedrichsen, 1978; Bowman, 1978; Nesibtt and Essene, 1982), and its usage should be considered in conjunction with other thermometers, especially in lower-grade rocks. For further information and tabulated fractionations the reader is referred to Friedman and O'Neil (1977).

PHASE EQUILIBRIA FOR PETROGENETIC GRIDS

Equilibria involving phases of fixed compositions are perhaps the simplest to apply to metamorphites. The experiments are often easy to interpret -- because inference of reaction direction is not complicated by solid solutions. It is equally simple to apply -- one just identifies the minerals in equilibrium. If the minerals have unique compositions then there is no need to analyze them. One just has to find enough critical assemblages to box out a restrictive range of P-T. There are several disadvantages with this approach. First, it is unusual to find univariant assemblages *because* of the lack of solid solution. This is the principle behind Goldschmidt's, or the mineralogical, phase rule which states that most rocks will have equilibrated under divariant conditions. Secondly, even where solid solution is limited, it may still have a significant effect on phase equilibria. Strens (1967) calculated the dramatic effect of even small amounts of Fe, Mn and Cr on phase relations of the Al_2SiO_5 system. Thirdly, minerals such as talc, phlogopite and tremolite are often assumed to be nearly stoichiometric in terms of the elements commonly analyzed on the microprobe, but may show extensive exchange of fluorine for hydroxyl, which strongly affects their stability (Valley et al., 1982). Fourthly, other minerals (such as magnetite, K-feldspar, and calcite), often appear to be nearly end-member upon analysis but have reset chemically by retrograde exchange from high-temperature solid solutions (Bowman, 1978; Bohlen, 1979; Nesbitt, 1979; Brown, 1980; Valley, 1980). Finally, some metamorphic minerals (such as aragonite in blueschists[4], pigeonite in high-temperature metamorphites, and periclase in high-grade marbles) back-react so easily that it is often difficult to be certain of their former presence. Sometimes textures (such as "onion-skin" brucite[5] after periclase or abundant exsolved clinopyroxene in an orthopyroxene host denoting former pigeonite) allow one to infer the former presence of the high-temperature phase. In other cases complete recrystallization may occur, making it difficult to distinguish retrograde from prograde assemblages.

----
[4] Isograds mapped for the reaction calcite-aragonite in all probability represent a retrograde locus of late access by H_2O rather than prograde pressure gradients.

[5] This texture can be mimicked by hydrotalcite after spinel.

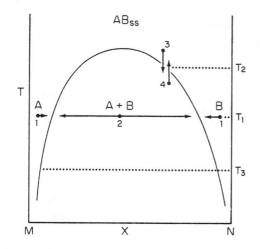

Figure 4. Compositional reversals and field boundary locations of the limbs of a solvus. See text for details.

Seldom is it possible to determine the P-T conditions of metamorphism using only phases of fixed compositions, and one must turn instead to reactions with phases of variable compositions requiring the use of the electron microprobe to evaluate metamorphic conditions. Solvus relations, solid-solid reactions and reactions involving a fluid phase will be discussed below.

Solvus thermometry

A compositional gap may exist between two coexisting, structurally related minerals, where the boundaries of the gap mark the mutual limits of miscibility (Fig. 4); such a miscibility gap will be loosely referred to as a solvus. With increasing temperatures mutual solution may become complete unless one or the other of the minerals first decomposes. Using phase diagrams such as Figure 4, the compositions of coexisting mineral pairs may be used directly to estimate temperature of equilibration. Experiments involving composition reversals are usually necessary to define the solvus location, and these experiments involve (1) increasing the mutual solution between the two phases, and (2) simultaneously unmixing a homogeneous phase (or a mixture of two phases with substantial mutual solid solution) at the same temperature (T_1 in Fig. 4). If disorder occurs in one or both minerals, then there will be an infinity of solvi, one for each degree of partial disorder. Many of the experiments on mineral solvi attempt to locate the solvus by identification of one phase versus two phases in experimental products. Reversals can be generated with this technique by a variety of experiments with variable temperatures (yielding T_2 in Fig. 4) but the compositional reversals generally

yield more accurate results. When performing unmixing experiments it is important *not* to rely solely on X-ray powder determinations of compositional
changes, because lattice parameters may shift by formation of coherent lamellae rather than by compositional change to the strain-free solvus. Microprobe determinations of compositional shifts should be obtained to demonstrate actual chemical changes. Even with tight reversals at high temperatures and with thermodynamic properties calculated from the solvus limbs
(Saxena, 1973), it is dangerous to extrapolate the miscibility limits to
lower temperatures (T_3 in Fig. 4) because the data often cannot be accurately
extrapolated and mineral compositions furthermore are relatively insensitive
to temperature. It is equally dangerous to apply an experimentally determined solvus in a simple compositional join to complex solid solutions.

Although many solvi are encountered in metamorphic rocks, few have been
well calibrated for geothermometry. Solvi among amphiboles and among micas
have not been generally established by experimental reversals and should not
be used. Even the paragonite-muscovite solvus has been determined only by
homogenization and not by unmixing experiments (Iiyama, 1964; Eugster et al.,
1972; Blencoe and Luth, 1973). The nepheline-kalsilite solvus is well determined (Ferry and Blencoe, 1978) but inapplicable to most metamorphites. Among
other solvi which have been well calibrated are calcite-dolomite, calcite-
siderite, diopside-enstatite, hedenbergite-ferrosilite, and K-feldspar-albite;
and several of these will be discussed below.

Albite - K-feldspar.   The alkali feldspar solvus was applied by Barth
(1934, 1957, 1962, 1969) based initially on field calibrations and later with
some experiments. The solvus was invesitgated experimentally by many workers,
including Orville (1963), Luth and Tuttle (1966), Parsons (1969), Morse (1970),
Bachinski and Muller (1971), Seck (1972), Goldsmith and Newton (1974), and
Lagache and Weisbrod (1977) (see summary in Parsons, 1978). The detailed
location of the limbs of the solvus is still somewhat uncertain due to experimental difficulties in unmixing a homogeneous feldspar and due to the
effect of Al-Si disorder, which may vary during an experiment. Application to
most metamorphites is made difficult by the effect of even small amounts of
calcium, and the two-feldspar exchange thermometer (below) is to be preferred
in quantitative work.

Ternary feldspars.   Most feldspars are well represented by the three
components $NaAlSi_3O_8$-$KAlSi_3O_8$-$CaAl_2Si_2O_8$. The plagioclase series and alkali
feldspar series show little mutual solution at T < 800°C and are well

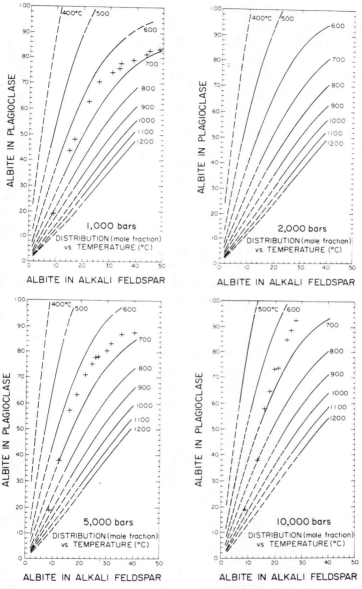

Figure 5. Stormer's (1975) calculated temperatures for coexisting alkali feldspar and plagioclase compared to Seck's (1972) experimental data. From Stormer (1975).

represented by the respective binaries, $NaAlSi_3O_8$-$CaAl_2Si_2O_8$ and $NaAlSi_3O_8$-$KAlSi_3O_8$. Barth (1957) and Stormer (1975) showed that the partitioning of $NaAlSi_3O_8$ between the two phases at equilibrium is controlled by temperature. Stormer (1975) calculated the partitioning considering the non-ideality in

169

alkali feldspars and considering the effect of pressure. His calculations fit
Seck's (1971) experiments well at 650°C for plagioclases between An_5 and An_{45},
but less well for more calcic compositions (Fig. 5). Powell and Powell (1977b)
modified Stormer's thermometer for the effect of $CaAl_2Si_2O_8$ in alkali feldspars but did not make an equivalent correction for $KAlSi_3O_8$ in plagioclases.
Whitney and Stormer (1977) have also produced a low-temperature thermometer
for ordered K-feldspar. Brown and Parsons (1981) pointed out shortcomings
in the treatments of Stormer (1975) and the Powells (1977) but were unable
to find an alternative procedure which gave a better fit to Seck's data.

Tests of the two-feldspar thermometers in metamorphic rocks are complicated by exsolution in the alkali feldspar upon cooling. Exsolved grains
must be reintegrated to obtain the composition of the original alkali feldspar solid solution. The only homogeneous grains of alkali feldspar which
survive high-grade metamorphism have compositions of $Or_{>90}$ and have reset at
T < 500°C by external granule exsolution (Bohlen and Essene, 1977; Valley and
Essene, 1980). Even with reintegration it is possible that some $NaAlSi_3O_8$
component has been lost at high temperatures, and two-feldspar thermometry
must always be regarded as providing minimum estimates of temperature (Bohlen and Essene, 1977). However, Bohlen and Essene showed that temperatures
of up to 800°C can be recovered with this thermometer, which deserves application in other high-grade terranes.

Applications of the Whitney-Stormer low-temperature feldspar thermometer
have been less successful. Nesibtt and Essene (1982) found that it gave 50°C
lower temperatures than calcite-dolomite thermometry conducted on adjacent
marbles for rocks regionally metamorphosed to 450-550°C. Some of the thermodynamic input parameters for the Whitney-Stormer model must be in error,
and the variable ordering and unmixing along the plagioclase binary should
be considered for ordered two-feldspar thermometry.

Calcite - dolomite. Thermometry relies on the temperature dependence
of the magnesium content of calcite in equilibrium with dolomite. Experimental determinations (Harker and Tuttle, 1955b; Graf and Goldsmith, 1955,
1958; Goldsmith and Heard, 1961; Goldsmith and Newton, 1969) include well-
reversed data. When reversals are plotted using the same X-ray determinative
curves for the Mg-content of calcite and the data are corrected for the small
effect of pressure (Goldsmith and Newton, 1969), most of the reversals are
consistent with a single solvus location between 400 and 800°C (Anovitz and
Essene, 1982) (Fig. 6). The solvus seems well calibrated for geothermometry
in this temperature range.

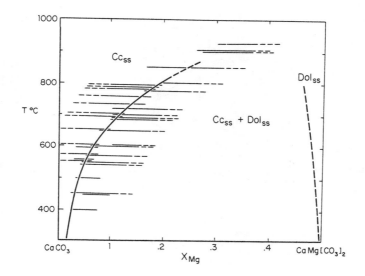

Figure 6. The calcite-dolomite solvus calculated by Anovitz and Essene (1982) with reversals taken from the literature. The dolomite limb of the solvus is based on analyses of coexisting dolomite and calcite by Bowman (1978) and Nesbitt (1979).

Unfortunately, application of the two-carbonate thermometer to metamorphic rocks is not simple because of retrogressive resetting. Magnesian calcites may exsolve dolomite lamellae which must be reintegrated during probe analysis (Perkins et al., 1982). More often the original high magnesian content is simply obliterated through diffusion, and yielding temperatures on the order of 300-400°C for high-grade marbles (Puhan, 1976; Valley and Essene, 1980). Occasionally, preserved magnesian calcites or cores of calcites yield temperatures approaching those from other thermometers (Bowman and Essene, 1982; Nesbitt and Essene, 1982). Even careful applications seldom yield temperatures above 600°C, which is taken to be the uppermost temperature obtainable for most high-grade regional metamorphic rocks. The thermometer seems to be more generally useful in low-grade regional metamorphic rocks (Bickle and Powell, 1977; Nesbitt and Essene, 1982) and in contact metamorphic rocks (Suzuki, 1977; Rice, 1977a,b; Bowman, 1978; Brown, 1980) where cooling times are presumably reduced.

Enstatite - diopside. This solvus has a distinguished history of experimentation (Boyd and Schairer, 1964; Davis and Boyd, 1966; Warner and Luth, 1974; Nehru and Wyllie, 1974; Lindsley and Dixon, 1976; Mori and Green, 1975, 1976) and thermodynamic modeling (Saxena and Nehru, 1975; Holland et al.,

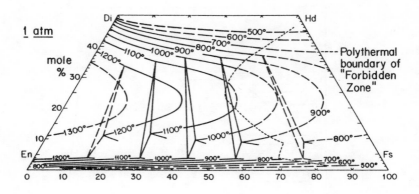

Figure 7. Solvus isotherms for coexisting pyroxenes in the system $CaMgSi_2O_6$-$CaFeSi_2O_6$-$Mg_2Si_2O_6$-$Fe_2Si_2O_6$. From Lindsley and Anderson (1982).

1979; Lindsley, 1980; Lindsley et al., 1981; Saxena, 1981). It is one variable in mantle geotherm thermobarometry of Boyd (1973), MacGregor (1974),
Mori (1977), and Herzberg (1978a). At high temperatures the diopside limb
of the solvus is temperature-dependent, but below 850°C it is insensitive to
temperature (Fig. 7). Although ignored in the applications by Boyd (1973)
and by MacGregor (1974), the major problem with the thermometer lies in correction for other components. Several approaches have been made to deal with
this problem: (1) performing experiments on actual rock compositions (Green
and Ringwood, 1967; Akella, 1976; Mori and Green, 1978); (2) adding additional
components to evaluate their effects separately (Akella and Boyd, 1973; Akella,
1976); and (3) correcting for the solid solutions by thermodynamic modeling
(Wood and Banno, 1973; Saxena and Nehru, 1975, Wells, 1977). Tests of the
thermometers applied to real systems suggest that the Wells thermometer works
well for ultramafic systems at temperatures >900°C (e.g., Wells, 1977; Ernst,
1981) but gives scattered and high values for iron-rich pyroxenes from granulites (Bohlen and Essene, 1979; Perkins et al., 1982). On the other hand,
Valley et al. (1982) found that a nearly pure enstatite-diopside rock from
the granulite facies of the Southern Adirondacks gave low temperatures (400-
600°C) consistent with resetting of the two-pyroxene solvus thermometer. This
solvus appears to provide no panacea for crustal granulites.

Quadrilateral pyroxenes. The quadrilateral pyroxene system is composed of diopside-hedenbergite-enstatite-ferrosilite. It provides a good
representation of the important solid solutions in pyroxenes from most crustal
metamorphic rocks (except for blueschists). If the clinopyroxene-orthopyroxene

172

solvus were well determined in this three-component ($CaSiO_3$-$MgSiO_3$-$FeSiO_3$) space, then it would be a useful thermometer for crustal pyroxenes composed of these major components. Ross and Huebner (1975) combined a variety of experimental and observational data in estimating a provisional ternary solvus at many temperatures. Lindsley and Anderson (1982) have revised the Ross-Huebner solvus for quadrilateral pyroxenes based on reversed experiments (Fig. 7). Applications of this diagram to most metamorphic clinopyroxenes coexisting with orthopyroxenes yields low temperatures (<600°C) consistent with Ca-Mg-Fe^{2+} exchange upon cooling.

Solid-solid reactions

These reactions are of the greatest use for geobarometry because their position in P-T space is not dependent on the compositions or presence of a fluid phase. Although solid-solid reactions generally have a temperature as well as a pressure dependence, most are still primarily used for geobarometry, with metamorphic temperatures determined by exchange or other thermometers. Application of these reactions usually requires chemical analysis and thermodynamic calculations to correct for additional solid solutions, but this is also advantageous because the additional components increase the variance of the subsystem under consideration and extend the stability of the assemblage allowing it to be found in a greater variety of rocks. The general equation for relating solid-solid mineral reactions to temperature and pressure is:

$$\Delta \bar{H}^O - T\Delta \bar{S}^O + (P-1)\Delta \bar{V}^O + RT \ln K = 0;$$

$\ln K = \Sigma \nu_i \ln a_i$, where ν_i is the stoichiometric coefficient of phase component i in the reaction (positive for products, negative for reactants) and a_i is the activity of component i in a participating mineral phase. Calibration of solid-solid reactions as thermobarometers involves determination of $\Delta \bar{H}^O$, $\Delta \bar{S}^O$, $\Delta \bar{V}^O$, and activity-composition relations from phase equilibrium experiments, calorimetry and/or measurements of heat capacity. For an assemblage with specified mineral compositions, the above equation can be used to define a line in P-T space along which the assemblage equilibrated.

Many of the most commonly applied reactions involve plagioclase, pyroxene, and garnet and require some evaluation of the thermodynamics of these solid solutions (i.e., activity-composition relationships). Developing these solution models is beyond the scope of this paper; the reader is referred to the Appendix for a brief, general discussion of activity-composition models and to Orville (1972), Saxena (1973), Ganguly and Kennedy (1974), Cressey et al. (1979), Lindsley (1981), and Ganguly (1982) for detailed treatments of solid solution models for these particular minerals. Reactions will be considered below in terms of mineral assemblages rather than in terms of chemical systems to help the reader identify critical assemblages and because a chemical-systems-approach would obscure the simple relationships between the end-member reactions.

Andalusite - sillimanite - kyanite. The Al_2SiO_5 system is comprised of the three polymorphs easily identified in hand-specimen and widely used as index minerals of metamorphism. Their P-T significance was recognized early although Harker (1932) led a generation of workers astray by the misplaced emphasis on stress as an important physiochemical variable of metamorphism. If reversed experiments obtained in a hydrostatic pressure apparatus are accepted, then the Al_2SiO_5 triple point is located at

> 6.5 kbar and 595°C (Althaus, 1967), or
> 5.5 kbar and 620°C (Richardson et al., 1969), or
> 3.8 kbar and 600°C (Holdaway, 1971).

The triple point of Holdaway (1975) is usually preferred by field petrologists who have tested Al_2SiO_5 occurrences against other thermobarometers. Likely errors in the triple point are probably ±0.5 kbar and ±50°C. The substantial experimental discrepancies may be related to the variation in composition and structure of the experimental versus the natural phases (Zen, 1969; Holdaway, 1971; Greenwood, 1972). These variations have a particularly large effect for the reaction andalusite = sillimanite due to its small free energy (Fyfe, 1971; Mueller and Saxena, 1977), and many of the discrepancies in the experimental determination of Al_2SiO_5 equilibria are reduced if one concedes that this reaction still has a 50-100°C uncertainty.

Despite remaining experimental disagreements, the Al_2SiO_5 system is widely used for estimating P-T in metamorphic rocks (e.g., Thompson and Norton, 1968; Carmichael, 1978). In many terranes only one isograd relating two Al_2SiO_5 minerals is found; pressures can still be estimated at the isograd if independent thermometry is available. Occasionally only one prograde Al_2SiO_5 mineral is found, but even this may give meaningful pressure limits with the use of other thermometers.

At high temperatures sillimanite reacts with corundum to form mullite, and at still higher temperatures it dissociates to mullite + tridymite (Holm and Kleppa, 1966). These reactions will be useful within the sanidinite facies where mullite is occasionally found, but the reactions need to be more exactly located. Holm and Kleppa's (1966) curves are the best available but

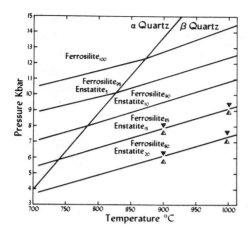

Figure 8. Stability limits of orthopyroxene solid solutions according to the reaction ferrosilite = fayalite + quartz. From Bohlen and Boettcher (1981).

do not consider potential effects of Al-Si order/disorder in mullite and sillimanite, or solid solutions of excess Al in sillimanite, excess Si in mullite or Al in tridymite. Careful experiments with both compositional and structural reversals in regard to the univariant equilibria are needed before these reactions should be used quantitatively.

Pyroxene/pyroxenoid - olivine - quartz. These assemblages are related by the reaction:

$$\text{pyroxene/pyroxenoid} = \text{olivine} + \text{quartz} \qquad (7)$$
$$2RSiO_3 = R_2SiO_4 + SiO_2 .$$

The olivine structure is stable with quartz at low pressures for iron-rich compositions. Lindsley et al. (1964) first synthesized ferrosilite and approximately located the equilibrium

$$\text{ferrosilite} = \text{fayalite} + \text{quartz} \qquad (8)$$
$$2FeSiO_3 = Fe_2SiO_4 + SiO_2 .$$

Smith (1971, 1972) experimentally determined the effect of enstatite solid solution upon the system at low pressures. Wood and Strens (1971) combined these data and proposed the system as a barometer. Bohlen et al. (1980) redetermined the P-T conditions for reaction (8) locating it 1-2 kbar lower than previously thought. Subsequently, Bohlen et al. (1980) and Bohlen and Boettcher (1981) have determined the effects of Mn and Mg (Fig. 8) on the location of reaction (8) and applied it as a barometer to many high-grade terranes. This barometer is useful only in iron-rich rocks such as fayalite granites and some charnockites.

The assemblage pyroxenoid-knebelite-quartz will also be a useful barometer limiting pressures in rhodonite- and pyroxmangite-bearing rocks. Once the activity of $FeSiO_3$ (orthopyroxene) can be calculated in pyroxenoid solid solutions, reaction (8) can be used as a barometer for pyroxenoids. Once calibrated, the occasionally-reported assemblage knebelite + pyroxenoid + quartz will fix a univariant curve in P-T space.

Pyroxene - plagioclase - quartz. Several reactions can be written balancing various end members in the pyroxene and plagioclase solid solutions (Gasparik and Lindsley, 1980), including:

$$albite = jadeite + quartz \qquad (9)$$
$$NaAlSi_3O_8 = NaAlSi_2O_6 + SiO_2$$

and

$$anorthite = tschermakite + quartz \qquad (10)$$
$$CaAl_2Si_2O_8 = CaAl_2SiO_6 + SiO_2 .$$

Reaction (9) has been evaluated by several workers, including Newton and Smith (1967), Johannes et al. (1971), Holland (1980), and Hemingway et al. (1981). It is complicated by the order-disorder transition in albite; at T < 500°C where ordered albite is stable, only a calculated reaction is available (Newton and Smith, 1967; Hemingway et al., 1981). Reaction (10) has been calibrated by several investigators including Gasparik and Lindsley (1980). In principle, any equilibrated pyroxene and plagioclase solid solution and quartz would be uniquely fixed in P-T space by these two reactions if the activities of jadeite, tschermakite, albite and anorthite could all be determined from chemical analysis of the pyroxene and plagioclase. Indeed, Gasparik and Lindsley give some P-T graphs with contours of mole fractions of pyroxene components for simplified synthetic systems. However, it is difficult to extract the activities of the components jadeite and tschermakite for applications to complex multicomponent natural clinopyroxenes, and it seems best to apply the reactions (9) and (10) to binary, or at most ternary, pyroxenes.

Pyroxene compositions are often close to the ternary system jadeite-aegirine-diopside in the blueschist facies (Essene and Fyfe, 1967). The jadeite-aegirine binary is close to an ideal solution (based on the experiments of Popp and Gilbert, 1972), and the stability of aegirine-jadeite + low albite + quartz rocks can be calculated without further ado. On the other hand, the cation ordering observed in omphacites lying along the diopside-jadeite binary (Clark and Papike, 1968) suggests that this system

176

is non-ideal. Ganguly (1973), Currie and Curtis (1976), and Gasparik and Lindsley (1980) all treated diopside-jadeite as a regular solution and calculated phase equilibria accordingly. While their calculations are certainly more accurate than the ideal solution model of Essene and Fyfe (1967), the system remains to be tested in nature against other barometers.

The tschermakite reaction (10) is also difficult to apply to natural aluminous pyroxene because of the substitution of titanium as $CaTiAl_2O_6$ or as $NaTiSiAlO_6$, and ferric iron as $CaFeSiAlO_6$ or as $NaFeSi_2O_6$; consequently, some of the tetrahedral or octahedral aluminum may be related to substitutions other than tschermakite. However, many ordinary granulite facies clinopyroxenes are close to the diopside-hedenbergite join with a small amount of tschermakite substitution. Wood (1976, 1979) has shown experimentally that the tschermakite substitution in hedenbergite is similar to that for diopside when equilibriated with anorthite + quartz at the same P-T. This suggests that the tschermakite activities are the same in both binaries and that one may use the tschermakite isopleths shown by Wood (1976) or by Gasparik and Lindsley (1980) for the granulite clinopyroxenes. Any calculation requires determination of anorthite activity for plagioclase solid solutions before P-T conditions of equilibrium may be determined. Application of this barometer by Martignole (pers. comm., 1981) and by the writer (unpublished) gave results scattered about the pressures given by other barometers. The tschermakite contents of such pyroxenes are sufficiently low as to represent substantial dilution, and the system should not be expected to be precise at crustal pressures when the activity of silica is one.

Garnet - plagioclase - Al_2SiO_5 - quartz. This reaction has become the most widely used barometer in medium- and high-grade rocks. It was developed by Ghent (1976) based on reaction (4), anorthite = grossular + sillimanite + quartz, and on various solution models for garnets and plagioclase. It has been modified somewhat by Newton and Haselton (1981) and by Perkins et al. (1982) based on different solution models for the activity of grossular in garnet. The advantage of this system is the widespread occurrence of garnet - Al_2SiO_5 - plagioclase-quartz assemblages in metapelites, mainly due to extensive solid solution of almandine in the garnet and of albite in plagioclase. Its very real disadvantages are related to: (1) the still uncertain $\Delta \bar{V}_s$ at P-T which is required to calculate the large shift in pressure from the end-member garnet; (2) the unknown effect of andradite solid solution on the activity of grossular; (3) the relatively large errors in analyzing a small

percentage of Ca in garnet; (4) the strong dependence of the barometer on the
assumed temperatures; and (5) the large uncertainties involved in the activity
coefficients for garnet and plagioclase for temperatures below 600°C. Despite
all these *caveats* this thermobarometer seems to work well in most terranes,
agreeing with Al_2SiO_5 constraints (Schmid and Wood, 1976; Ghent et al., 1979)
and with garnet-pyroxene and other barometers (Perkins et al., 1982; Martignole and Nantel, 1982). This system should be tested for internal consistency
in a terrane where plagioclase and garnet have a wide range of anorthite- and
grossular-contents, respectively. If it passes this test, then it would truly
deserve the place it has already assumed among crustal barometers.

Garnet - plagioclase - olivine. Wood and Banno (1973) and Wood (1975)
used the reaction:

$$\text{plagioclase}_{ss} + \text{olivine}_{ss} = \text{garnet}_{ss} \qquad (11)$$

for barometry based on the reaction

$$\text{anorthite} + \text{fayalite} = \text{grossular}_{1/3}\text{almandine}_{2/3} \qquad (12)$$
$$CaAl_2Si_2O_8 + Fe_2SiO_4 = CaFe_2Al_2Si_3O_{12}$$

first calibrated by Green and Hibberson (1970). Bohlen et al. (1982) redetermined this reaction in a special low-friction cell piston-cylinder apparatus
and also in an internally heated apparatus. They obtained a somewhat different curve for the reaction than did Green and Hibberson (1970). The equivalent magnesium-rich reaction

$$\text{anorthite} + \text{forsterite} = \text{grossular}_{1/3}\text{pyrope}_{2/3} \qquad (13)$$
$$CaAl_2Si_2O_8 + Mg_2SiO_4 = CaMg_2Al_2Si_3O_{12}$$

may be calculated from experiments of Kushiro and Yoder (1966) and of Herzberg (1978a) to yield a potential barometer for garnetiferous-olivine metagabbros (Johnson et al., 1979). To apply it one must correct all three phases
for the natural solid solutions; reactions (12) and (13) should then cross or
give equivalent answers. In many garnetiferous metabasites garnet-olivine-
plagioclase is metastable with respect to aluminous orthopyroxene and aluminous clinopyroxene (Sack, 1980a) as might be expected by the experiments of
Kushiro and Yoder (1966). Nevertheless, for some iron-rich and albite-rich
solid solutions the garnet-plagioclase-olivine assemblage is stable (Sack,
1980a) and may be used in barometry. Reaction (12) has one additional important use. It may be combined with reaction (8), fayalite + quartz = ferrosilite, to locate one end member of the next assemblage (Bohlen et al., 1982).

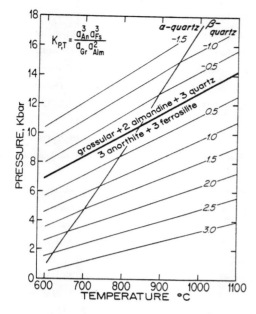

$$K_{P,T} = \frac{a_{An}^3 \, a_{Fs}^3}{a_{Gr} \, a_{Alm}^2}$$

a-quartz / β-quartz

grossular + 2 almandine + 3 quartz

3 anorthite + 3 ferrosilite

Figure 9. $Log_{10}K$ contoured on a P-T diagram for the reaction ferrosilite + anorthite = grossular$_{1/3}$, almandine$_{2/3}$ + quartz. From Bohlen et al. (1982a).

Garnet - plagioclase - orthopyroxene - quartz.   This assemblage is related by the reaction:

$$\text{plagioclase} + \text{orthopyroxene} = \text{garnet} + \text{quartz} \qquad (14)$$

which can be calculated from:

$$\text{anorthite} + \text{ferrosilite} = \text{grossular}_{1/3}\text{almandine}_{2/3} + \text{quartz} \quad (15)$$
$$CaAl_2Si_2O_8 + 2FeSiO_3 = CaFe_2Al_2Si_3O_{12} + SiO_2$$

or from:

$$\text{anorthite} + \text{enstatite} = \text{grossular}_{1/3}\text{pyrope}_{2/3} + \text{quartz} \quad (16)$$
$$CaAl_2Si_2O_8 + 2MgSiO_3 = CaMg_2Al_2Si_3O_{12} + SiO_2 \,.$$

This reaction has been calculated by Wood (1975) from the experiments of Kushiro and Yoder (1966) and was calibrated by Perkins and Newton (1981) and Newton and Perkins (1982) who found systematic errors in an attempted calorimetric calibration. Perkins and Newton (1981) found excellent agreement between their barometer and other available barometers, but this is not surprising because their barometer was calibrated to yield reasonable results when compared to Al_2SiO_5 assemblages. An independent test of their model would consist of an assemblage of much more iron-rich minerals or one that equilibrated at far different pressures.

Bohlen et al. (1982) have experimentally located reaction (15), and therefore it should stand as an independent barometer (Fig. 9). Use of the diagram requires calculation of the activities of the garnet $(Gr_{1/3}Alm_{2/3})$, anorthite, and ferrosilite from probe analyses of the natural garnet, plagioclase and hypersthene in the mineral assemblage. This entails use of activity models and some estimate of temperature (generally required in calculation of the activities). The formulations recommended for this system are Perkins (1979) for garnet, Seil and Blencoe (1979) for plagioclase, and Saxena (1973) for orthopyroxene. Initial applications of the data of Bohlen et al. (1982) yield generally reasonable pressures in granulite terranes, and this barometer seems to offer much promise in P-T estimation for garnet granulites and some garnet amphibolites.

Garnet - cordierite - sillimanite - quartz. This reaction has attracted a great deal of attention because of its potential for cordierite granulites. Calibrations have been attempted by Currie (1971), by Hensen and Green (1973), by Weisbrod (1973), by Hensen (1977), by Holdaway and Lee (1977), by Newton and Wood (1979), by Martignole and Sisi (1981), and by Lonker (1981), and others. All of these calibrations disagree with each other as to (1) the P-T location of a given reaction; (2) the importance of P_{H_2O}/P_{solid}; (3) the activity of anhydrous cordierite in a hydrous cordierite; and (4) the locations and slopes of the end-member reactions:

$$\text{iron-cordierite} = \text{almandine} + \text{sillimanite} + \text{quartz} \qquad (17)$$
$$3Fe_2Al_4Si_5O_{18} = 2Fe_3Al_2Si_3O_{12} + 4Al_2SiO_5 + 5SiO_2$$

$$\text{cordierite} = \text{pyrope} + \text{sillimanite} + \text{quartz}$$
$$3Mg_2Al_4Si_5O_{18} = 2Mg_3Al_2Si_3O_{12} + 4Al_2SiO_5 + 5SiO_2 .$$

Direct experiments on these reactions and intermediate solid solutions are difficult to interpret. Currie (1971) failed to duplicate the garnet/cordierite Mg-Fe^{2+} exchange K_D observed in rocks in his relatively short-term experiments. Hensen and Green (1973) used glasses as part of their starting materials so that true reversals might not have been attained, but Hensen (1977) confirmed that their data were in agreement with carefully attained compositional reversals at 9 kbar and 900°C. Application of Hensen and Green's (1973) thermobarometer to natural occurrences gives kyanite P-T for sillimanite-bearing rocks (Perkins et al., 1982), and it is possible that the necessary pressure corrections were underestimated for a 5/8" piston-cylinder device at the relatively low pressures of the experiments. It is also possible that synthetic cordierites grown at high temperatures (>800°C?) become substantially disordered, thus affecting phase equilibria and extrapolations to geologic P-T. The models of Holdaway and Lee (1977), Newton and Wood

(1979), and Martignole and Sisi (1981) all give approximately correct pressures for intermediate values of water pressure. This system needs more carefully documented experiments independently evaluating the effects of cordierite order/disorder and variable water activity.

Garnet - spinel - sillimanite - quartz. This system is related by the reaction:

$$\text{garnet} + \text{sillimanite} = \text{spinel} + \text{quartz} \qquad (19)$$
$$R_3Al_2Si_3O_{12} + 2Al_2SiO_5 = 3RAl_2O_4 + 5SiO_2 \; .$$

It is unclear whether magnesium and manganese spinels are ever stable with quartz, and the only reaction that has been calibrated is

$$\text{almandine} + \text{sillimanite} = \text{hercynite} + \text{quartz} \qquad (20)$$
$$Fe_3Al_2Si_3O_{12} + 2Al_2SiO_5 = 3FeAl_2O_4 + 5SiO_2 \; .$$

Richardson (1968) and Holdaway and Lee (1977) estimated the placement of this reaction but had few constraints on its slope. Wall and England (1980) obtained tight reversals on this reaction in an internally heated gas apparatus with direct control of hydrogen fugacity by a Shaw membrane. However, all hydrothermal experiments on hercynite at high temperatures yield spinels with significant magnetite solution, and one must correct experiments on reaction (20) for reduced hercynite activity by modeling the hercynite-magnetite solvus (determined by Turnock and Eugster, 1962). Since the amount of magnetite dissolved in the hercynite after the experiment is temperature dependent (but not buffered with excess magnetite), this serves to generate larger corrections at higher temperatures, thereby increasing the slope of the ideal reaction. Many high-grade metapelites carry spinel-bearing assemblages where the spinel is dominantly a gahnite-hercynite solid solution, and reaction (20) should prove to be an excellent thermobarometer for these rocks. However, no quantitative applications of this reaction have been published.

Garnet - rutile - ilmenite - sillimanite - quartz. The appropriate reaction is balanced as:

$$\text{ilmenite} + \text{sillimanite} + \text{quartz} = \text{almandine} + \text{rutile} \quad (21)$$
$$3FeTiO_3 + Al_2SiO_5 + 2SiO_2 = Fe_3Al_2Si_3O_{12} + 3TiO_2 \; .$$

Bohlen et al. (1982a) have tightly reversed this reaction experimentally, finding it to be metastable with respect to an equivalent kyanite-bearing reaction (Fig. 10) at the P-T conditions of their experiments. The reaction is balanced here in terms of sillimanite because the common garnet-rutile-ilmenite-quartz occurrences in high-grade rocks are with sillimanite.

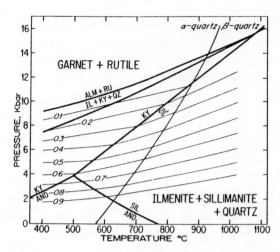

Figure 10. $Log_{10}K$ contoured on a P-T diagram for the reaction ilmenite + sillimanite + quartz = almandine + rutile. From Bohlen et al. (1982b).

This system should prove to be one of the best high-grade metapelite barometers available for the following reasons: (1) Its location is almost independent of temperature; (2) it is tightly reversed in the vicinity of P-T for which it will generally be used; (3) only one phase, garnet (or at most two phases, garnet and ilmenite) shows significant solid solution in ordinary metamorphic rocks with this five-phase assemblage; (4) analysis of the garnet is for its major component, almandine, and long compositional extrapolations are unnecessary. Application of this barometer by Bohlen et al. (1982) and by Edwards and Essene (1982) in metapelites from the western Adirondacks gives pressures of 7-8 kbar, in good agreement with estimates based on other barometers.

   Sphalerite - pyrrhotite - pyrite.   Calculations (Scott and Barnes, 1971) and experiments (Scott, 1973) show that the iron content of sphalerite in equilibrium with pyrrhotite and pyrite provides a barometer in the temperature range 300-650°C. Subsequent calculations and experiments by Lusk and Ford (1978) have led to minor modifications at high pressures, but Scott's (1973) calibrations remain in essence unchanged (Fig. 11).

   Applications of this barometer to metamorphic rocks have yielded excellent to mediocre results. Metamorphosed ore deposits seem to be a locally "juicy" environment during retrograde metamorphism with extensive resetting of sulfide compositions at lower pressures. However, if sphalerite re-equilibrates with monoclinic pyrrhotite at T < 300°C, the iron content of the sphalerite drops to 11-12 percent, indicating to the unwary user that pressures

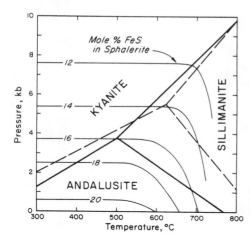

Figure 11. Isopleths of X_{FeS} in sphalerite equilibrated with pyrrhotite + pyrite on a P-T diagram. From Scott (1976).

were near 10 kbar. This double-edged resetting has led to considerable confusion in applications of the barometer to metamorphosed ore deposits, and any pressure of >8 kbar obtained with this system should be viewed with considerable skepticism (Nesbitt and Essene, 1982). However, reasonable pressures may be attained by using disseminated, unexsolved, coexisting, three-phase assemblages (DeWitt and Essene, 1974; Scott, 1976; Brown et al., 1980).

Reactions involving fluid species

Univariant equilibria involving a fluid are generally less directly applicable for thermobarometry because placement of the reaction in P-T space requires prior knowledge of the fugacity of the fluid species involved in the reaction at the time of metamorphism. One may assume that the pressure of a fluid species equals solid pressure for a given P-T, but this is demonstrably erroneous in some rocks where $P_f < P_s$ or where the fluid is composed of several species. One generally cannot analyze the peak metamorphic fluid directly (except occasionally through fluid inclusions) to constrain fugacities of individual fluid species. So what can be done? If enough independent reactions are represented by minerals in the same rock, then sometimes a unique solution of the composition of the fluid and hence the fugacities of each component in the fluid can be calculated (Ghent et al., 1979; Ferry, 1980; Nesibtt and Essene, 1982). In other cases, only one additional assumption need be made; namely, that $P_f = P_s$ (e.g., Valley et al., 1982). In yet other metamorphic rocks, there are too many degrees of freedom for an exact determination of fluid composition, and the best one can do is to use mineral-fluid equilibria to restrict the ranges of possible fugacities of fluid species.

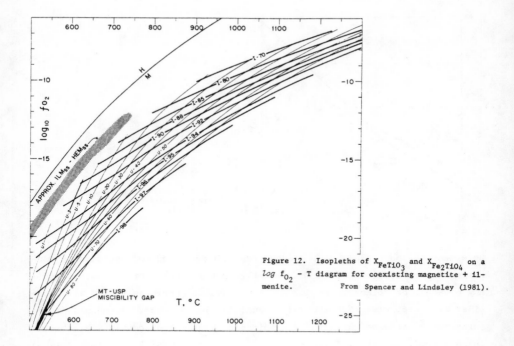

Figure 12. Isopleths of X_{FeTiO_3} and $X_{Fe_2TiO_4}$ on a $Log\ f_{O_2}$ - T diagram for coexisting magnetite + ilmenite. From Spencer and Lindsley (1981).

In order to apply reactions involving volatile components, one must learn to recognize calibrated univariant equilibria fixed or limited by a given assemblage in metamorphic rocks. Some simplified yet widely applicable chemical systems are presented below. Phase equilibria are also available for many other systems of potential interest and use, including $Na_2O-K_2O-Al_2O_3-SiO_2-H_2O$, $MgO-SiO_2-H_2O-CO_2$, $MgO-Al_2O_3-SiO_2-H_2O$, $CaO-MgO-Al_2O_3-SiO_2-H_2O-CO_2$, $CaO-MgO-FeO-SiO_2-H_2O-O_2$, etc., but limited space precludes their discussion. Good starting points for analysis of equilibria in some of these systems are given by Mueller and Saxena (1977), Winkler (1978), and Turner (1979).

$FeO - TiO_2 - O_2$. This system includes the famous magnetite-ilmenite thermometer of Buddington and Lindsley (1964). It also contains more oxidized equilibria involving pseudobrookite and hematite, as well as more reduced equilibria with wüstite, ferropseudobrookite, and native iron. The bulk of crustal metamorphic rocks are in the appropriate f_{O_2} range for magnetite-ilmenite and only this subsystem will be discussed, but the reader should be aware that grids similar to the one presented for magnetite-ilmenite will exist for assemblages like magnetite-pseudobrookite and ilmenite-ferropseudo-brookite.

184

Buddington and Lindsley (1964) performed experiments on coexisting titanian magnetite and ferrian ilmenite, and contoured isopleths of ilmenite
(X_{FeTiO_3}) and ulvöspinel ($X_{Fe_2TiO_4}$) on a $log f_{O_2}$-T diagram. They showed that
intersection of these isopleths yields f_{O_2} and T, and applied this thermometer
to a variety of igneous and metamorphic rocks containing magnetite and ilmenite. Powell and Powell (1977a) divided magnetite-ilmenite equilibria into
a reaction dependent on f_{O_2}:

$$4[Fe_3O_4]_{ss}^{mt} + O_2 = 6[Fe_2O_3]_{ss}^{ilm} \tag{22}$$

and a temperature-dependent exchange reaction:

$$[Fe_2TiO_4]_{ss}^{mt} + [Fe_2O_3]_{ss}^{ilm} = [Fe_3O_4]_{ss}^{mt} + [FeTiO_3]_{ss}^{ilm} . \tag{23}$$

Their calculated curves agreed well with Buddington and Lindsley's experiments. Spencer and Lindsley (1981) revised the Buddington/Lindsley curves
with new experimental data and with improved thermodynamic models, generating
the curves in Figure 12.

Application of these phase equilibria to metamorphic magnetite-ilmenite
pairs is not straightforward because both minerals re-equilibrate, oxidize,
and exsolve on cooling. If the metamorphic ilmenites have more than 10-15
percent hematite they exsolve to a fine intergrowth of ilmenite-hematite which
may oxidize to rutile-ilmenite-hematite or even rutile-hematite. If magnetites
have more than 10-20 percent ulvöspinel, they oxidize upon cooling to magnetite-ilmenite (Buddington and Lindsley, 1964; Bohlen and Essene, 1977). These
exsolution/oxidation processes may lead to complete recrystallization of the
original ferrian hematite or titanian magnetite -- the external granule process
of Buddington and Lindsley. In this case, all earlier thermal information is
lost because the minerals have reset totally. If the process is arrested with
the low-temperature assemblage still contained as lamellae and host within
single grains, the high-temperature information is still available for recovery
by reintegrating the phases and calculating the former composition of the
high-temperature titanian magnetite and/or ferrian ilmenite (Bohlen and
Essene, 1977). Application of this system is quite analogous to applications
of the two-feldspar thermometer in high-grade metamorphites in that: (1)
Reintegration of exsolved material is often required; (2) the systems are
presently inapplicable below 600°C; (3) formation of exsolved materials in
low-temperature chemical equilibrium is rapid and, once formed, is often preserved. Applications of magnetite-ilmenite thermometry to metamorphic rocks
are still uncommon because of retrogression, tediousness of reintegration,

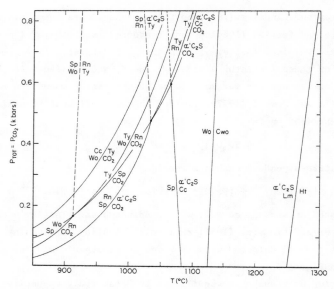

Figure 13. Phase diagram calculated for CaO-SiO$_2$-CO$_2$ from available experimental data: 'C$_2$S = 'Ca$_2$SiO$_4$; Cc = Calcite; Cwo = Cyclowollastonite; Ht = Hatrurite; Rn = Rankinite; Sp = Spurrite; Ti = Tilleyite; Wo = Wollastonite. From Treiman and Essene (1982).

temperature restrictions, and rarity of primary metamorphic ilmenite equilibrated with magnetite.

CaO - SiO$_2$ - CO$_2$. Equilibria for this system, based on a variety of experimental and thermodynamic data of a large number of workers, were calculated by Treiman and Essene (1982) and are presented in Figure 13. The reactions in this system apply to sanidinite facies metamorphism, most often as contact metamorphism of siliceous limestones by basalts, but occasionally by ignition of carbon in shallow sedimentary strata. It may be seen from the figure that the calc-silicates, tilleyite, spurrite, rankinite, α'Ca$_2$SiO$_4$ (which inverts to larnite upon cooling), cyclowollastonite, lime, and hartrurite, are generated successively at temperatures of 900–1300°C depending on total and/or CO$_2$ pressures. Even for the extreme thermal metamorphism of these rocks, water may circulate into the system lowering P$_{CO_2}$ and reducing the temperatures needed for the formation of these calc-silicates by as much as 100° to 150°C (Treiman and Essene, 1982). However, the solid-solid reactions remain unaffected and minimum temperatures are required by certain assemblages such as: rankinite-tilleyite (>920°C); larnite-tilleyite (>1020°C); larnite-calcite (>1060°C); cyclowollastonite (>1125°C); and hartrurite (>1250°C). These assemblages are occasionally found and mark the extremes of thermal metamorphism.

186

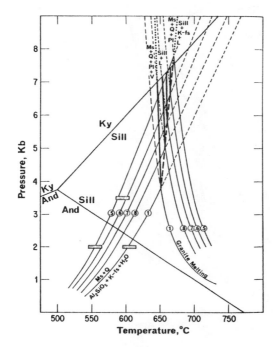

Figure 14. Phase diagram for K_2O-Al_2O_3-SiO_2-H_2O with variable X_{H_2O}. From Kerrick (1972).

$K_2O - Al_2O_3 - SiO_2 - H_2O$. This system contains reactions relating to the stability of muscovite, e.g.,

$$\text{muscovite} + \text{quartz} = Al_2SiO_5 + \text{K-feldspar} + H_2O \qquad (24)$$
$$KAl_3Si_3O_8(OH)_2 + SiO_2 = Al_2SiO_5 + KAlSi_3O_8 + H_2O \ .$$

Many workers have experimented on this system, and the more recent data are in reasonable agreement. The results of Kerrick (1972) are presented in Figure 14 and show the effect of reduced water pressures ($X_{H_2O} < 1$) on the equilibrium. This is important because geothermometry in the vicinity of the sillimanite + K-feldspar isograd may indicate lower temperatures than that predicted for $X_{H_2O} = 1$ (Evans and Guidotti, 1966; Schmid and Wood, 1976). Careful application of the equilibrium to the isograd requires correction for additional components in the solids, particularly Na in both K-feldspar and muscovite (Thompson, 1974). The decrease in water activity in the vicinity of the isograd may be explained by loss of water to migmatites and to ascending water-undersaturated granites or possibly by the influx of other deep-seated non-H_2O fluid species into the rocks. Studies which treat the affect on mineral equilibria of adding Na_2O to the system K_2O-Al_2O_3-SiO_2-H_2O include

187

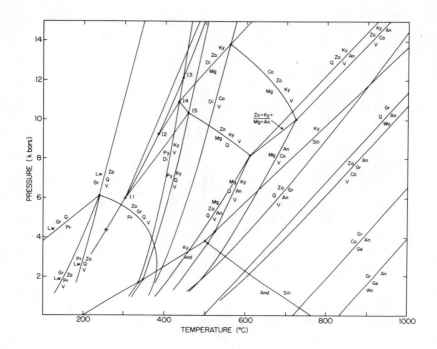

Figure 15. Phase diagram calculated for $CaO-Al_2O_3-SiO_2-H_2O$ from available experimental and thermodynamic data. An = Anorthite; And = Andalusite; Co = Corundum; Di = Diaspore; Ge = Gehlenite; Gr = Grossular; Ky = Kyanite; Lw = Lawsonite; Mg = Margarite; Pr = Prehnite; Py = Pyrophyllite; Q = Quartz; V = H_2O; Wo = Wollastonite; Zo = Zoisite. From Perkins et al.(1980).

those by Thompson (1974), Chatterjee and Froese (1975), and Thompson and Thompson (1976).

$CaO - Al_2O_3 - SiO_2 - H_2O.$ This system has been the object of many experimental and thermodynamic investigations. Because there are dozens of phases stable at metamorphic conditions in this system, there are hundreds of possible univariant reactions which may be considered. Perkins et al. (1980) obtained new heat capacity data on a half-dozen calc-aluminum silicates, combined these data with available reversed experiments, and calculated the phase equilibria in Figure 15. Many of these curves offer great potential in terms of thermobarometry and have been so used. For instance, Nesibtt and Essene (1982) applied equilibria involving zoisite, plagioclase, margarite, quartz, calcite and kyanite to calculate P-T at Ducktown, Tennessee, in good agreement with other thermobarometers. Low-temperature reactions involving prehnite, zoisite, lawsonite, and laumontite (not shown in Fig. 15) will closely constrain conditions in the prehnite-pumpellyite and other low grade facies. Of course, many other four-, five-, ... and n-component systems offer greater complexity than this system and provide considerable potential in locating still other metamorphic equilibria. Only one such system will be discussed briefly below.

$K_2O - FeO - Al_2O_3 - SiO_2 - H_2O - O_2$. This may be termed the *haplo-pelite system* because so many of the minerals and mineral reactions in metapelites can be modeled by this system. Although experiments are complicated by possible oxidation reactions, this iron-bearing system is to be preferred over the equivalent magnesian system (Seifert, 1970) because minerals like chloritoid, staurolite, and garnet are dominantly iron-aluminum-silicates in metapelites. The pelitic minerals stilpnomelane, chlorite, biotite, chloritoid, staurolite, iron-cordierite, and ferrogedrite are well represented by this system, which also contains muscovite, K-feldspar, grunerite, ferrosilite, Al_2SiO_5 minerals, quartz, hercynite, magnetite, hematite, corundum, and H_2O.

Most equilibria in this system have not been located accurately, although there are many estimates of the locations of their analogs in pelitic schists (which could be taken to represent this system), based mostly on limited experiments and on field observations (Albee, 1965, 1972; Hess, 1969; Grant, 1973; Carmichael, 1978). There have been a number of experiments on the stability of staurolite, iron-cordierite, and almandine with sillimanite (e.g., Richardson, 1968; Ganguly, 1972; Weisbrod, 1973; Holdaway and Lee, 1977); and on biotite (Wones and Eugster, 1965; Wones et al., 1971). Thermodynamic calculations which could generate a complete P-T net have been hampered by the lack of entropy data on most iron silicates. Once a good experimental and thermodynamic data base is available, the equilibria for this halplo-pelitic system can be calculated. Then accurate solution models must be developed for the complex phases biotite, staurolite, cordierite, and stilpnomelane in order to apply these equilibria to metamorphic assemblages. This task will probably consume many years before it is adequately completed.

Studies which treat the affect on mineral equilibria of adding MgO to the system $K_2O-FeO-Al_2O_3-SiO_2-H_2O-O_2$ include those by Albee (1965), Hess (1969), Thompson (1976), and Carmichael (1978).

OTHER METHODS

A large number of other direct and indirect methods have occasionally been used for thermobarometry. These range from: direct measurements of $P-T-X_f$ of active metamorphism in hot springs or geothermal areas; to determination of the change in fabric of oriented quartz in quartzites in comparison with laboratory studies; to stratigraphic reconstructions with calculations of burial depths; to observations of the P-T path of zero strain birefringence (piezothermometry) caused by an inclusion in a host mineral (Adams et al., 1975a,b); to fluid inclusions, illite crystallinity, vitrinite reflectance, and conodont color indices. The last four techniques will be briefly discussed because of their wide usage.

Fluid inclusions

The literature on studies of fluid inclusions is extensive although largely composed of applications to veins and ore deposits. Metamorphic fluid inclusions tend to be very small and difficult to work with. Furthermore, necking and leaking of inclusions may change the fluid composition and the inferences made from it (Roedder, 1981). Most two-phase inclusions tend to fill at low temperatures and give meaningless minimum temperature estimates (except in very low-grade rocks). If the composition of the fluid can be determined, an isochore which traces out a line from the minimum filling temperature to higher P-T may be calculated. In some cases, this line will pass through the peak metamorphic P-T range, giving permissive evidence (but not proof) that the fluid was trapped at or near the peak of metamorphism (Crawford, 1981; Touret, 1981). More often the isochore gives information on events occurring after the peak of metamorphism and may be used to trace retrograde P-T paths. Fluid inclusions often give more information on fluid compositions and their evolution than metamorphic P-T. For further information, the reader is directed to the MAC Short Course on *Fluid Inclusions: Applications to Petrology* (Hollister and Crawford, 1981).

Illite crystallinity, vitrinite reflectance, and the conodont color index

These three techniques are generally applied to very low-grade metamorphic rocks between the diagenetic zone and the prehnite-pumpellyite facies, sometimes called the anchizone. All three are presumably rate-dependent, i.e., the longer an illite remains at a constant temperature the better crystallized it will be (up to some limit). In order to compare different field areas with these techniques one must assume that they were subjected to the same temperatures for about the same times, because the phenomena measured can give the same results for a short period of time at relatively high temperatures or a longer period of time at somewhat lower temperatures. This behavior is typical of metastable materials tending towards equilibrium, and many other factors besides time and temperature may affect such transformations. Thus while the techniques may qualitatively distinguish different temperatures

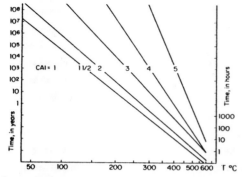

Figure 16. Experiments on Conodont Color Index (CAI) versus time and temperature. Figure modified from Epstein et al. (1977).

in ordinary rocks with similar burial and thermal histories, they are unlikely to ever yield quantitative thermometry.

Illite crystallinity is a measure of the peak width compared to the peak height for the 060 X-ray powder diffraction peak (Kisch, 1966, 1980; Frey, 1970; Kubler et al., 1974; Frey et al., 1980). Because this line broadening is known to increase with decreased grain size as well as crystal imperfection, and may be affected by chemical zoning in the illite, the significance of the measured parameter is unclear. Direct electron microscopic observations are needed on these illites to systematically evaluate crystallite size versus crystal perfection versus chemical zonation. Nevertheless, the "crystallinity" does change systematically across areas of increasing metamorphic grade, and provides some qualitative measure of changing grade.

Vitrinite reflectance is a measure of the conversion of coal from complex hydrocarbons to graphite, and reflectivity of polished specimens has been correlated with this process. Here, too, the grain size of the graphite crystals and/or other intermediate hydrocarbons should influence the measurement, and one has to assume that processes of grain growth have occurred at the same rate in order to correlate measurements from different areas. This process is certainly a "one-way" or a kinetically controlled reaction, and various factors such as P_{CO_2}, P_{H_2}, etc. should be expected to alter the rates. Its best use is as a qualitative thermometer.

Conodont color index is the most quantitatively developed of these techniques. Epstein et al. (1977) measured the rate of change of conodont colors by heating in the laboratory and constructed the graph shown in Figure 16. Experimental data measured up to 10^3 hours were extrapolated to 10^{12} hours (100 million years). Even slight errors over the laboratory time scale generate large errors on the geologic time scale. The exact reaction(s) leading to the color changes have not been determined but presumably involve oxidation of organic compounds. Such a process, like coal metamorphism, should be

influenced by P_{H_2O}, P_{CO_2}, P_{H_2}, etc. Indeed, Epstein et al. (1977) experimentally demonstrated the retardation of the color changes at high P_{H_2O}.
It is also possible that the mechanisms of color loss are different at low
temperatures over geologic times, so that the color function could consist
of two or more straight-line segments or an infinity of such segments resulting in a curved line. In either case, the straight line extrapolation would
yield erroneous time/temperature constraints for geological processes. Organic geochemists need to study the processes involved in the color changes.

These three techniques should not be used for quantitative thermometry
or even as time/temperature indicators until the physical and chemical changes
involved are directly observed and fully understood.

CONCLUSIONS

Thermobarometry is a valid technique today. However, the careful worker
should do much more than plug in the numbers or read off the graphs. One
should test individual thermobarometers in the field by: (1) repeated
measurements on the same assemblage in variable bulk compositions to test
reproducibility and the effect of variable composition; (2) calibrations
against other thermobarometers; (3) measurements over a terrane known from
isograds or other data to have variable P and/or T to ensure that the system
tested reveals an appropriate gradient; and (4) evaluation of the effect of
varying the analytical input data for expected errors and for expected variation in Fe^{2+}/Fe^{3+} (if appropriate) to make sure these do not affect the system seriously.

ACKNOWLEDGMENTS*

The writer thanks the present and former Ph.D. students with whom he has
worked, including L.M. Anovitz, S.R. Bohlen, J.W. Bowman, P.E. Brown, J.W.
Geissman, B.E. Nesibtt, D. Perkins III, E.U. Peterson, J. Ruiz, A.H. Treiman,
and J.W. Valley for educating him more about the diversity of metamorphic
rocks and their problems. Without their help -- and questions -- the writer
would not have begun to pursue the breadth and depth of studies that he has
managed with them. Dr. A.H. Treiman is particularly thanked for helping in
time-consuming preparation of the final stages of the manuscript. Drs. J.R.
Bowman, J.M. Ferry, D.R. Peacor, J. Ruiz and A.H. Treiman, are thanked for
their helpful reviews, and Drs. S.R. Bohlen, D.M. Kerrick, D.H. Lindsley,

----

*Contribution 383 from the Mineralogical Laboratory, The University of
 Michigan.

S.D. Scott, and J.C. Stormer for the use of their diagrams. Mary Schatz is gratefully thanked for her patient typing and retyping of various versions of the manuscript.

APPENDIX

ACTIVITY-COMPOSITION RELATIONS FOR MINERAL SOLID SOLUTIONS

Activity-composition relations for minerals are required for evaluation of the terms $RT \ln K_\gamma$ and $RT \ln K$ and hence for evaluation of metamorphic temperature and pressure. A few naturally-occurring minerals (e.g., quartz and the Al_2SiO_5 polymorphs) are sufficiently close to pure substances that their activities are commonly taken as one. Most minerals in metamorphic rocks, however, are clearly solid solutions and activity-composition relations must either be estimated or measured directly by laboratory experiment. Some of the most commonly used approaches are summarized here.

Ideal activity-composition relations

The mixing properties of solid solutions have only been experimentally determined for a few binary and ternary mineral solutions. For lack of an alternative, many mineral solid solutions therefore are modeled as ideal solutions. Kerrick and Darken (1975) outline simple procedures, based in statistical mechanics, by which activities of components in an ideal mineral solid solution may be estimated from a knowledge of mineral composition and of the atomic substitution mechanisms in the mineral. As an example, ideal activity-composition relations for olivine, derived in this manner, are:

$$a_{M_2SiO_4,\text{olivine}} = X^2_{M_2SiO_4,\text{olivine}} \tag{A1}$$

where M = Fe, Mg, Mn, Ca, etc.; X = M/(Fe + Mg + Mn + Ca + ...), and the exponent refers to two sites of mixing for the divalent cations per standard olivine mineral formula. For some substitutions in some minerals, the ideal solution model appears to adequately describe experimentally-determined activity-composition relations (e.g., pyroxenes: Wood, 1976; olivines: Bradley, 1962).

Non-ideal activity-composition relations

Many experimentally-determined activity-composition relations indicate that mineral solid solutions are non-ideal. The non-ideality is most commonly described with a regular solution model (Thompson, 1967). Less often-used models, described by Saxena (1973), Grover (1977), and Blencoe (1977) for binary solid solutions, are the van Laar model and the quasi-chemical model. The regular solution model represents the activity coefficient of a component in a binary solution, λ_1, by an expression which involves one or two adjustable empirical parameters:

194

$$RT\ln\lambda_1 = X_2^2 \, W_G \tag{A2}$$

for the case of one adjustable parameter, and

$$RT\ln\lambda_1 = X_2^2 \, [W_{G_1} + 2(W_{G_2} - W_{G_1})X_1] \, , \tag{A3}$$

for the case to two adjustable parameters. The W_G are constants dependent on pressure and temperature but not on composition. Direct experiment dictates whether equation (A2) or (A3) is more appropriate for a particular solid solution. The W_G are usually, in turn, modelled as a linear function of P and T:

$$W_G = W_H + PW_V - TW_S \, . \tag{A4}$$

Activity-composition relations for a number of common binary mineral solid solutions have been characterized with the regular solution model: alkali feldspars (Waldbaum and Thompson, 1969; Bachinski and Muller, 1971), diopside-enstatite (Warner and Luth, 1974; Lindsley et al., 1981), muscovite (Chatterjee and Froese, 1975), nepheline (Ferry and Blencoe, 1978), Fe-Ti oxides (Spencer and Lindsley, 1981), olivine (O'Neill and Wood, 1979; Sack, 1980b), orthopyroxene (Sack, 1980b), garnet (Newton and Perkins, 1982); plagioclase (Newton and Perkins, 1982; Newton and Haselton, 1981), spinel (Zen, 1973), and calcite (Gordon and Greenwood, 1970).

Formulation of a regular solution model for ternary and higher order solid solutions is considerably more complex (Saxena, 1973). One simplified model that relies only on binary interaction parameters (W_G, equation A2), which are relatively easy to measure experimentally, is the "simple mixture" model of Prigogine and DeFay (1954). For a four-component solid solution, for example, the model represents the activity coefficient for component 1 by

$$RT\ln\lambda_1 = W_{12}X_2^2 + W_{13}X_3^2 + W_{14}X_4^2 + (W_{12} + W_{13} - W_{23})X_2X_3$$
$$(W_{13} + W_{14} - W_{34})X_3X_4 + (W_{12} + W_{14} - W_{24})X_2X_4 \tag{A5}$$

where the X_i are mole fractions of the four components. The W_{ij} are binary interaction parameters (as in equation A2), and are derived from experimental data on each of the binary subsystems. Equation (A5), for example, has been applied to non-ideal $M_3Al_2Si_3O_{12}$ garnet solid solutions by Ganguly and Kennedy (1974), Newton and Haselton (1981), and Newton and Perkins (1982).

Adams, H.G., L.H. Cohen and J.L. Rosenfeld (1975a) Solid inclusion piezothermometry I: Comparison dilatometry. Am. Mineral. 60, 574-583.

_____, _____, _____ (1975b) Solid inclusion piezothermometry II. Geometric basis, calibration for the association quartz-garnet, and application to some pelitic schists. Am. Mineral. 60, 584-598.

Akella, J. (1976) Garnet-pyroxene equilibria in the system $CaSiO_3$-$MgSiO_3$-Al_2O_3 and in a natural mineral mixture. Am. Mineral. 61, 589-598.

_____ and F.R. Boyd (1973) Effect of pressure on the composition of coexisting pyroxenes and garnet in the system $CaSiO_3$-$MgSiO_3$-$FeSiO_3$-$CaAlTi_2O_6$. Carneg. Inst. Wash. Year Book 72, 523-526.

Albee, A.L. (1965) A petrogenetic grid for the Fe-Mg silicates of pelitic schists. Am. J. Sci. 263, 512-536.

_____ (1972) Metamorphism of pelitic schists: Reaction relations of chloritoid and staurolite. Geol. Soc. Am. Bull. 83, 3249-3268.

Anderson, P.A.M., R.C. Newton and O.J. Kleppa (1977) The enthalpy change of the andalusite-sillimanite reaction and the Al_2SiO_5 diagram. Am. J. Sci., 277, 585-593.

Anovitz, L.M. and E.J. Essene (1982) Phase relations in the System $CaCO_3$-$MgCO_3$-$FeCO_3$. Trans. Am. Geophys. Union 63, 464.

Bachinski, S.W. and G. Muller (1971) Experimental determinations of the microcline - low albite solvus. J. Petrol. 12, 329-356.

Banno, S. (1970) Classification of eclogites in terms of physical conditions of their origin. Phys. Earth Planet. Interiors 3, 405-421.

_____ and Y. Matsui (1965) Eclogite types and partitioning of Mg, Fe, and Mn between clinopyroxene and garnet. Japan Academy Proc. 41, 716-721.

Barth, T.F.W. (1934) Polymorphic phenomena and crystal structure. Am. J. Sci. 232, 273-313.

_____ (1957) The feldspar geological thermometers. N. Jahrb. Mineral. Abh. 82, 143-154.

_____ (1962) The feldspar geologic thermometer. Norsk Geol. Tidsskr. B42, 330-339.

_____ (1969) Additional data for the two-feldspar geothermometer. Lithos 1, 305-306.

Bell, P.M. (1963) Aluminum silicate system: Experimental determination of the triple point. Science 139, 1055-1056.

Besancon, J.R. (1981) Rate of cation disordering in orthopyroxene. Am. Mineral. 66, 965-973.

Bickle, M.J. and R. Powell (1977) Calcite-dolomite geothermometry for iron-bearing carbonates. The Glockner area of the Tauern window, Austria. Contrib. Mineral. Petrol. 59, 281-289.

Bishop, F.C. (1980) The distribution of Fe^{2+} and Mg between coexisting ilmenite and pyroxene with application to geothermometry. Am. J. Sci. 280, 46-77.

Black, P.M. (1973) Mineralogy of New Caledonian metamorphic rocks. III. Pyroxenes, and major element partitioning between coexisting pyroxenes, amphiboles and garnets from Ouegoa District. Contrib. Mineral. Petrol. 45, 281-303.

Blencoe, J.G. (1977) Computation of thermodynamic mixing parameters for isostructural, binary crystalline solutions using solvus experimental data. Computers Geosci. 3, 1-18.

_____ and W.C. Luth (1973) Muscovite-paragonite solvi at 2,4, and 8 kb pressure. Geol. Soc. Am. Abstr. Prog. 5, 553-554.

Bohlen, S.R. (1979) *Pressure, temperature, and fluid composition of Adirondack metamorphism as determined in orthogneisses, Adirondack Mountains, New York.* Ph.D. thesis, University of Michigan, Ann Arbor.

_____, and A.L. Boettcher (1981) Experimental investigations and geological applications of orthopyroxene geobarometry. Am. Mineral. 66, 951-964.

_____, _____, W.R. Dollase and E.J. Essene (1980) The effect of manganese on olivine-quartz-orthopyroxene stability. Earth Planet. Sci. Lett. 47, 11-20.

_____ and E.J. Essene (1977) Feldspar and oxide thermometry of granulites in the Adirondack Highlands. Contrib. Mineral. Petrol. 62, 153-169.

_____ and _____ (1978) Igneous pyroxenes from metamorphosed anorthosite massifs. Contrib. Mineral. Petrol. 65, 433-442.

_____ and _____ (1979) A critical evaluation of two-pyroxene thermometry in Adirondack granulites. Lithos 12, 335-345.

_____ and _____ (1980) Evaluation of coexisting garnet-biotite, garnet-clinopyroxene, and other thermometers in Adirondacks granulites. Geol. Soc. Am. Bull. 91, 685-719.

- - - - - - - - - -

*In addition to references cited in the text, this list contains a number of other references to geothermometry/geobarometry based on mineral equilibria in metamorphic rocks.

_____, _____ and A.L. Boettcher (1980) Reinvestigation and application of olivine-quartz-orthopyroxene barometry. Earth Planet. Sci. Lett. 47, 1-10.

_____, _____ and K.S. Hoffman (1980) Feldspar and oxide thermometry in the Adirondacks: An update. Geol. Soc. Am. Bull. 91, 110-113.

_____, V.J. Wall and A.L. Boettcher (1982a) Experimental investigations and geological applications of equilibria in the system $FeO-TiO_2-Al_2O_3-SiO_2-H_2O$, manuscript in preparation.

_____, _____, _____ (1982b) Experimental investigation of model garnet granulite equilibria. Contrib. Mineral. Petrol. (in press).

Bostwick, N.H. (1974) Phytoclasts as indicators of thermal metamorphism, Franciscan assemblage and Great Valley sequence (upper Mesozoic), California, *In* Dutcher, R.R., Ed., *Carbonaceous Materials as Indicators of Metamorphism*, Geol. Soc. Am. Special Paper 153, 1-17.

Bowen, N.L. (1940) Progressive metamorphism of siliceous limestone and dolomite. J. Geol. 48, 225-274.

_____ and O.F. Tuttle (1949) The system $MgO-SiO_2-H_2O$. Geol. Soc. Am. Bull. 60, 439-460.

Bowman, J.R. (1978) *Contact Metamorphism, Skarn Formation, and Origin of C-O-H Skarn Fluids in the Black Butte Aureole, Elkhorn, Montana*. Ph.D. thesis, University of Michigan, Ann Arbor.

_____ and E.J. Essene (1982) $P-T-X(CO_2)$ conditions of contact metamorphism in the Black Butte aureole, Elkhorn, Montana. Am. J. Sci. 282, 311-340.

Boyd, F.R. (1973) The pyroxene geotherm. Geochim. Cosmochim. Acta 37, 2533-2546.

_____ and J.F. Schairer (1964) The system $MgSiO_3-CaMgSi_2O_6$. J. Petrol. 5, 275-309.

Bradley, R.S. (1962) Thermodynamic calculations on fused salts. Part II. Solid solutions and application to the olivines. Am. J. Sci. 260, 550-554.

Brown, P.E. (1980) *A petrologic and stable isotopic study of skarn formation and mineralization at the Pine Creek, California Tungsten Mine*. Ph.D. thesis, University of Michigan, Ann Arbor.

_____, E.J. Essene and W.C. Kelly (1978) Sphalerite geobarometry in the Balmat-Edwards district, New York. Am. Mineral. 63, 250-257.

Brown, W.L. and I. Parsons (1981) Towards a more practical two-feldspar geothermometer. Contrib. Mineral. Petrol 76, 369-377.

Buddington, A.F. and D.H. Lindsley (1964) Iron-titanium oxide minerals and synthetic equivalents. J. Petrol. 5, 310-357.

Carmichael, D.M. (1978) Metamorphic bathozones and bathograds: A measure of the depth of post-metamorphic uplift and erosion on the regional scale. Am. J. Sci. 278, 769-797.

Carswell, D.A. and F.G.F. Gibb (1980) The equilibrium conditions and petrogenesis of European crustal garnet lherzolites. Lithos 13, 19-29.

Chatterjee, N.D. (1971) The upper stability limit of the assemblage paragonite + quartz and its natural occurrences. Contrib. Mineral. Petrol. 34, 288-303.

_____ and E. Froese (1975) A thermodynamic study of the pseudobinary join muscovite-paragonite in the system $KAlSi_3O_8-NaAlSi_3O_8-Al_2O_3-SiO_2-H_2O$. Am. Mineral. 60, 985-993.

_____ and W. Johannes (1974) Thermal stability and standard thermodynamic properties of synthetic $2M_1$ muscovite, $KAl_2(Si_3Al)O_{10}(OH)_2$. Contrib. Mineral. Petrol. 48, 89-114.

Clark, J.R. and J.J. Papike (1968) Crystal-chemical characterization of omphacites. Am. Mineral. 53, 840-868.

Clark, S.P., Jr., E.C. Robertson and F. Birch (1957) Experimental determination of kyanite-sillimanite equilibrium relations at high pressures and temperatures. Am. J. Sci. 255, 628-640.

Crawford, M.L. (1981) Fluid inclusions in metamorphic rocks - low and medium grade. *In* Hollister, L.S. and M.L. Crawford, Eds., *Short Course in Fluid Inclusions: Applications to Petrology*. V. 6, Short Course Notes. Mineral. Assoc. Canada, Calgary, pp. 157-181.

Cressey, G., R. Schmid, and B.J. Wood (1979) Thermodynamic properties of almandine-grossular garnet solid solutions. Contrib. Mineral. Petrol 67, 397-448.

Currie, K.L. and L.W. Curtis (1976) An application of multicomponent solution theory to jadeitic pyroxenes. J. Geol. 84, 179-194.

Dahl, P.S. (1980) The thermal-compositional dependence of Fe^{2+}-Mg distributions between coexisting garnet and pyroxene: Applications to geothermometry. Am. Mineral. 65, 852-866.

Dallmeyer, R.D. (1974) The role of crystal structure in controlling the partitioning of Mg and Fe^{2+} between coexisting garnet and biotite. Am. Mineral. 59, 201-203.

De Witt, D.B. and E.J. Essene (1974) Sphalerite geobarometry applied to Grenville marbles. Geol. Soc. Am. Abs. Prog. 6, 709-710.

Edwards, R.L. and E.J. Essene (1981) Zoning patterns and their effect on biotite garnet K_D thermometry. Trans. Am. Geophys. Union 62, 411.

_____ and _____ (1982) Barometry in metapelitic rocks of the N.W. Adirondacks, New York. Contrib. Mineral. Petrol. (submitted).

Eitel, W. (1925) Neues Jahrbuch fur Mineralogie, Beilage Band II, 477-493. In Bowen (1940) q.v.

Ellis, D.J. and D.H. Green (1979) An experimental study of the effect of Ca upon garnet-clinopyroxene exchange equilibria. Contrib. Mineral. Petrol 71, 13-22.

Engi, M. (1978) *Mg-Fe Exchange Equilibria Among Spinel, Olivine, Orthopyroxene and Cordierite.* Ph.D. thesis, No. 6256, Eidgenossische Technische Hochschule, Zürich, Switzerland.

Epstein, A.G., J.B. Epstein and L.D. Harris (1977) Conodont color alteration - an index to organic metamorphism. U.S. Geol. Surv. Prof. Paper 995, 1-27.

Ernst, W.G. (1981) Petrogenesis of eclogites and peridotites from the Western and Ligurian Alps. Am. Mineral. 66, 443-472.

Essene, E.J. and W.S. Fyfe (1967) Omphacite in Franciscan metamorphic rocks. Contrib. Mineral. Petrol. 15, 1-23.

Eugster, H.P., A.L. Albee, A.E. Bence, J.B. Thompson, Jr. and D.R. Waldbaum (1972) The two-phase region and excess mixing properties of paragonite-muscovite crystalline solutions. J. Petrol. 13, 147-179.

Evans, B.W. (1965a) Application of a reaction rate method to the breakdown equilibria of muscovite and muscovite + quartz. Am. J. Sci. 263, 647-667.

_____ (1965b) Pyrope garnet-barometer or thermometer? Geol. Soc. Am. Bull. 76, 1265-1300.

_____ and C.V. Guidotti (1966) The sillimanite-potash feldspar isograd in western Maine, U.S.A. Contrib. Mineral. Petrol. 12, 25-62.

_____ and V. Trommsdorff (1978) Petrogenesis of garnet lherzolite, Cima di Gagnone, Lepontine Alps. Earth Planet. Sci. Lett. 40, 333-348.

Ferry, J.M. (1980) A comparative study of geothermometers and geobarometers in pelitic schists from south-central Maine. Am. Mineral. 65, 113-117.

_____ and J.G. Blencoe (1978) Subsolidus phase relations in the nepheline-kalsilite system at 0.5, 2.0 and 5.0 kbar. Am. Mineral. 63, 1225-1240.

_____ and F.S. Spear (1978) Experimental calibration of the partitioning of Fe and Mg between biotite and garnet. Contrib. Mineral. Petrol. 66, 113-117.

Fleet, M.E., C.T. Herzberg, G.M. Bancroft and L.P. Aldridge (1978) Omphacite studies I. The $P2/n \rightarrow C2/c$ transformation. Am. Mineral. 63, 1100-1106.

Frey, M. (1970) The step from diagenesis to metamorphism in pelitic rocks during Alpine orogenesis. Sedimentology 15, 261-279.

_____, M. Teichmueller, R. Teichmueller, J. Mullis, B. Kunzi, A. Breitschmid, U. Gruner and B. Schwizer (1980) Very low-grade metamorphism in external parts of the Central Alps: Illite crystallinity, coal rank, and fluid inclusion data. Eclog. Geol. Helv. 73, 173-203.

Friedman, I. and J.R. O'Neil (1977) Compilation of stable isotopic fractionation factors of geochemical interest. Data of Geochemistry, 6th ed. Fleischer, M., Ed., U.S. Geol. Surv. Prof. Paper 440-KK.

Fujii, T. (1977) Fe-Mg partitioning between olivine and spinel. Carnegie Inst. Wash. Year Book 76, 563-569.

Fyfe, W.S. (1960) Hydrothermal synthesis and determination of equilibrium between minerals in the subliquidus region. J. Geol. 68, 553-566.

_____ and F.J. Turner (1966) Reappraisal of the mineral facies concept. Contrib. Mineral. Petrol. 12, 354-364.

Ganguly, J. (1972) Staurolite stability and related parageneses: Theory, experiments and applications. J. Petrol. 13, 335-365.

_____ (1973) Activity-composition relation of jadeite in omphacite pyroxene. Earth Planet. Sci. Lett. 19, 145-153.

_____ (1979) Garnet and clinopyroxene solid solutions and geothermometry based on Fe-Mg distribution coefficient. Geochim. Cosmochim. Acta 43, 1021-1029.

_____ (1982) Mg-Fe order-disorder in ferromagnesian silicates. II. Thermodynamics, kinetics and geological application. *In* Saxena, S.K., Ed., *Advances in Physical Geochemistry,* 2, Springer-Verlag, New York, pp. 58-100.

_____ and G.C. Kennedy (1974). The energetics of natural garnet solution. I. Mixing of the aluminosilicate endmembers. Contrib. Mineral. Petrol., 48, 137-148.

Gasparik, T. and D.H. Lindsley (1980) Phase equilibria at high pressures of pyroxenes containing monovalent and trivalent ions. *In* Prewitt, C.T., Ed., *Pyroxenes.* Reviews in Mineralogy 7, 309-340.

Ghent, E.D. (1976) Plagioclase-garnet-Al_2SiO_5-quartz: A potential geobarometer-geothermometer. Am. Mineral. 61, 710-714.

_____, D.B. Robbins and M.Z. Stout (1979) Geothermometry, geobarometry, and fluid compositions of metamorphosed calc-silicates and pelites, Mica Creek, British Columbia. Am. Mineral. 64, 874-885.

_____, P.S. Simony and C.C. Knitter (1980) Geometry and pressure-temperature significance of the kyanite-sillimanite isograd in the Mica Creek area, British Columbia. Contrib. Mineral. Petrol. 74, 67-73.

_____ and M.Z. Stout (1981) Geobarometry and geothermometry of plagioclase-biotite-garnet-muscovite assemblages. Contrib. Mineral. Petrol. 76, 92-97.

Ghose, S. (1982) Mg-Fe order-disorder in ferromagnesian silicates. I. Crystal chemistry. *In* Saxena, S.K., Ed., *Advances in Physical Geochemistry*, 2, Springer-Verlag, New York, pp. 4-57.

Goldman, D.S. and A.L. Albee (1977) Correlation of Mg/Fe partitioning between garnet and biotite with $^{18}O/^{16}O$ partitioning between quartz and magnetite. Am. J. Sci. 277, 750-767.

Goldschmidt, V.M. (1912) Die Gesetze der Gesteinsmetamorphose mit Beispielen aus der Geologie des Sudlichen Norwegens, Kristina Vidensk Skriften I Math - Naturu. KI. 22.

Goldsmith, J.R. and H.C. Heard (1961) Subsolidus phase relations in the system $CaCO_3-MgCO_3$. J. Geol. 69, 45-74.

_____ and R.C. Newton (1969) P-T-X relations in the system $CaCO_3-MgCO_3$ at high temperatures and pressures. Am. J. Sci. 267A, 160-190.

_____ and _____ (1974) An experimental determination of the alkali feldspar solvus. *In* Mackenzie, W.S. and J. Zussman, Eds., *The Feldspars*, Manchester Univ. Press, Manchester, pp. 337-359.

Gordon, T.M. and H.J. Greenwood (1971) The stability of grossularite in H_2O-CO_2 mixtures. Am. Mineral. 56, 1674-1688.

Graf, D.F. and J.R. Goldsmith (1955) Dolomite-magnesian calcite relations at elevated temperatures and CO_2 pressures. Geochim. Cosmochim. Acta 7, 109-128.

_____ and _____ (1958) The solid solubility of $MgCO_3$ in $CaCO_3$: A revision. Geochim. Cosmochim. Acta 13, 218-219.

Grant, J.A. (1973) Phase equilibria in high-grade rocks and partial melting of pelitic rocks. Am. J. Sci. 273, 289-317.

Green, D.H. and W. Hibberson (1970) The instability of plagioclase in peridotite at high pressure. Lithos 3, 209-221.

_____ and A.E. Ringwood (1967a) The stability fields of aluminous pyroxene peridotite and garnet peridotite and their relevance in upper mantle structure. Earth Planet. Sci. Lett. 3, 151-160.

_____ and _____ (1967b) An experimental investigation of the gabbro to eclogite transformation and its petrological applications. Geochim. Cosmochim. Acta 31, 767-833.

_____ and _____ (1972) A comparison of recent experimental data on the gabbro-garnet granulite-eclogite transition. J. Geol. 80, 277-288.

Greenwood, H.J. (1969) Wollastonite: Stability in H_2O-CO_2 mixtures and occurrence in a contact-metamorphic aureole near Salmo, British Columbia, Canada. Am. Mineral. 52, 1669-1680.

_____ (1972) $Al^{IV}-Si^{IV}$ disorder in sillimanite and its effect on phase relations of the aluminum silicate minerals. Geol. Soc. Am. Mem. 132, 553-571.

Grover, J. (1977) Chemical mixing in multicomponent solutions: An introduction to the use of Margules and other thermodynamic excess functions to represent non-ideal behaviour. *In* Fraser, D.G., Ed., *Thermodynamics in Geology*, Reidel, Boston, 67-98.

Hariya, Y. and G.C. Kennedy (1968) Equilibrium study of anorthite under high pressure and high temperature. Am. J. Sci. 266, 196-203.

Harker, A. (1932) *Metamorphism*. Methuen, London.

Harker, R.I. (1959) The synthesis and stability of tilleyite, $Ca_5Si_2O_7(CO_3)_2$. Am. J. Sci. 257, 656-667.

_____ and O.F. Tuttle (1955a) Studies in the system $CaO-MgO-CO_2$. Part I. Am. J. Sci. 253, 209-224.

_____ and _____ (1955b) Studies in the system $CaO-MgO-CO_2$. Part 2: Limits of solid solutions along the binary join, $CaCO_3-MgCO_3$. Am. J. Sci. 253, 274-282.

_____ and _____ (1956) Experimental data on the P_{CO_2}-T curve for the reaction: calcite + quartz = wollastonite + carbon dioxide. Am. J. Sci. 267, 729-804.

Helgeson, H.C., J.M. Delany, H.W. Nesbitt and D.K. Bird (1978) Summary and critique of the thermodynamic properties of rock-forming minerals. Am. J. Sci. 278A, 229.

Hemingway, B.S., K.M. Krupka and R.A. Robie (1981) Heat capacities of the alkali feldspars between 350 and 1000K from differential scanning calorimetry, the thermodynamic functions of the alkali feldspars from 298.15 to 1400 K, and the reaction quartz + jadeite = analbite. Am. Mineral. 66, 1202-1215.

Henry, D.J. and L.G. Medaris, Jr. (1980) Application of pyroxene and olivine-spinel geothermometers to spinel peridotites in S.W. Oregon. Am. J. Sci. The Jackson Volume,
211-231.

Henson, B.J. (1977) Cordierite-garnet bearing assemblages as geothermometers and barometers
in granulite facies terranes. Tectonophys. 43, 73-88.

_____ and D.H. Green (1973) Experimental study of the stability of cordierite and garnet in
pelitic compositions at high pressures and temperatures. Contrib. Mineral. Petrol. 38,
151-166.

Herzberg, C.T. (1978a) Pyroxene geothermometry and geobarometry: Experimental and thermodynamic evaluation of some subsolidus phase relations involving pyroxenes in the system
$CaO-MgO-Al_2O_3-SiO_2$. Geochim. Cosmochim. Acta 42, 945-957.

_____ (1978b) The bearing of phase equilibria in simple and complex systems on the origin
and evolution of some garnet websterites. Contrib. Mineral. Petrol. 66, 375-382.

Hess, P.C. (1969) The metamorphic paragenesis of cordierite in metamorphic rocks. Contrib.
Mineral. Petrol. 24, 191-207.

Hietanen, A. (1969) Distribution of Fe and Mg between garnet, staurolite and biotite in an
aluminum-rich schist in various metamorphic zones north of the Idaho Batholith. Am. J.
Sci. 267, 422-456.

Hoernes, S. and H. Friedrichsen (1978) Oxygen and hydrogen isotope study of the polymetamorphic
area of the northern Otzal-Stubai Alps (Tyrol). Contrib. Mineral. Petrol. 67, 305-315.

Hoffman, K.S. and E.J. Essene (1982) A re-evaluation of the orthopyroxene isograd, in the
N.W. Adirondacks, New York. Contrib. Mineral. Petrol. (submitted).

Holdaway, M.J. (1971) Stability of andalusite and the aluminum silicate phase diagram. Am.
J. Sci. 271, 97-131.

_____ and S.M. Lee (1977) Fe-Mg cordierite stability in high-grade pelitic rocks based on
experimental, theoretical, and natural observations. Contrib. Mineral. Petrol. 63,
175-193.

Holland, T.J.B. (1980) The reaction albite = jadeite + quartz determined experimentally in
the range 600-1200°C. Am. Mineral. 65, 129-134.

_____, A. Navrotsky and R.C. Newton (1979) Thermodynamic parameters of $CaMgSi_2O_6-Mg_2Si_2O_6$
pyroxenes based on regular solution and cooperative disordering models. Contrib. Mineral.
Petrol. 69, 337-350.

Hollister, L.S. and M.L. Crawford (1981) *Short Course in Fluid Inclusions: Applications to
Petrology.* Short Course Notes 6, Mineral. Assoc. Canada, Calgary.

Holm, J.L. and O.J. Kleppa (1966) The thermodynamic properties of the aluminum silicates.
Am. Mineral. 51, 1608-1622.

Hutcheon, I. (1978) Calculation of metamorphic pressure using the sphalerite-pyrrhotite-
pyrite equilibrium. Am. Mineral. 63, 87-95.

Iiyama, J.T. (1964) Etude des reactions d'exchange d'ions Na-K dans la serie muscovite-
paragonite. Bull. Soc. fran. Mineral. Cristallog. 87, 532-541.

Irving, A.J. (1974) Geochemical and high-pressure experimental studies of garnet pyroxenite
and granulite xenoliths from the Delegate basaltic pipes, Australia. J. Petrol. 15, 1-40.

Jacobs, G.K. and D.M. Kerrick (1981) Devolatilization equilibria in H_2O-CO_2 and H_2O-CO_2-
NaCl fluids: An experimental and thermodynamic evaluation at elevated pressures and
temperatures. Am. Mineral. 66, 1135-1153.

Jenkins, D.M. (1981) Experimental phase relations of hydrous peridotites modelling the system $H_2O-CaO-MgO-Al_2O_3-SiO_2$. Contrib. Mineral. Petrol. 77, 166-176.

_____ and R.C. Newton (1979) Experimental determination of the spinel peridotite to garnet
peridotite inversion at 900°C and 1000°C in the system $CaO-MgO-Al_2O_3-SiO_2$, and at 900°C
with natural garnet and olivine. Contrib. Mineral. Petrol. 68, 407-419.

Johannes, W., P.M. Bell, H.K. Mao, A.L. Boettcher, D.W. Chipman, J.F. Hays, R.C. Newton and
F. Seifert (1971). An interlaboratory comparison of piston-cylinder pressure calibration
using the albite-breakdown reaction. Contrib. Mineral. Petrol. 32, 24-38.

Johnson, C.A., E.J. Essene and S.R. Bohlen (1979) Garnet formation in olivine-bearing
metagabbros from the Adirondack Mtns., N.Y. Geol. Soc. Am. Abs. Prog. 11, 451.

Kase, H. and P. Metz (1980) Experimental investigation of the metamorphism of siliceous
dolomites. IV. Equilibrium data for the reaction diopside + 3 dolomite = 2 forsterite
+ 4 calcite + $2CO_2$. Contrib. Mineral. Petrol. 73, 151-159.

Kawasaki, T. and Y. Matsui (1977) Partitioning of Fe^{2+} and Mg^{2+} between olivine and garnet.
Earth Planet. Sci. Lett. 37, 159-166.

Kerrick, D.M. (1972) Experimental determination of muscovite + quartz stability with $P_{H_2O} <
P_{total}$. Am. J. Sci. 272, 946-958.

_____ and L.S. Darken (1975) Statistical thermodynamic models for ideal oxide and silicate
solid solutions, with application to plagioclase. Geochim. Cosmochim. Acta 39,
1431-1442.

Khitarov, N.I., V.A. Pugin, Chzabo-Bin and A.B. Slutsky (1963) Relations among andalusite, kyanite and sillimanite under conditions of moderate temperature and pressures. Geochimiya. 3, 235-249.

Kisch, J.J. (1966) Zeolite facies and regional rank of bituminous coals. Geol. Mag. 103, 414-422.

_____ (1980) Illite crystallinity and coal rank associated with lowest grade metamorphism of the Taveyanne greywacke in the Helvetic zone of the Swiss Alps. Eclog. Geol. Helv. 73, 753-777.

Kretz, R. (1959) Chemical study of garnet, biotite, and hornblende from gneisses of S.W. Quebec, with emphasis on distribution of elements in coexisting minerals. J. Geol. 67, 371-402.

_____ (1964) Analysis of equilibrium in garnet-biotite-sillimanite gneisses from Quebec. J. Petrol. 5, 1-20.

Kubler, B., J. Martini and M. Vuagnat (1974) Very low grade metamorphism in the western Alps. Schweiz. Mineral. Petrograph. Mitt. 54, 461-469.

Kushiro, I. and H.S. Yoder, Jr. (1966) Anorthite-forsterite and anorthite-enstatite reactions and their bearing on the basalt-eclogite transformation. J. Petrol. 7, 337-362.

Labotka, T.C. (1980) Petrology of a medium-pressure regional metamorphic terrane, Funeral Mountains, California. Am. Mineral 65, 670-689.

_____, J.J. Papike, D.T. Vaniman, G.B. Morey (1981) Petrology of contact metamorphosed argillite from the Rove formation, Gunflint trail, Minnesota. Am. Mineral. 66, 70-86.

Lagache, M. and A. Weisbrod (1977) The system: Two alkali feldspars - $KCl-NaCl-H_2O$ at moderate to high temperatures and low pressures. Contrib. Mineral. Petrol. 62, 77-102.

Lindsley, D.H. (1980) Phase equilibria of pyroxenes at pressures >1 atmosphere. *In* Prewitt, C.T., Ed., *Pyroxenes*, Reviews in Mineralogy 7, 289-308.

_____ and D.J. Anderson (1982) A two-pyroxene thermometer. Proc. 13th Lunar Planet. Sci. Conf. (submitted).

_____, B.T.C. Davis and I.D. MacGregor (1964) Ferrosilite ($FeSiO_3$) synthesis at high pressures and temperatures. Science 144, 73-75.

_____ and S.A. Dixon (1976) Diopside-enstatite equilibria at 850-1400°C, 5-35 kbar. Am. J. Sci. 276, 1285-1301.

_____, J.E. Grover and P.M. Davidson (1981) The thermodynamics of the $Mg_2Si_2O_6-CaMgSi_2O_6$ join: A review and an improved model. *In* S.K. Saxena, Ed., *Advances in Physical Geochemistry*, Springer-Verlag, New York, 146-172.

Lonker, S.W. (1981) The P-T-X relations of the cordierite-garnet-sillimanite-quartz equilibrium. Am. J. Sci. 281, 1056-1090.

Lusk, J. and C.E. Ford (1978) Experimental extension of the sphalerite geobarometer to 10 kbar. Am. Mineral. 63, 516-519.

Luth, W.C. and O.F. Tuttle (1966) The alkali feldspar solvus in the system $Na_2O-K_2O-Al_2O_3-SiO_2-H_2O$. Am. Mineral. 51, 1359-1373.

MacGregor, I.D. (1974) The system $MgO-Al_2O_3-SiO_2$: Solubility of Al_2O_3 in enstatite for spinel and garnet peridotite compositions. Am. Mineral. 59, 110-119.

Maresch, W.V. and A. Mottana (1976) The pyroxmangite-rhodonite transformation for the $MnSiO_3$ composition. Contrib. Mineral. Petrol 55, 69-79.

Martignole, J. and S. Nantel (1982) Geothermobarometry of cordierite-bearing metapelites near the Morin Anorthosite complex (Grenville Province). Canadian Mineral. (submitted).

_____ and J.C. Sisi (1981) Cordierite-garnet-H_2O equilibrium: A geological thermometer, barometer and water fugacity indicator. Contrib. Mineral. Petrol. 77, 38-46.

Matsui, Y. and O. Nishizawa (1974) Iron-magnesium exchange equilibrium between olivine and orthopyroxene over a temperature range 800-1300°C. Bull. Soc. franc. Mineral. Crystallogr. 97, 122-130.

Medaris, L.G., Jr. (1969) Partitioning of Fe^{2+} and Mg^{2+} between coexisting synthetic olivine and orthopyroxene. Am. J. Sci. 267, 945-968.

Metz, P. (1967a) Die obere stabilitatsgrenze aus Tremolit bei der metamorphose von kieseligen Karbonaten. Contrib. Mineral. Petrol., 15, 272-280.

_____ (1967b) Experimentelle Bildung von Forsterit und Calcit aus Tremolit und Dolomit. Geochim. Cosmochim. Acta 31, 1517-1532.

_____ (1970) Experimental investigation of the metamorphism of siliceous dolomites, 2. The conditions of diopside formation. Contrib. Mineral. Petrol. 28, 221-250.

_____ (1977) Temperature, pressure, and H_2O-CO_2 gas composition during metamorphism of siliceous dolomitic limestones deduced from experimentally determined equilibria of forsterite-forming reactions. Tectonophys. 43, 163-167.

_____ and D. Puhan (1970) Experimentelle Untersuchung der Metamorphose von kieselig dolomitischen Sedimenten. Contrib. Mineral. Petrol. 26, 302-314.

_____ and V. Trommsdorff (1968) On phase equilibria in metamorphosed siliceous dolomites. Contrib. Mineral. Petrol. 18, 305-309.

_____ and H.G.F. Winkler (1964) Experimentelle Intersuchung der Diopsid Bildung aus Tremolit, Calcit and Quartz. Naturwiss. 51, 460.

Miyashiro, A. (1961) Evolution of metamorphic belts. J. Petrol. 2, 277-318.

Mori, T. (1977) Geothermometry of spinel lherzolites. Contrib. Mineral. Petrol. 59, 261-279.

_____ (1978) Experimental study of pyroxene equilibria in the system $CaO-MgO-FeO-SiO_2$. J. Geol. 19, 45-65.

_____ and D.H. Green (1975) Pyroxenes in the system $Mg_2Si_2O_6-CaMgSi_2O_6$ at high pressure. Earth Planet. Sci. Lett. 26, 277-286.

_____ and _____ (1976) Subsolidus equilibria between pyroxenes in the $CaO-MgO-SiO_2$ system at high pressures and temperatures. Am. Mineral. 61, 616-625.

_____ and _____ (1978) Laboratory duplication of phase equilibria observed in natural garnet lherzolites. J. Geol. 86, 83-97.

Morse, S.A. (1970) Alkali feldspars with water at 5 kb pressure. J. Petrol. 11, 221-253.

_____ (1975) High-Al pyroxene in anorthosite: Barometer or speedometer? Int'l Conf. Geothermometry and Geobarometry, Extended Abstracts, Pennsylvania State University.

Mueller, R.F. (1961) Analysis of relations among Mg, Fe, Mn in certain metamorphic minerals. Geochim. Cosmochim. Acta 25, 267-296.

_____ and S.K. Saxena (1977) *Chemical Petrology*. Springer-Verlag, New York.

Mysen, B.O. and K.S. Heier (1972) Petrogenesis of eclogites in high grade metamorphic gneisses, exemplified by the Hareidland eclogite, Western Norway. Contrib. Mineral. Petrol. 36, 73-87.

Nafziger, R.H. and A. Muan (1967) Equilibrium phase compositions and thermodynamic properties of olivines and pyroxenes in the system $MgO-"FeO"-SiO_2$. Am. Mineral. 52, 1364-1385.

Nehru, C.E. and P.J. Wyllie (1974) Electron microprobe measurements of pyroxenes coexisting with H_2O-undersaturated liquids on the join $CaMgSi_2O_6-Mg_2Si_2O_6-H_2O$ at 30 kb with application to geothermometry. Contrib. Mineral. Petrol. 48, 221-228.

Nesbitt, B.E. (1979) *Regional Metamorphism of the Ducktown, Tennessee Massive Sulfides and Adjoining Portions of the Blue Ridge Province*. Ph.D. thesis, University of Michigan, Ann Arbor.

_____ and E.J. Essene (1982) Metamorphic thermometry and barometry of a portion of the Southern Blue Ridge Province. Am. J. Sci. 282, 701-729.

Newton, R.C. (1966) Kyanite-sillimanite equilibrium at 750°C. Science 151, 1222-1225.

_____ and H.T. Haselton (1981) Thermodynamics of the garnet-plagioclase-Al_2SiO_5-quartz geobarometer. *In* Newton, R.C., A. Navrotsky and B.J. Wood, Eds., *Thermodynamics of Minerals and Melts*, Springer-Verlag, New York, 129-145.

_____ and D. Perkins III (1982) Thermodynamic calibration of geobarometers based on the assemblages garnet-plagioclase-orthopyroxene (clinopyroxene)-quartz. Am. Mineral. 67, 203-222.

_____ and J.V. Smith (1967) Investigations concerning the breakdown of albite at depth in the earth. J. Geol. 75, 268-286.

_____ and B.J. Wood (1979) Thermodynamics of water in cordierite and some petrologic consequences of cordierite as a hydrous phase. Contrib. Mineral. Petrol. 68, 391-405.

Nishizawa, O. and S. Akimoto (1973) Partition of magnesium and iron between olivine and spinel, and between pyroxene and spinel. Contrib. Mineral. Petrol. 41, 217-240.

Novak, J.M. and M.J. Holdaway (1981) Metamorphic petrology, mineral equilibria, and polymetamorphism in the Augusta quadrangle, south-central Maine. Am. Mineral. 66, 51-69.

Obata, M. (1980) The Ronda peridotite: garnet, spinel-, and plagioclase-lherzolite facies and the P-T trajectories of a high-temperature mantle intrusion. J. Petrol. 21, 533-572.

O'Hara, M.J. and G. Yarwood (1978) High pressure-temperature point on an Archaean geotherm, implied magma genesis by crustal anatexis and consequences for garnet-clinopyroxene thermometry and barometry. Trans. Roy. Soc. of Edin. 65, 251-314.

Oka, Y. and T. Matsumoto (1974) Study on the compositional dependence of the apparent partition coefficient of iron and magnesium between coexisting garnet and clinopyroxene solid solutions. Contrib. Mineral. Petrol. 48, 115-121.

O'Neill, H. St. C. (1981) The transition between spinel and garnet lherzolite and its use as a barometer. Contrib. Mineral. Petrol. 77, 185-194.

_____ and B.J. Wood (1979) An experimental study of Fe-Mg partitioning between garnet and olivine and its calibration as a geothermometer. Contrib. Mineral. Petrol. 70, 59-70.

O'Neill, J.R. and R.N. Clayton (1964) Oxygen isotope geothermometry. *In* Craig, H., S.L. Miller, and G.J. Wasserburg, Eds., *Isotopic and Cosmic Chemistry*, North Holland Publishing Company, Amsterdam, 157-168.

Orville, P.M. (1963) Alkali ion exchange between vapor and feldspar phases. Am. J. Sci. 261, 201-237.

_____ (1972) Plagioclase cation exchange equilibria with aqueous chloride solution at 700°C and 2000 bars in the presence of quartz. Am. J. Sci. 272, 234-272.

Oterdoom, W.H. (1978) Tremolite and diopside-bearing assemblages in the CaO-MgO-SiO$_2$-H$_2$O multisystem. Schweiz. Mineral. Petrograph. Mitt. 58, 127-138.

Parsons, I. (1969) Subsolidus crystallization behavior in the system KAlSi$_3$O$_8$-NaAlSi$_3$O$_8$. Mineral. Mag. 37, 173-180.

_____ (1978) Alkali feldspars: Which solvus? Phys. Chem. Minerals 2, 199-213.

Perchuk, L.L. (1967) Biotite-garnet geothermometer. Dokl. Akad. Nauk SSSR 177, 411-414.

_____ (1968) Pyroxene-garnet equilibrium and the depth facies of eclogites. Inter. Geol. Rev. 10, 280-318.

_____ (1977) Thermodynamic control of metamorphic processes. In Saxena, S.K. and Batacharji, Eds., Energetics of Geological Processes, Springer-Verlag, New York, 283-352.

_____, K.K. Podlesski and L.Y. Aranovich (1981) Calculation of thermodynamic properties of end-member minerals from natural parageneses. In Newton, R.C., A. Navrotsky, and B.J. Wood, Eds., Advances in Physical Geochemistry, 1, Springer-Verlag, New York, 110-129.

Perkins, D. III, (1979) Application of New Thermodynamic Data to Mineral Equilibria. Ph.D. thesis, University of Michigan, Ann Arbor, Michigan.

_____, E.J. Essene and L.A. Marcotty (1982) Thermometry and barometry of some amphibolite-granulite facies rocks from the Otter Lake area, S. Quebec. Canadian J. Earth Sci. in press.

_____, T.J.B. Holland and R.C. Newton (1981) The Al$_2$O$_3$ contents of enstatite in equilibrium with garnet in the system MgO-Al$_2$O$_3$-SiO$_2$ at 15-40 kbar and 900-1600°C. Contrib. Mineral. Petrol. 78, 99-109.

_____ and R.C. Newton (1981) Charnockite barometers based on coexisting garnet-pyroxene-plagioclase-quartz. Nature 292, 144-146.

_____, E.F. Westrum, Jr., and E.J. Essene (1980) The thermodynamic properties and phase relations of some minerals on the system CaO-Al$_2$O$_3$-SiO$_2$-H$_2$O. Geochim. Cosmochim. Acta 44, 61-84.

Phillips, M.W. and P.H. Ribbe (1973) The structures of monoclinic potassium-rich feldspars. Am. Mineral. 58, 263-270.

Pitcher, W.S. and G.W. Flinn Eds. (1965) Controls of Metamorphism. Geol. J. Special Issue No. 1. Oliver and Boyd, Edinburgh.

Popp, R.K. and M.C. Gilbert (1972) Stability of acmite-jadeite pyroxenes at low pressure. Am. Mineral. 57, 1210-1231.

Prigogine, I. and R. Defay (1954) Chemical Thermydynamics, Longmans Green, London.

Powell, M. and R. Powell (1977b) Plagioclase-alkali feldspar thermometry revisited. Mineral. Mag. 41, 253-256.

Powell, R. and M. Powell (1977a) Geothermometry and oxygen barometry using iron-titanium oxides: A reappraisal. Mineral. Mag. 41, 257-263.

Puhan, D. (1976) Metamorphic temperature determined by means of the dolomite-calcite solvus thermometer - examples from the central Damara Orogen (Southwest Africa). Contrib. Mineral. Petrol. 58, 23-28.

Raheim, A. (1975) Mineral zoning as a record of P, T history of Precambrian metamorphic rocks in W. Tasmania. Lithos 8, 221-236.

_____ (1976) Petrology of eclogites and surrounding schists from the Lyell Highway-Collingwood River area, W. Tasmania. Geol. Soc. of Australia J. 23, 313-327.

_____ and D.H. Green (1974) Experimental determination of the temperature and pressure dependence of the Fe-Mg partition coefficient for coexisting garnet and clinopyroxene. Contrib. Mineral. Petrol. 48, 179-203.

_____ and _____ (1975) P, T paths of natural eclogites during metamorphism, a record of subduction. Lithos 8, 317-328.

Ramberg, H. (1952) The Origin of Metamorphic and Metasomatic Rocks. University of Chicago Press, Chicago.

_____ and G. DeVore (1951) The distribution of Fe^{2+} and Mg in coexisting olivines and pyroxenes. J. Geol. 59, 193-210.

Rice, J.M. (1977a) Progressive metamorphism of impure dolomitic limestone in the Marysville aureole, Montana. Am. J. Sci. 277, 1-24.

_____ (1977b) Contact metamorphism of impure dolomitic limestone in the Boulder aureole, Montana. Contrib. Mineral. Petrol. 59, 237-259.

Richardson, S.W. (1968) Staurolite stability in a part of the system Fe-Al-Si-O-H. J. Petrol. 9, 467-488.

Robie, R.A., B.S. Hemingway and J.R. Fisher (1978) Thermodynamic properties of minerals and related substances at 298.15K and 1 bar (10^5Pascals) and at higher temperatures. U.S. Geol. Surv. Bull. 452, 1456 pp.

Roedder, E. (1981) Origin of fluid inclusions and changes that occur after trapping. *In* Hollister, L.S. and M.L. Crawford, Eds., *Short Course in Fluid Inclusions: Applications to Petrology*. Mineral. Assoc. Canada, 101-137.

Rollinson, H.R. (1980) Garnet-pyroxene thermometry and barometry in the Scourie granulites, N.W. Scotland. Lithos 14, 225-238.

Ross, M. and J.S. Huebner (1975) A pyroxene geothermometer based on composition-temperature relationships of naturally occurring orthopyroxene, pigeonite, and augite. Extended Abstract, Int'l Conf. Geothermometry and Geobarometry, Pennsylvania State University.

Sack, R.O. (1980a) Adirondack mafic granulites and a model lower crust. Geol. Soc. Am. Bull 91, 349-442.

_____ (1980b) Some constraints on the thermodynamic mixing properties of Fe-Mg orthopyroxenes and olivines. Contrib. Mineral. Petrol. 71, 237-246.

Saxena, S.K. (1968) Distribution of iron and magnesium between coexisting garnet and clinopyroxene in rocks of varying metamorphic grade. Am. Mineral. 53, 2018-2021.

_____ (1969a) Silicate solid solution and geothermometry. 3. Distribution of Fe and Mg between coexisting garnet and biotite. Contrib. Mineral. Petrol. 22, 259-267.

_____ (1969b) Silicate solid solutions and geothermometry. 4. Statistical study of chemical data on garnets and clinopyroxenes. Contrib. Mineral. Petrol. 23, 140-156.

_____ (1973) *Thermodynamics of Rock-Forming Crystalline Solutions*. Minerals, Rocks and Inorganic Materials 8, Springer-Verlag, New York.

_____ (1976) Two pyroxene geothermometer: A model with an approximate solution. Am. Mineral. 61, 643-652.

_____ (1979) Garnet-clinopyroxene geothermometer. Contrib. Mineral. Petrol. 70, 229-235.

_____ (1981) Fictive component model of pyroxenes and multicomponent phase equilibria. Contrib. Mineral. Petrol. 78, 345-351.

_____ and S. Ghose (1971) Mg^{2+}-Fe^{2+} order disorder and the thermodynamics of the orthopyroxene-crystalline solution. Am. Mineral. 56, 532-559.

_____ and C.E. Nehru (1975) Enstatite-diopside solvus and geothermometry. Contrib. Mineral. Petrol. 49, 259-267.

Schmid, R. and B.J. Wood (1976) Phase relationships in granulitic metapelites from the Ivren-Verbavo zone (Northern Italy). Contrib. Mineral. Petrol. 54, 255-279.

Schuiling, R.D. (1957) A geo-experimental phase diagram of Al_2SiO_5 (sillimanite, kyanite, andalusite). Koninkl. Ned. Akad. Wetenschap. B60, #3, 220-226.

_____ (1962) Die petrogenetische Deutung der drei Modificationen von Al_2SiO_5. N. Jahrb. Mineral. Mh. 1962, 200-214.

Scott, S.D. (1973) Experimental calibration of the sphalerite geobarometer. Econ. Geol. 68, 466-474.

_____ (1976) Application of the sphalerite geobarometer to regionally metamorphosed terrains. Am. Mineral. 61, 661-670.

_____ and H.L. Barnes (1971) Sphalerite geothermometry and geobarometry. Econ. Geol. 66, 653-669.

Seck, H.A. (1972) The influence of pressure on the alkali-feldspar solvus from peraluminous and persilicic materials. Fortschr. Mineral. 49, 31-49.

Seifert, F. (1970) Low temperature compatibility relations of cordierite in haplopelites of the system K_2O-MgO-Al_2O_3-SiO_2-H_2O. J. Petrol. 11, 73-99.

Seil, M.K. and J.G. Blencoe (1979) Activity-composition relations of $NaAlSi_3O_8$-$CaAl_2Si_2O_8$ feldspars at 2kb, 600-800°C. Geol. Soc. Am. Abstr. Prog. 11, 513.

Sheppard, S.M.F. and H.P. Schwarcz (1970) Fractionation of carbon and oxygen isotopes and magnesium between metamorphic calcite and dolomite. Contrib. Mineral. Petrol., 26, 161-198.

Skippen, G.B. (1971) Experimental data for reactions in siliceous marbles. J. Geol. 79, 457-481.

Slaughter, J., D.M. Kerrick and V.J. Wall (1975) Experimental and thermodynamic study of equilibria in the system CaO-MgO-SiO_2-CO_2-H_2O. Am. J. Sci. 275, 143-162.

Slavinskiy, V.V. (1976) The clinopyroxene-garnet geothermometer. Dokl. Akad. Nauk SSSR 231, 181-184.

Smith, D. (1971) Stability of the assemblage iron-rich orthopyroxene-olivine-quartz. Am. J. Sci. 271, 370-382.

_____ (1972) Stability of iron-rich pyroxene in the system $CaSiO_3$-$FeSiO_3$-$MgSiO_3$. Am. Mineral. 57, 1413-1438.

204

Speidel, D.H. and E.F. Osborn (1967) Element distribution among coexisting phases in the system $MgO-FeO-Fe_2O_3-SiO_2$ as a function of temperature and oxygen fugacity. Am. Mineral. 52, 1139-1152.

Spencer, K.J. and D.H. Lindsley (1981) A solution model for coexisting iron-titanium oxides. Am. Mineral. 66, 1189-1201.

Stewart, D.B. and P.H. Ribbe (1969) Structural explanation for variations in cell parameters of alkali feldspars with Al/Si ordering. Am. J. Sci. 267A, 444-462.

Stoddard, E.F. (1976) Granulite facies metamorphism in the Colton-Rainbow Falls area, northwest Adirondacks, New York. Ph.D. thesis, University of California, Los Angeles.

_____ (1979) Metamorphic conditions at the northern end of the N.W. Adirondack lowlands. Geol. Soc. Amer. Bull. 91, 589-614.

Stormer, J.C., Jr. (1965) A practical two feldspar thermometer. Am. Mineral. 60, 667-674.

Strens, R.G.J. (1967) Stability of Al_2SiO_5 solid solutions. Mineral. Mag. 31, 839-849.

Suzuki, K. (1977) Local equilibrium during the contact metamorphism of siliceous dolomites in Kasuga-Mura, Gifu-ken, Japan. Contrib. Mineral. Petrol. 61, 79-89.

Thompson, A.B. (1974) Calculation of muscovite-paragonite-alkali feldspar phase relations. Contrib. Mineral. Petrol. 44, 173-194.

_____ (1976) Mineral reactions in pelitic rocks II: Calculations of some P-T-X (Fe-Mg) phase relations. Am. J. Sci. 276, 425-454.

Thompson, J.B., Jr. (1967) Thermodynamic properties of simple solutions. *In* P.H. Abelson, Ed., *Researches in Geochemistry* II, John Wiley and Sons, New York, 340-361.

_____ and S.A. Norton (1968) Paleozoic regional metamorphism in New England and adjacent areas. *In* Zen, E-an et al., Eds., *Studies of Appalachian Geology: Northern and Maritime*, John Wiley and Sons, New York, 319-327.

_____ and A.B. Thompson (1976) A model system for mineral facies in pelitic schists. Contrib. Mineral. Petrol. 58, 243-277.

Treiman, A.H. and E.J. Essene (1982) Phase equilibria in the system $CaO-SiO_2-CO_2$, Am. J. Sci., Orville Volume (in press).

Touret, J. (1981) Fluid inclusions in high grade metamorphic rocks. *In* Hollister, L.S. and M.L. Crawford, Eds., *Short Course in Fluid Inclusions: Application to Petrology*, Short Course Notes, Mineral. Assoc. Canada, 182-208.

Trommsdorff, V. and B.W. Evans (1972) Progressive metamorphism of antigorite schist in the Bergell tonalite aureole (Italy). Am. J. Sci. 272, 423-437.

Turncock, A.C. and H.P. Eugster (1962) Fe-Al oxides: phase relationships below 1000°C. J. Petrol. 3, 533-565.

Turner, F.J. (1979) *Metamorphic Petrology-Mineralogical and Field Aspects*, 2nd ed. McGraw-Hill, New York.

Valley, J.W. (1980) *The Role of Fluids During Metamorphism of Marbles and Associated Rocks in the Adirondack Mountains, New York*. Ph.D. thesis, University of Michigan, Ann Arbor.

_____ and E.J. Essene (1980) Calc-silicate reactions in Adirondack marbles: The role of fluids and solid solutions. Geol. Soc. Am. Bull. 91, 114-117, 720-815.

_____, J. McLelland, E.J. Essene and W. Lamb (1982) Metamorphic fluids in the deep crust: Evidence from the Adirondacks, N.Y., U.S.A. Nature (submitted).

_____, E.U. Petersen, E.J. Essene and J.R. Bowman (1982) Fluorphlogopite and fluortremolite in Adirondack marbles and calculated C-O-H-F fluid compositions. Am. Mineral. 67, 545-557.

Virgo, D. and S.S. Hafner (1969) Fe^{2+}, Mg order-disorder in heated orthopyroxenes. Mineral. Soc. Am. Special Paper 2, 67-81.

Waldbaum, D.R. (1965) Thermodynamic properties of mullite, andalusite, kyanite and sillimanite. Am. Mineral. 50, 186-195.

_____ and J.B. Thompson, Jr. (1969) Mixing properties of sanidine crystalline solutions: IV. Phase diagrams from equations of state. Am. Mineral. 54, 1274-1298.

Wall, V.J. and R.N. England (1979) Zn-Fe spinel-silicate-sulphide reactions as sensors of metamorphic intensive variables and processes. Geol. Soc. Am. Abstr. Prog. 11, 534.

Warner, R.D. and W.C. Luth (1974) The diopside-orthoenstatite two-phase region in the system $CaMgSi_2O_6-Mg_2Si_2O_6$. Am. Mineral. 59, 98-109.

Weill, D.F. (1966) Stability relations in the $Al_2O_3-SiO_2$ system calculated from solubilities in the $Al_2O_3-SiO_2-Na_3AlF_6$ system. Geochim. Cosmochim. Acta 30, 223-237.

_____ and W.S. Fyfe (1961) A preliminary note on the relative stability of andalusite, kyanite and sillimanite. J. of the Am. Chem. Soc. 72, 4742-4743.

Weisbrod, A. (1973) Refinements of the equilibrium conditions of the reaction Fe-cordierite = almandine + quartz + sillimanite + H_2O. Carnegie Inst. Wash. Year Book 72, 515-522.

Wells, P.R.A. (1976) Late Archean metamorphism in the Buksefjorden region, East Greenland. Contrib. Mineral. Petrol. 56, 229-243.

_____ (1977) Pyroxene thermometry in simple and complex systems. Contrib. Mineral. Petrol. 62, 129-139.

Whitney, J.A. and J.R. Stormer (1977) The distribution of $NaAlSi_3O_8$ between coexisting microcline and plagioclase and its effect on geothermometric calculations. Am. Mineral. 62, 687-691.

Winkler, H.G.F. (1978) *Petrogenesis of Metamorphic Rocks.* 5th ed. Springer-Verlag, New York.

Winter, G.A., E.J. Essene and D.R. Peacor (1981) Carbonates and pyroxenoids from the manganese deposit near Bald Knob, North Carolina. Am. Mineral. 66, 278-279.

Wones, D.R., R.G. Burns and B.M. Caroll (1971) Stability and properties of synthetic annite. Trans. Am. Geophys. Union 52, 369-370.

_____ and H.P. Eugster (1965) Stability of biotite: Experiment, theory and application. Am. Mineral. 50, 1228-1272.

Wood, B.J. (1975) The influence of pressure, temperature and bulk composition on the appearance of garnet in orthogneisses - an example from South Harris, Scotland. Earth Planet. Sci. Lett. 26, 299-311.

_____ (1976) Mixing properties of tschermakitic pyroxenes. Am. Mineral. 61, 599-602.

_____ (1979) Activity-composition relations in $Ca(Mg,Fe)Si_2O_6-CaAl_2SiO_6$ clinopyroxene solid solutions. Am. J. Sci. 279, 854-87.

_____ and S. Banno (1973) Garnet-orthopyroxene and orthopyroxene-clinopyroxene relationships in simple and complex systems. Contrib. Mineral. Petrol. 42, 109-124.

_____ and R.G.J. Strens (1971) The orthopyroxene geobarometer. Earth Planet. Sci. Lett. 11, 1-6.

Wright, T.L. and D.B. Stewart (1968a) X-ray and optical study of alkali feldspar: I. Determination of composition and structural state from refined unit-cell parameters and 2V. Am. Mineral. 53, 38-87.

_____ and _____ (1968b) X-ray and optical study of alkali feldspar: II. An X-ray method for determining the composition and structural state from measurement of 2θ values for three reflections. Am. Mineral. 53, 88-104.

Zen, E-an (1969) The stability relations of the polymorphs of aluminum silicate: A survey and some comments. Am. J. Sci. 267, 677-710.

_____ (1973) Thermochemical parameters of minerals from oxygen-buffered hydrothermal equilibrium data: Method, application to annite and almandine. Contrib. Mineral. Petrol. 39, 65-80.

Ziegenbein, P. and W. Johannes (1974) Wollastonitbildung aus Quarz und Calcit bei P_f = 2, 4 und 6 kb. Fortschr. Mineral. 52, 77-79.

Chapter 6

CHARACTERIZATION of METAMORPHIC FLUID COMPOSITION through MINERAL EQUILIBRIA

J.M. Ferry and D.M. Burt

INTRODUCTION

In this chapter, we discuss the volatile-rich phase which is present at some times and in some places in the Earth's crust during metamorphism. We refer to the volatile-rich phase as a fluid (as opposed to a liquid, a vapor, or a gas) for two reasons. First, the pressure-temperature conditions of metamorphism are above the critical point for major species constituting the phase. At metamorphic P-T conditions, the distinction between a liquid and vapor is meaningless -- the phase is a supercritical fluid. Second, the density of the volatile-rich phase is on the order of 1 g/cm^3 at metamorphic conditions. Such a substance has properties unlike those of what we normally think of as a gas. The term "fluid" therefore is also used to emphasize this distinction.

Evidence for a fluid phase during metamorphism

There are numerous lines of evidence for the existence of a volatile-rich fluid phase in rocks during metamorphism. (a) Minerals in metamorphic rocks commonly contain fluid inclusions. While it is sometimes uncertain whether a particular fluid inclusion is synmetamorphic or postmetamorphic, at least some of these inclusions are believed to contain trapped metamorphic fluids (Crawford, 1981; Touret, 1981). (b) Whole-rock chemical analyses indicate that high-grade metamorphic rocks are depleted in volatile components relative to their low-grade equivalents. Shaw (1955), for example, compiled analyses of pelitic schists and observed that "low-grade" schists contain an average of 4.34 wt % H_2O and 2.31 wt % CO_2 while "high-grade" schists contain an average of 2.42 wt % H_2O and 0.22 wt % CO_2. The difference in H_2O and CO_2 contents is interpreted as a loss of CO_2 and H_2O from the schists during progressive metamorphism. The CO_2 and H_2O were probably lost as a fluid phase. (c) Prograde mineral reactions commonly involve devolatilization. The devolalitization reactions are the mechanisms which account for the decrease in whole-rock contents of H_2O and CO_2 during progressive metamorphism. The evolved CO_2 and H_2O are believed to form a fluid phase. (d) The presence of hydrates, carbonates, sulfides, etc. in metamorphic rocks requires non-zero values for the partial pressure of H_2O, CO_2, S_2, etc. during metamorphism.

For example, the presence of muscovite, as opposed to its anhydrous equivalent, K-feldspar + corundum, requires a certain finite p_{H_2O} during the metamorphic event. When partial pressures of the abundant volatile species are calculated from mineral equilibria, their sum often is essentially the same as the estimated lithostatic pressure during metamorphism (Ferry, 1976b). The condition of $\Sigma p_i \sim P_{lithostatic}$ implies the presence of a fluid phase during metamorphic events. (e) Active metamorphism appears to be taking place in some modern geothermal fields such as the one at the Salton Sea, California (Muffler and White, 1969; McDowell and McCurry, 1977; McDowell and Elders, 1980). Fluids are, of course, present at depth in these geothermal fields and the composition of fluid, sampled by wells, appears to be appropriate for equilibrium with the metamorphic minerals brought up in rock fragments during drilling of the wells (Bird and Norton, 1981). If modern geothermal fields are at least generally analogous to some kinds of metamorphism, then a fluid phase exists in chemical equilibrium with rocks during metamorphic events.

These five lines of evidence indicate that fluids are generated by mineral reactions during metamorphism and that a fluid phase is present at least at some points in space and time during metamorphic events. A fundamental goal in metamorphic petrology is therefore to characterize the composition of metamorphic fluids. An understanding of metamorphic fluid compositions is a prerequisite to understanding the mechanisms by which many metamorphic mineral reactions proceed. The systematics of fluid composition may also lead to the identification of fundamental controls on intensive variables during metamorphism (e.g., Rice and Ferry, Chapter 7, this volume).

The composition of metamorphic fluids: general statement

Before the methods for determining fluid composition are discussed, it is useful to identify what to look for. What are the major species in metamorphic fluids? The composition of metamorphic fluids will be reviewed in detail later, but a few general remarks are appropriate here. Recent reviews of this subject conclude that metamorphic fluids principally are composed of species in the system C-O-H-S as well as various associated and disassociated chloride complexes (Roedder, 1972; Fyfe et al., 1978; Holloway, 1981).

The system C-O-H-S. Under most metamorphic P-T conditions in the crust those C-O-H-S fluid species present in concentrations $\gtrsim 0.1$ mol % are: H_2O, CO_2, CH_4, H_2S, H_2, CO, and SO_2. The relation between speciation and elemental composition of a C-O-H-S fluid is illustrated in Figure 1. The diagram is constructed for f_{S_2} defined by pyrrhotite of composition $Fe_{0.905}S$,

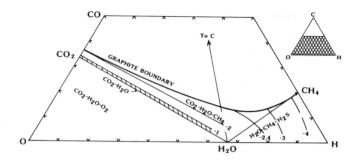

Figure 1. Speciation of C-O-H-S fluids with compositions projected through S onto C-O-H. Physical conditions are T = 1100°C; P = 5000 bars; f_{S_2} is defined by pyrrhotite with composition $Fe_{0.905}S$. Numbered lines are isobars of f_{O_2} relative to f_{O_2} defined by quartz + fayalite + magnetite. From Holloway (1981).

and fluid compositions are projected through sulfur. Although the diagram refers to a temperature greater than most which are attained during metamorphism, it does illustrate some general relations. The diagram is divided into a number of regions of bulk composition and the major species which constitute the fluid differ between regions. Along the CO_2-H_2O join fluids are binary H_2O-CO_2 mixtures. Fluids with bulk compositions bounded by the H_2O-CO_2 join, the H_2O-CH_4 join, and the graphite boundary are principally H_2O-CO_2-CH_4 mixtures with proportions varying as a function of bulk composition. Fluids with bulk compositions bounded by the joins H_2O-CH_4, CH_4-H, and H_2O-H are principally H_2O-CH_4-H_2S mixtures with proportions varying with bulk composition. Fluids with compositions in the region bounded by the joins CO_2-O, O-H_2O, and H_2O-CO_2 are probably not found in the crust during metamorphism. Fluids with carbon contents greater than that defined by the graphite boundary also do not occur: Such bulk compositions are represented by mixtures of graphite and fluid, with fluid composition along the graphite boundary. It is evident from Figure 1 that the presence of graphite in a rock places a substantial limit on the composition of a coexisting C-O-H-S fluid. The presence of graphite in metamorphic rocks is therefore a very useful tool in characterizing metamorphic fluid compositions.

The general relationship between speciation of a C-O-H-S fluid and temperature is shown in Figure 2. The diagram illustrates speciation of a fluid with f_{S_2} defined by equilibrium with pyrrhotite $Fe_{0.905}S$ and at f_{O_2} two log units below that defined by quartz + magnetite + fayalite. The fluid is almost entirely composed of CH_4 at temperatures below ∿500°C and is dominated by H_2O and CO_2 at temperatures above ∿900°C. At metamorphic temperatures

209

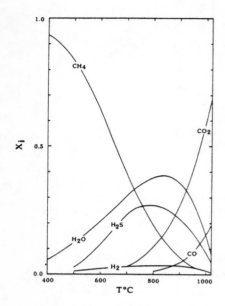

Figure 2. Speciation of C-O-H-S fluid coexisting with graphite as a function of temperature at $P = 2000$ bars. f_{S_2} is defined as in Figure 1. f_{O_2} is fixed at two *log* units less than that defined by quartz + fayalite + magnetite. From Holloway (1981).

($T \simeq 400°-800°C$), the principal C-O-H-S fluid species are CH_4, H_2O, CO_2, and H_2S. As a rough approximation we may look for CH_4, H_2O, CO_2 and H_2S as the species which compose most of the C-O-H-S fraction of metamorphic fluids. The concentration of sulfur-bearing species (principally H_2S or SO_2) is usually substantially less than implied by Figure 2: It is almost always less than 10 mol % and often less than 1 mol %.

Dissolved chlorides. Natural metamorphic fluids contain chloride complexes in addition to C-O-H-S species. At the elevated pressures and temperatures of metamorphism, the chlorides are probably nearly completely associated. The most abundant chloride species is NaCl; KCl, $FeCl_2$, and $CaCl_2$ occur in less concentrations but may be the dominant chloride in some metamorphic fluids. Metamorphic fluids may contain very large concentrations of chlorides. Rich (1979), for example, reported fluid inclusions in metamorphic rocks from Vermont which contain more than 25 wt % NaCl. In general, however, the concentration of dissolved chlorides is one of the less well-characterized aspects of metamorphic fluids. Except for some metamorphic waters collected from hot-springs, Cl appears to be the chief anion in metamorphic fluids.

Species whose concentrations serve as monitors of process. Although the principal metamorphic fluid species are H_2O, CO_2, CH_4, H_2S, and various chlorides, petrologists often find it useful to characterize the concentration of other species which serve as monitors or "operators" of certain metamorphic processes. Such monitor species include O_2, S_2, and F_2 among others. The species commonly occur in vanishingly small concentrations (e.g., a value of $X_{O_2,fluid} \simeq 10^{-20}$ would not be unusual). Consequently, their concentrations are most often presented in $logf$ units where f is fugacity (we will define and discuss fugacity later). Although the species O_2, S_2, F_2, etc. are an insignificant contribution to metamorphic fluids, their concentrations serve as valuable monitors of oxidation/reduction, sulfidation/desulfidation, fluorination/defluorination, etc. For example, a fluid with $logf_{O_2}$ = -15 will oxidize a mineral assemblage which defines $logf_{O_2}$ = -20. (The mineral-fluid redox reaction, however, will involve species mostly *other than* O_2). We may also be interested in variables such as HK_{-1} ("H, K minus 1") and $FeMg_{-1}$ ("Fe, Mg minus 1") which have no physical meaning at all as actual fluid species. Such quantities are referred to as exchange components (Thompson, Chapter 1, this volume) and serve as monitors of processes involving exchange of atoms between minerals and fluid. For example, HK_{-1} monitors exchange of H for K (acid metasomatism); $FeMg_{-1}$ monitors exchange of Fe for Mg (iron metasomatism); etc. Many diagrams which refer to metamorphic fluid composition involve variables such as O_2 and HK_{-1} whose principal use is to monitor a process rather than to characterize the concentration of a major species.

Methods of investigating the composition of metamorphic fluids

There are three general ways in which petrologists characterize the composition of metamorphic fluid. First, some waters from hotsprings and geothermal fields are believed to be direct samples of metamorphic fluid (Barnes and O'Neil, 1969; Barnes, 1970; Barnes et al., 1972; White et al., 1973, Bird and Norton, 1981). The chemical composition of these metamorphic fluids may then be determined by direct chemical analysis. One drawback, of course, is an uncertainty as to the extent that the metamorphic component of the fluids has been diluted by non-metamorphic components (e.g., drilling fluids, meteoric water). Furthermore, waters from springs and wells can never represent samples of metamorphic fluid which has come from depths of more than a few kilometers. Contemporary metamorphic fluids from depths of 20-30 km probably could never be directly sampled. Direct chemical analysis of contemporary metamorphic fluids nevertheless gives important information about the nature and concentration of dissolved charged species which is difficult to obtain by other means.

Second, some fluid inclusions are believed to contain samples of metamorphic fluid. In some cases these fluids may be extracted and analyzed directly (Roedder, 1972). More commonly, the composition of fluids in fluid inclusions is inferred through heating and freezing experiments. The study of fluid inclusions and their application to petrology were the subjects of a short course sponsored by the Mineralogical Association of Canada: See Hollister and Crawford (1981), which contains an exhaustive and up-to-date summary of the characterization of metamorphic fluids through fluid inclusions in metamorphic rocks.

Third, the composition of metamorphic fluids may be inferred from equilibria involving minerals which contain those elements which constitute the fluid phase. Because of the theme of this book (and because fluid inclusions have been discussed in the M.A.C. handbook), the remainder of this chapter concentrates on the characterization of metamorphic fluids by mineral equilibria in metamorphic rocks. The fundamental concept which permits estimation of fluid composition from mineral equilibria is the relationship among the chemical potentials of components in a system at chemical equilibrium. Consider a collection of phases at equilibrium: *For every balanced stoichiometric relation that can be written among independently variable components of the phases, there exists a corresponding relation among the chemical potentials of those components* (see Spear, Ferry & Rumble, this volume). For example, the following stoichiometric relation can be written among components of wollastonite, calcite, quartz, and fluid:

$$CaCO_3 + SiO_2 = CaSiO_3 + CO_2 \ . \tag{1}$$

If the mineral phases and fluid phase were in equilibrium during metamorphism, then

$$\mu_{CaCO_3,calcite} + \mu_{SiO_2,quartz} = \mu_{CaSiO_3,wollastonite} + \mu_{CO_2,fluid} \ . \tag{2}$$

[See Appendix 1 for all notation.] If the chemical potentials of components in the minerals can be quantitatively evaluated, equation (2) allows calculation of $\mu_{CO_2,fluid}$. If chemical potential-composition relations (i.e., equation of state) are known for the fluid, then the estimation of $\mu_{CO_2,fluid}$ allows, in turn, calculation of the concentration of CO_2 in the metamorphic fluid in equilibrium with calcite, quartz, and wollastonite. The concentrations of other species in the fluid may be estimated from relations like (2) which involve the other species. For example, the concentration of H_2O could be estimated from the occurrence of zoisite, anorthite, wollastonite, and

212

quartz in a rock according to:

$$2\mu_{Ca_2Al_3Si_3O_{12}(OH),zoisite} + \mu_{SiO_2,quartz} =$$

$$3\mu_{CaAl_2Si_2O_8,plagioclase} + \mu_{CaSiO_3,wollastonite} + \mu_{H_2O,fluid} \qquad (3)$$

etc.

Mineral equilibria are also useful in characterizing the concentration of dissolved chlorides and other non-C-O-H-S species in metamorphic fluid. The following stoichiometric relation may be written among components in muscovite, quartz, K-feldspar, and coexisting fluid:

$$3KAlSi_3O_8 + 2HCl = KAl_3Si_3O_{10}(OH)_2 + 6SiO_2 + 2KCl \qquad (4)$$

and if the minerals and fluid are in equilibrium:

$$3\mu_{KAlSi_3O_8,feldspar} + 2\mu_{HCl,fluid} =$$

$$\mu_{KAl_3Si_3O_{10}(OH)_2,muscovite} + 6\mu_{SiO_2,quartz} + 2\mu_{KCl,fluid} . \qquad (5)$$

Equation (5) would permit the evaluation of $\mu_{KCl,fluid} - \mu_{HCl,fluid}$ and, in turn, the ratio of the concentration of KCl to HCl in the fluid. If the disassociation constants of HCl and KCl are known at the P-T conditions of mineral-fluid equilibrium, the ratio of the concentration of KCl to HCl can be used to estimate the ratio of the concentration of the charged species K^+ to H^+.

In the following sections we explore in more detail the use of equilibrium relations among chemical potentials in characterizing the composition of metamorphic fluids, beginning with a consideration of qualitative constraints on fluid composition exerted by mineral equilibria. Next we consider the quantitative determination of metamorphic fluid composition from mineral assemblages and mineral equilibria.

QUALITATIVE CHARACTERIZATION OF METAMORPHIC FLUID COMPOSITION FROM MINERAL EQUILIBRIA

In the absence of precise thermochemical data, metamorphic fluid compositions can commonly be qualitatively characterized from a variety of diagrams on which one or both coordinates refers to fluid composition.

Isobaric, isothermal $\mu_1-\mu_2$ diagrams

The application of $\mu_1-\mu_2$ diagrams to petrology and geochemistry was pioneered by D.S. Korzhinskii (1959 and earlier publications) and has been more widely used in the Soviet Union than in the United States. The diagrams

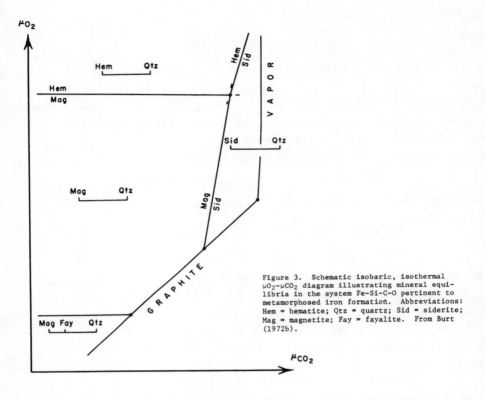

Figure 3. Schematic isobaric, isothermal μ_{O_2}–μ_{CO_2} diagram illustrating mineral equilibria in the system Fe-Si-C-O pertinent to metamorphosed iron formation. Abbreviations: Hem = hematite; Qtz = quartz; Sid = siderite; Mag = magnetite; Fay = fayalite. From Burt (1972b).

are easy to construct and are topologically related to the types of diagrams more commonly used by petrologists. Specifically, μ_{H_2O}–μ_{CO_2} diagrams generally are similar to T-X_{CO_2} or P-X_{CO_2} diagrams (Korzhinskii, 1959; Rose and Burt, 1979); μ_{O_2}–μ_{CO_2} diagrams are similar to $logf_{O_2}$-T diagrams (Burt, 1971a); and μ_{H_2O}–$\mu_{HK_{-1}}$ diagrams are similar to $loga_{K^+}/loga_{H^+}$ diagrams (Burt, 1976).

The principal advantage of isobaric, isothermal μ_1–μ_2 diagrams is that the slopes of isobaric isothermal univariant reaction boundaries are fixed by the stoichiometric coefficients of the reaction (which in turn are fixed solely by the compositions of the participating phases). These compositions can easily be measured, approximated from optical or X-ray data, or assumed. The topology of the complete diagram (intersections of the various univariant lines at invariant points and arrays of invariant points) can be determined most directly using thermochemical data. Although such data are often unavailable for the construction of μ–μ diagrams, natural mineral assemblages can be used instead to determine the topology of the diagrams. In doing so, the most conservative procedure is to look for mineral assemblages that *never* occur in nature and to assume that these are unstable in the simplified model system that is chosen. Simply using observed mineral assemblages can lead to results at variance with experimental work because solid solution of unaccounted-for components can stabilize mineral assemblages in nature that are unstable in the simplified model system.

An example of a qualitative isobaric, isothermal μ_{CO_2}-μ_{O_2} diagram is Figure 3 which refers to mineral equilibria in the system Fe-Si-C-O (e.g., metamorphosed iron formations). Straight lines correspond to assemblages in which a balanced stoichiometric relation can be written among compositions of coexisting minerals ($\pm O_2 \pm CO_2$). For example, along the hematite-siderite reaction boundary,

$$2Fe_2O_3 + 4CO_2 = 4FeCO_3 + O_2 \ . \tag{6}$$

If the minerals coexist in equilibrium with a fluid, the corresponding relation among chemical potentials is:

$$2\mu_{Fe_2O_3,hematite} + 4\mu_{CO_2,fluid} = 4\mu_{FeCO_3,siderite} + \mu_{O_2,fluid} \ . \tag{7}$$

The slope of the hematite-siderite boundary on the μ_{CO_2}-μ_{O_2} diagram can be derived by differentiating equation (7), noting that $d\mu_{Fe_2O_3} = d\mu_{FeCO_3} = 0$ because of the presence of hematite + siderite, and rearranging terms:

$$(\partial\mu_{O_2}/\partial\mu_{CO_2})_{P,T,siderite+hematite} = 4 \ . \tag{8}$$

In the general case, the slope of an isobaric, isothermal univariant reaction on a μ_1-μ_2 diagram is:

$$(\partial\mu_2/\partial\mu_1)_{P,T,equilibrium} = -\nu_1/\nu_2 \ , \tag{9}$$

where ν_1 and ν_2 are the stoichiometric coefficients of species 1 and 2 in stoichiometric relations such as (6). Values of ν_i are positive for products and negative for reactants.

Qualitative μ_1-μ_2 diagrams can be used to make qualitative statements about the nature of metamorphic fluid that coexisted with mineral assemblages. For example, with reference to Figure 3, siderite + hematite records both higher μ_{O_2} *and* μ_{CO_2} than coexisting siderite + magnetite. Fluids in equilibrium with siderite + hematite during metamorphism contain higher concentrations of CO_2 and O_2 than fluids in equilibrium with siderite + magnetite at the same pressure and temperature.

Qualitative μ_1-μ_2 diagrams are topologically identical to qualitative isobaric, isothermal $logf_1$-$logf_2$ and $loga_1$-$loga_2$ diagrams because:

$$\mu_i^{P,T} = (\mu_i^o)^{1,T} + 2.303RTlogf_i \tag{10}$$

215

and

$$\mu_i^{P,T} = (\mu_i^o)^{P,T} + 2.303RT\log a_i .$$ (11)

Because the μ_i^o terms are constant at constant P and T, $\log f_1 - \log f_2$, $\log a_1 - \log a_2$, and $\mu_1 - \mu_2$ diagrams differ only by a constant scale factor. Slopes of univariant boundaries are identical on all three types of diagrams and may be determined by equation (9).

In Figure 3 (and in others which follow), univariant lines meet at invariant points. The geometrical arrangement of univariant curves about the invariant points is governed by strict rules which are referred to as *Schreinemakers analysis* (Zen, 1966a). Arrays of multiple invariant points are termed *multisystems* (Korzhinskii, 1959). Schreinemakers analysis and the topological properties of multisystems have been the subject of numerous studies (Korzhinskii, 1959; Zen, 1966a,b, 1967; Zen and Roseboom, 1972; Day, 1972; Guo, 1980; Roseboom and Zen, 1982). These studies treat the purely geometrical properties of multisystems which must exist regardless of any quantitative characterization of the thermodynamic properties of participating phases (e.g., volume, entropy, enthalpy, etc.). The use of purely geometrical considerations to draw multisystems arrays, however, is seldom practical or necessary. Usually one has additional information provided by natural mineral assemblages, molar volumes and compositions of minerals, or partial thermochemical or experimental data. The reader should consult references cited above for discussions of Schreinemakers analysis and multisystems which do the subject more justice than our brief remarks here.

Isothermal $P_s - \mu_i$ diagrams

A type of diagram closely related to $\mu - \mu$, $\log f - \log f$, and $\log a - \log a$ diagrams is the isothermal $P_s - \mu_i$ diagram (Marakushev, 1973; Zen, 1974; Burt, 1978). The only additional data needed to construct such a diagram are the $\Delta \bar{V}_s$ of the crystalline phases. Because molar volume data on minerals are almost always retrievable from measured densities and unit cell volumes, $P_s - \mu_i$ diagrams are in practice almost as easy to construct as $\mu - \mu$ diagrams. The slope of isothermally univariant boundaries on a $P_s - \mu_i$ is calculated using a relation analogous to equation (9):

$$(\partial P_s / \partial \mu_i)_{T,equilibrium} = -\nu_i / \Delta \bar{V}_s ,$$ (12)

where ν_i is the stoichiometric coefficient of species i in the reaction and $\Delta \bar{V}_s$ is the change in volume of the solids in a reaction relationship such as

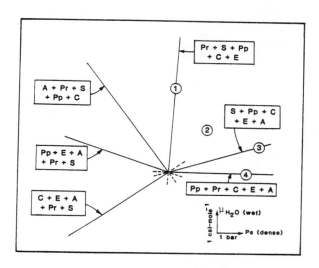

Figure 4. Schematic isothermal $P_s-\mu H_2O$ diagram illustrating mineral equilibrium pertinent to metamorphosed mafic igneous rocks in the prehnite-pumpellyite facies. All assemblages with K-mica + albite + quartz + sphene. Abbreviations: Pp = pumpellyite; Pr = prehnite; C = chlorite; A = actinolite; E = epidote; S = stilpnomelane. From Zen (1974).

(1) (expressed in appropriate units). A $P_s-(-\mu H_2O)$ diagram must be topologically the same as a P-T diagram (Burt, 1978).

Figure 4 is an example of a $P_s-\mu_i$ diagram in which i is H_2O. The diagram refers to mineral equilibria in mafic rocks metamorphosed under conditions of the prehnite-pumpellyite facies (Zen, 1974). Circled numbers refer to different assemblages collected at various localities in the northern Appalachians. Figure 4 permits characterization of differences in the composition of fluid that coexisted with minerals of the four assemblages. For example, μH_2O was less in assemblage 4 than in the other three assemblages.

Isobaric T-X$_i$ diagrams

If metamorphic fluids can be approximated by an i-k binary solution, then qualitative T-X$_i$ diagrams [X$_i$ = molar i/(i+k)] are useful in characterizing the composition of fluid that coexisted with minerals during metamorphism. The most common of this type of diagram is a T-X$_{CO_2}$ diagram where $X_{CO_2} = CO_2/(CO_2+H_2O)$. T-X$_{CO_2}$ diagrams were devised by Greenwood (1962, 1967). In his theoretical treatment Greenwood demonstrated that isobaric univariant curves have one of six possible shapes which solely depend on the way in which CO_2 and H_2O enter into balanced stoichiometric relations among mineral formulas such as (1). The six geometries are illustrated in Figure 5, and they are: (1) $\nu_{CO_2} = \nu_{H_2O} = 0$; (2) $\nu_{CO_2} > 0$, $\nu_{H_2O} = 0$; (3) $\nu_{CO_2} = 0$, $\nu_{H_2O} > 0$;

217

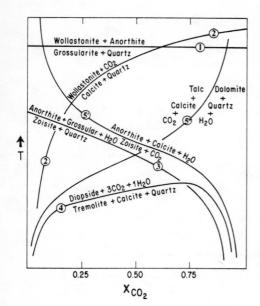

Figure 5. The six possible topologies of isobaric univariant mineral equilibria on an isobaric T-X_{CO_2} diagram. From Kerrick (1974).

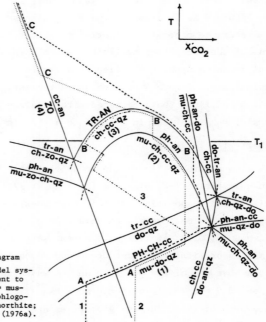

Figure 6. Qualitative isobaric T-X_{CO_2} diagram illustrating mineral equilibria in the model system $CaO-MgO-Al_2O_3-SiO_2-K_2O-H_2O-CO_2$ pertinent to metamorphosed marls. Abbreviations: mu = muscovite; do = dolomite; qz = quartz; ph = phlogopite; ch = chlorite; cc = calcite; an = anorthite; tr = tremolite; zo = zoisite. From Ferry (1976a).

218

(4) ν_{CO_2}, $\nu_{H_2O} > 0$; (5') $\nu_{CO_2} < 0$, $\nu_{H_2O} > 0$; and (5'') $\nu_{CO_2} > 0$, $\nu_{H_2O} < 0$. Qualitative $T-X_{CO_2}$ diagrams therefore may be constructed from (a) knowledge of the shapes of isobaric univariant curves based on the ν_{CO_2} and ν_{H_2O} of appropriate stoichiometric relations and (b) Schreinemakers analysis. Often simple observations about mineral stabilities based either on experimental studies or on mineral assemblages in metamorphic rocks are useful in constraining the topology of qualitative $T-X_{CO_2}$ diagrams.

Figure 6 is a qualitative $T-X_{CO_2}$ diagram which illustrates equilibria among minerals in metamorphosed marls and a CO_2-H_2O metamorphic fluid. Different mineral assemblages obviously act as useful qualitative records of differences in metamorphic fluid composition. For example, at temperature T_1, X_{CO_2} of coexisting fluid increases from the assemblage chlorite-calcite-quartz-tremolite-anorthite to muscovite-chlorite-calcite-phlogopite-anorthite-dolomite to chlorite-calcite-dolomite-tremolite-anorthite.

Exchange operators as components of the fluid

The components whose chemical potentials are used as variables on diagrams may range from volatiles (e.g., H_2O, CH_4, O_2, S_2, F_2), acid volatiles (CO_2, HF, HCl, H_2S, H_2SO_4), acid oxides (SO_2, P_2O_5, SiO_2), basic oxides (CaO, MgO, K_2O, Na_2O), soluble salts (NaCl, KCl, NaF, Na_2CO_3), to just about any component one might choose. It is sometimes convenient to choose components that have no physical significance as actual species in a fluid but whose use allows a reduction in the total number of components. For example, in problems involving the exchange of cations or anions between minerals and fluids, components called *exchange operators* may be used (Burt, 1974, 1979; see also Thompson, Chapter 1, this volume). As an example, muscovitization of K-feldspar involves the component HK_{-1} because the process may be represented by the reaction

$$3KAlSi_3O_8 + 2HK_{-1} = KAl_3Si_3O_{10}(OH)_2 + 6SiO_2 . \qquad (13)$$

Reaction (13) is represented by the "Ksp-Qtz-Mus" boundary in Figure 6 which is an isobaric, isothermal $\mu_1-\mu_2$ diagram in which one of the components is the exchange operator HK_{-1}. The slope of the "Ksp-Qtz-Mus" curve is derived from equation (9): $(\partial\mu_{HK_{-1}}/\partial\mu_{H_2O})_{P,T} = 0/(-2) = 0$. Use of the exchange operator permits reduction of three fluid components (e.g., HCl, KCl, H_2O) to two (HK_{-1}, H_2O) and hence representation of equilibrium relations on a two-dimensional diagram. Diagrams involving exchange operators do not contain information about the absolute concentration of species in fluid coexisting with minerals but information about the *difference* in concentration

219

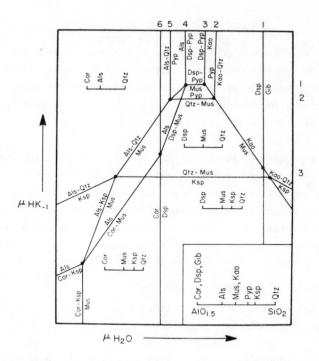

Figure 7. Schematic isobaric, isothermal $\mu H_2O-\mu HK_{-1}$ diagram illustrating mineral equilibria in the system $SiO_2-Al_2O_3-H_2O-HK_{-1}$ pertinent to the muscovitization of K-feldspar. Abbreviations: Cor = corundum; Als = Al_2SiO_5; Qtz = quartz; Dsp = diaspore; Gib = gibbsite; Mus = muscovite; Kao = kaolinite; Pyp = pyrophyllite; Ksp = K-feldspar. From Burt (1976).

between two species (or about a ratio of concentrations). Mineral equilibria in Figure 7, for example, may be used to make qualitative statements about the difference between the concentrations of HCl and KCl in fluid coexisting with mica and feldspar during metamorphism (or the difference between the concentrations of H_2O and K_2O or of H^+ and K^+, etc.).

Other processes to which exchange operators are pertinent include: dolomitization of limestone which involves $MgCa_{-1}$; fluorination of andalusite to topaz which involves F_2O_{-1}; and F-OH substitution in micas which involves $F(OH)_{-1}$. Other applications of exchange operators are given by Thompson et al. (1982) and in Chapter 1, this volume.

Exchange operators must be intrinsically acidic or basic in the Lewis sense because the different ions have intrinsically different attractions for electrons (Burt, 1974). We also note that an exchange operator such as F_2O_{-1} can be thought of as the "acid anhydride" of HF (that is, $F_2O_{-1} + H_2O$

220

= 2HF) and that it makes more sense to use F_2O_{-1} as a component in discussing
fluorine equilibria than to use F_2 and O_2 separately (Valley et al., 1982)
because doing so (1) eliminates a component in cases not involving oxidation
and (2) points out the analogy between reactions involving calcite + CO_2 and
those involving fluorite + F_2O_{-1} (Burt, 1972a). Thus the reaction analogous
to (1) is

$$CaF_2 + SiO_2 = CaSiO_3 + F_2O_{-1} . \tag{14}$$

Here "F_2O_{-1}" is being used as a convenient shorthand or "operator" for the
species actually present $[F_2O_{-1} = F_2 - \frac{1}{2}O_2 = \frac{1}{2}(SiF_4 - SiO_2) = 2HF - H_2O,$ etc.].
In general, we might not know what these species actually are. A familiar
analog is the use of O_2 as an "operator" to perform oxidation in reactions
in metamorphic rocks, even though its abundance as a species is negligible
(Eugster, 1959).

QUANTITATIVE CHARACTERIZATION OF METAMORPHIC FLUID COMPOSITION FROM MINERAL EQUILIBRIA

General statement

The quantitative characterization of metamorphic fluid composition is
developed directly from basic equilibrium relations among chemical potentials
of components in an equilibrium set of minerals coexisting with fluid. We
will begin by making the fundamental assumption that fluid pressure equals
lithostatic pressure during metamorphism. The thermodynamics of mineral-
fluid systems in which fluid pressure is less than lithostatic pressure are
more complicated and beyond the scope of this chapter. The strategy for determining fluid composition is presented through the specific example of
coexisting calcite, quartz, wollastonite, and fluid which defines the equilibrium relation (2) among chemical potentials. Because chemical potentials
are equivalent to the partial molar Gibbs energy of a component in a phase,
equation (2) may be rewritten:

$$\bar{G}^{P,T}_{CaCO_3,\text{calcite}} + \bar{G}^{P,T}_{SiO_2,\text{quartz}} = \bar{G}^{P,T}_{CaSiO_3,\text{wollastonite}} + \bar{G}^{P,T}_{CO_2,\text{fluid}} \tag{15}$$

which may be expanded to

$$\{(\bar{G}^o_{CaCO_3})^{P,T} + RT\ln a_{CaCO_3,\text{calcite}}\} + \{(\bar{G}^o_{SiO_2})^{P,T} + RT\ln a_{SiO_2,\text{quartz}}\} =$$
$$\{(\bar{G}^o_{CaSiO_3})^{P,T} + RT\ln a_{CaSiO_3,\text{wollastonite}}\} + \{(\bar{G}^o_{CO_2})^{P,T} + RT\ln a_{CO_2,\text{fluid}}\} \tag{16}$$

Because

$$(\bar{G}^o_i)^{P,T} = (\bar{G}^o_i)^{1,T} + \int_1^P \bar{V}^o_i dP \tag{17}$$

equation (16) may be rewritten (with some rearrangement) as:

$$\{(\bar{G}^o_{CaSiO_3})^{1,T} + (\bar{G}^o_{CO_2})^{1,T} - (\bar{G}^o_{CaCO_3})^{1,T} - (\bar{G}^o_{SiO_2})^{1,T}\} +$$

$$\int_1^P (\bar{V}^o_{CaSiO_3} - \bar{V}^o_{CaCO_3} - \bar{V}^o_{SiO_2})dP +$$

$$RT\ln\{a_{CaSiO_3,wollastonite}/(a_{CaCO_3,calcite})(a_{SiO_2,quartz})\} +$$

$$\int_1^P \bar{V}^o_{CO_2} dP + RT\ln a_{CO_2,fluid} = 0 . \qquad (18)$$

Using the standard Δ notation equation (18) becomes

$$(\Delta\bar{G}^o)^{1,T} + \int_1^P \Delta\bar{V}^o_s dP + RT\ln\{a_{CaSiO_3,wollastonite}/(a_{CaCO_3,calcite})(a_{SiO_2,quartz})\} +$$

$$\int_1^P \bar{V}^o_{CO_2} dP + RT\ln a_{CO_2,fluid} = 0 . \qquad (19)$$

Because minerals are relatively incompressible, $\Delta\bar{V}^o_s$ may be approximated as a constant independent of P and T within the range of metamorphic pressures and temperatures. The term $a_{CO_2,fluid}$ may be expanded to $\lambda_{CO_2}X_{CO_2,fluid}$. Equation (19) then becomes

$$(\Delta\bar{G}^o)^{1,T} + \Delta\bar{V}^o_s(P-1) + RT\ln\{a_{CaSiO_3,wollastonite}/(a_{CaCO_3,calcite})(a_{SiO_2,quartz})\} +$$

$$\int_1^P \bar{V}^o_{CO_2} dP + RT\ln\lambda_{CO_2}X_{CO_2,fluid} = 0 . \qquad (20)$$

The mole fraction of CO_2 in metamorphic fluid at metamorphic pressure P and temperature T can be calculated from equation (20) provided all terms other than $X_{CO_2,fluid}$ are known or estimated.

In general, for a stoichiometric relation among j components in minerals (ν_j stoichiometric coefficient on component j) and i components in fluid (ν_i stoichiometric coefficient on component i), the relation for mineral-fluid equilibria like (20) is:

$$(\Delta\bar{G}^o)^{1,T} + \Delta\bar{V}^o_s P + \sum_j RT\nu_j \ln a_j + \sum_i (\nu_i \int_1^P \bar{V}_i dP) + \sum_i RT\nu_i \ln a_{i,fluid} = 0 \qquad (21)$$

($P \sim P-1$ for metamorphic pressures). Replacing the a_i,

$$(\Delta\bar{G}^o)^{1,T} + \Delta\bar{V}^o_s P + \sum_j RT\nu_j \ln a_j + \sum_i (\nu_i \int_1^P \bar{V}^o_i dP) + \sum_i RT\nu_i \ln(\lambda_i X_{i,fluid}) = 0 \qquad (22)$$

Equation (22) highlights the information required to quantitatively evaluate the concentrations of components in metamorphic fluids from mineral equilibria: (a) temperature and pressure at the time of mineral-fluid equilibrium; (b) thermochemical data for the pure components ($\Delta\bar{G}^o$, $\Delta\bar{V}^o_s$); (c) equations of state for the participating mineral phases (activity-composition relations);

222

and (d) an equation of state for the fluid phase (a model for evaluating $\int_1^P \bar{V}_i^o dP$ and activity-composition relations). In the following sections we discuss how each of the kinds of required information may be obtained.

Pressure and temperature

Accurate estimates of pressure and temperature are needed in order to estimate metamorphic fluid composition. Chapter 5 of this volume is devoted to techniques which petrologists may use to estimate metamorphic pressures and temperatures from mineral equilibria, and the subject therefore is not repeated here.

Thermochemical data

Thermochemical data are required to evaluate $(\Delta \bar{G}^o)^{1,T}$ and $\Delta \bar{V}_s^o$ in equation (22). The molar volumes of end-member components of minerals are quite well characterized and do not significantly vary over the range of P-T conditions normally associated with metamorphism. The value of $\Delta \bar{V}_s^o$ therefore may be calculated from molar volume data tabulated for conditions of 25°C, 1 atm (Robie et al., 1967, 1978; Helgeson et al., 1978).

The term $(\Delta \bar{G}^o)^{1,T}$ in equation (22) is commonly expanded to:

$$(\Delta \bar{G}^o)^{1,T} = (\Delta \bar{H}^o)^{1,T} - T(\Delta \bar{S}^o)^{1,T} \; ; \tag{23}$$

$$(\Delta \bar{G}^o)^{1,T} = (\Delta \bar{H}^o)^{1,298} + \int_{298}^T (\Delta c_p^o) dT - T(\Delta \bar{S}^o)^{1,298} - T \int_{298}^T (\Delta c_p^o/T) dT \; . \tag{24}$$

Estimation of $(\Delta \bar{G}^o)^{1,T}$ may involve various degrees of simplification depending on the accuracy required in a calculation of metamorphic fluid composition. Three methods of determining $(\Delta \bar{G}^o)^{1,T}$ are commonly used. (1) Values of $(\Delta \bar{H}^o)^{1,T}$ and $(\Delta \bar{S}^o)^{1,T}$ are extracted from results of hydrothermal phase equilibrium experiments. The method was pioneered by Orville and Greenwood (1965) and has been applied in numerous studies (e.g., Skippen, 1971, 1974; Chatterjee and Johannes, 1974; Ferry, 1976b, 1980b; Hoy, 1976; Crawford et al., 1979; Ghent et al., 1979). The method assumes that $\Delta \bar{H}^o$ and $\Delta \bar{S}^o$ do not change substantially with temperature (i.e., $\Delta c_p^o \sim 0$) and hence works best in applications where metamorphic conditions are similar to those of the phase equilibrium experiments used to derive $\Delta \bar{H}^o$ and $\Delta \bar{S}^o$. (2) Values of $(\Delta \bar{H}^o)^{1,T}$ and $(\Delta \bar{S}^o)^{1,T}$ are derived from tabulated values of $(\bar{H}^o)^{1,298}$, $(\bar{S}^o)^{1,298}$, and heat capacity. The usual sources of the tabulated values are either Helgeson et al. (1978) or Robie et al. (1978). (3) Values of $(\Delta \bar{S}^o)^{1,T}$ are derived from tabulated values of $(\bar{S}^o)^{1,298}$ and heat capacity; $(\Delta \bar{H}^o)^{1,T}$ values are extracted from results of hydrothermal phase equilibrium experiments. The method was developed by D.M. Kerrick and co-workers (Slaughter et al., 1975; Kerrick and Slaughter, 1976; Jacobs and Kerrick, 1981) and is probably the most accurate of the three methods for determining $(\Delta \bar{H}^o)^{1,T}$.

Equations of state for participating mineral phases

Equations of state (activity-composition relations) for minerals are required for evaluation of the term $RT(v_j \mathit{l}na_j)$ in equation (22) and hence for evaluation of metamorphic fluid composition. Activity-composition relations for mineral solid solutions are considered in the appendix to Chapter 5 by Essene and are not reconsidered here.

Equation of state for the fluid phase

An equation of state for the fluid phase is required to evaluate the terms $\int_1^P \bar{V}_i^o dP$ and λ_i in equation (22) and hence metamorphic fluid composition. We examine a number of models of metamorphic fluid which have been or are now widely used by petrologists for this purpose. The models involve various degrees of simplification; which particular model is chosen depends on the accuracy required in a calculation of metamorphic fluid composition.

Bulk fluids versus intergranular films. The fluid phase occupies pores, fractures, and grain boundaries in rocks during metamorphism. The sizes of these features may be on the order of 100Å in their shortest dimension, judging from direct observations of microcracks in crystalline rocks (Batzle and Simmons, 1976; Padovani et al., 1978). Consequently, the fluid probably occurs as a thin film. It is appropriate to question whether the thermodynamic properties of metamorphic fluids are the same as those of bulk fluids or whether, as films, they are "anomalous." The interest in anomalous properties of a fluid occupying small cavities in part stems from the controversy over "polywater" in the mid-1960's. Some investigators measured anomalously low partial pressures of H_2O over water held in thin capillaries. The low partial pressures of H_2O were initially interpreted as the formation of a new form of H_2O with order intermediate between that for liquid and solid. The formation of this new form of H_2O (polywater) was supposed to be induced by its constriction to a small cavity. The low partial pressures were later shown to be simply the result of solution of material in the H_2O occupying the capillaries. Although hard data are still scant, it appears that the properties of H_2O in small cavities may not be greatly different from those of bulk H_2O. For example, Churayev et al. (1970) showed that the viscosity of H_2O in capillaries with diameters of 500Å is nearly the same as that of bulk H_2O at temperatures above 70°C. The diffusivity of H_2O in films of $H_2O \geq 100$Å thick on clays is not much less than in bulk H_2O (Olejnik and White, 1972). It appears that treatment of metamorphic fluid as a bulk fluid may not be inappropriate. In any case, the thermodynamic properties of thin aqueous films are so poorly characterized that petrologists are left with little alternative; we now, therefore, discuss several models for bulk fluids.

Ideal C-O-H-S-N fluids which are pure substances. With reference to equation (22), we will begin by first considering fluids which are pure substances (e.g., pure CO_2, pure H_2O, etc.). For these, $\lambda_i = X_{i,fluid} = 1$ and $\sum_i RT\nu_i ln(\lambda_i X_{i,fluid}) = 0$; we need only consider the integral. At low pressures, the behavior of many fluids may be adequately described by the ideal gas law: $P\bar{V}_i^o = RT$. Substituting RT/P for $\bar{V}_i^o$ we obtain:

$$\int_1^P (\partial \bar{G}_i^o / \partial P)_T dP = \int_1^P \bar{V}_i^o dP = \int_1^P (RT/P) dP = \int_1^P RT d ln P = RT ln P - RT ln(1) = RT ln P .$$
$$(25)$$

Equation (25) is referred to as the *ideal gas model.* Equation (22) now becomes (for ideal gases which are pure substances):

$$(\Delta \bar{G}^o)^{1,T} + \Delta \bar{V}_s^o P + \sum_j \nu_j RT ln a_j + \nu_i RT ln P = 0 .$$
$$(26)$$

Real C-O-H-S-N fluids which are pure substances. At the elevated pressures of metamorphism, equation (25) greatly oversimplifies the behavior of fluids. To adequately represent *real fluids* at metamorphic conditions, we introduce the concept of fugacity. The fugacity of a pure fluid at pressure P and temperature T, $(f_i^o)^{P,T}$, is a measurable quantity defined by the equation:

$$\int_1^P (\partial \bar{G}_i^o / \partial P)_T dP = \int_1^P \bar{V}_i^o dP = \int_1^P RT d ln f_i^o = RT ln(f_i^o)^{P,T} - RT ln(f_i^o)^{1,T} .$$
$$(27)$$

Note the similarity between the role of f_i^o in equation (27) and of P in equation (25). In fact, fugacity is defined by equation (27) precisely in order to preserve the formalism of the ideal gas model as expressed by equation (25). At low pressures (e.g., 1 bar), the behavior of real fluids (gases) is sufficiently ideal that $(f_i^o)^{1,T} = P = 1$; at higher pressures, $(f_i^o)^{P,T} \neq P$. Consequently, equation (27) reduces to

$$\int_1^P \bar{V}_i^o dP = RT ln(f_i^o)^{P,T} .$$
$$(28)$$

The relation between the concepts of an ideal fluid and of a real fluid may be further explored. At a specified pressure and temperature, consider the difference between the molar Gibbs energy of a real fluid and a hypothetical ideal fluid of the same composition:

$$(\bar{G}^o_{real})^{P,T} - (\bar{G}^o_{ideal})^{P,T} = \int_1^P RTd\ln f^o_i - \int_1^P RT\ln P =$$
$$RT\{\ln(f^o_i)^{P,T} - \ln(f^o_i)^{1,T} - \ln P + \ln(1)\} \ . \tag{29}$$

A pressure of 1 bar is sufficiently low that real fluids may be approximated by the ideal gas model (i.e., $(f^o_i)^{1,T} = P = 1$) and

$$(\bar{G}^o_{real})^{P,T} - (\bar{G}^o_{ideal})^{P,T} = RT\{\ln(f^o_i)^{P,T} - \ln P\} = RT\ln\{(f^o_i)^{P,T}/P\} \ . \tag{30}$$

The ratio $(f^o_i)^{P,T}/P$ therefore is of fundamental importance because it serves as a measure of the departure from ideality that real gases display at elevated P-T conditions. The ratio is referred to as the *fugacity coefficient*, γ^o_i:

$$\gamma^o_i = (f^o_i)^{P,T}/P \ . \tag{31}$$

Fugacity coefficients are species-specific and functions of pressure and temperature. In practice, the fugacity of species i at P and T is calculated from tabulations of fugacity coefficients as a function of pressure and temperature: $(f^o_i)^{P,T} = \gamma^o_i P$. Equation (22) now becomes (for real fluids which are pure substances):

$$(\Delta\bar{G}^o)^{1,T} + \Delta\bar{V}^o_s P + \sum_j \nu_j RT\ln a_j + \nu_i RT\ln\gamma^o_i P = 0 \ . \tag{32}$$

Tabulations of γ^o_i may be derived from P-V-T relations because of the relationship between fugacity and the molar volume of fluids at P and T:

$$(\partial\ln f^o_i/\partial P)_T = \bar{V}^o_i/RT \ . \tag{33}$$

Equation (31) then allows calculation of fugacity coefficients from fugacities derived by fitting equation (33) to P-V-T data. P-V-T relations have been experimentally determined for only a few fluid species (H_2: Presnall, 1969; H_2O: Burnham et al., 1969; CO_2: Burnham and Wall, pers. comm.; Shmonov and Shmulovich, 1974; N_2: Malbrunot and Vodar, 1973). Consequently, fugacity coefficients which are based on direct experiment are available only for H_2, H_2O, and CO_2 (see above references as well as Shaw and Wones, 1964). Fortunately, fugacity coefficients may be quite accurately estimated by the method of reduced variables. The method of reduced variables is based on the empirical observation that many fluid species have the same equation of state when pressure, temperature, and volume are expressed as the reduced variables P/P_c, T/T_c, and V/V_c, where P_c, T_c, and V_c correspond to the critical pressure, temperature, and volume of the fluid species, respectively. Ryzhenko and

Volkov (1971) used the method of reduced variables to calculate γ_i^o for H_2O, CO_2, N_2, H_2, O_2, H_2S, CO, and CH_4 over essentially the entire range of metamorphic temperatures and pressures. Their results compare favorably to values of γ_i^o determined from direct experiment on H_2, H_2O, and CO_2 fluids.

Ideal mixing in ideal C-O-H-S-N fluids. When species mix ideally in a solution, $\bar{V}_i = \bar{V}_i^o$ and $\lambda_i = 1$. Consequently, for equilibrium between minerals and an ideal fluid which displays ideal mixing, equation (22) becomes

$$(\Delta\bar{G}^o)^{1,T} + \Delta\bar{V}_s^o P + \sum_j \nu_j RT\ln a_j + \sum_i \nu_i RT\ln P + \sum_i \nu_i RT\ln X_{i,fluid} = 0 \qquad (34)$$

or

$$(\Delta\bar{G}^o)^{1,T} + \Delta\bar{V}_s^o P + \sum_j \nu_j RT\ln a_j + \sum_i \nu_i RT\ln(PX_{i,fluid}) = 0 \qquad (35)$$

or

$$(\Delta\bar{G}^o)^{1,T} + \Delta\bar{V}_s^o P + \sum_j \nu_j RT\ln a_j + \sum_i \nu_i RT\ln P_{i,fluid} = 0 \ , \qquad (36)$$

where $P_{i,fluid}$ is the partial pressure of component i in the fluid solution. Equation (36) was used by French (1966) to calculate equilibria between minerals and a C-O-H fluid, but is now considered unsatisfactory for these purposes.

Ideal mixing in real C-O-H-S-N fluids. If species in a real fluid mix ideally then equation (22) becomes

$$(\Delta\bar{G}^o)^{1,T} + \Delta\bar{V}_s^o P + \sum_j \nu_j RT\ln a_j + \sum_i \nu_i RT\ln f_i^o + \sum_i \nu_i RT\ln X_{i,fluid} = 0 \qquad (37)$$

or

$$(\Delta\bar{G}^o)^{1,T} + \Delta\bar{V}_s^o P + \sum_j \nu_j RT\ln a_j + \sum_i \nu_i RT\ln(f_i^o X_{i,fluid}) = 0 \ . \qquad (38)$$

Equation (38) is often written as

$$(\Delta\bar{G}^o)^{1,T} + \Delta\bar{V}_s^o P + \sum_j \nu_j RT\ln a_j + \sum_i \nu_i RT\ln f_i = 0 \ , \qquad (39)$$

where

$$f_i = f_i^o(X_{i,fluid}) = \gamma_i^o P(X_{i,fluid}) = \gamma_i^o P_{i,fluid} \ , \qquad (40)$$

and f_i is referred to as the fugacity of component i in a fluid solution. Equation (40) is the Lewis and Randall rule (Lewis and Randall, 1961).

Ideal mixing in a real fluid solution (equation 40) has served as an adequate model for C-O-H-S-N fluids in experimental studies of mineral-fluid equilibria (e.g., Greenwood, 1967; Skippen, 1971, 1974; Hewitt, 1973b, 1975; Gordon and Greenwood, 1970), in theoretical studies (e.g., Ohmoto and Kerrick, 1977), and in studies which characterized the composition of natural metamorphic fluids from mineral equilibria in metamorphic rocks (e.g., Ghent, 1975;

Ghent et al., 1979; Ferry, 1976b, 1981; Hoy, 1976; Valley and Essene, 1980).

Non-ideal mixing in real C-O-H-S-N fluids. Limited P-V-T data for CO_2-H_2O mixtures at 450°C $\leq$ T $\leq$ 800°C and P = 0-500 bars indicate that CO_2 and H_2O mix non-ideally at least below $\sim$500°C, i.e., $\lambda_i \neq 1$ in equation (22) (Greenwood, 1973). Ryzhenko and Malinin (1971) have also analyzed P-V-T data for CO_2-CH_4, CO_2-N_2, and CO_2-H_2 mixtures at low pressures (P < 700 bars) and low temperatures (40° $\leq$ T < 238°C) and found values of λ_i which range from $\sim$1 to $\sim$2 depending on temperature, pressure, and composition. There exists too little P-V-T data on C-O-H-S mixtures for any comprehensive non-ideal mixing models to be developed solely from experimental data. Fortunately, a Redlich-Kwong type equation of state appears to be a satisfactory semi-empirical model which can be formulated from existing P-V-T data for pure fluid species H_2O, CO_2, H_2, N_2, etc. (Holloway, 1977, 1981; Flowers, 1979). The Redlich-Kwong equation can be used to model non-ideal mixing in real C-O-H-S-N fluids. We begin with the van der Waals equation (which is simply a modification of the ideal gas law, $P = RT/\bar{V}$):

$$P = RT/(\bar{V}-b) - a/\bar{V}^2 , \qquad (41)$$

where "a" is a constant which measures the cohesion among molecules in a fluid and "b" is a measure of the volume of the fluid occupied by the molecules themselves. The Redlich-Kwong equation modifies equation (41), in turn, by:

$$P = RT/(\bar{V}-b) - a/\{(\bar{V}^2+b\bar{V})\sqrt{T}\}, \qquad (42)$$

where the constants "a" and "b" have the same significance as in the van der Waals equation (41). Because of the meaning of "b," it is taken as a species-specific constant which is independent of pressure and temperature. The "a" parameter is species-specific and dependent on temperature in most formulations. Kerrick and Jacobs (1981) present a Redlich-Kwong equation of state which utilizes an "a" parameter dependent on both pressure and temperature.

The adjustable parameters "a" and "b" are derived by fitting experimentally determined P-V-T data to equation (42). Holloway (1977) gives values of "b" and temperature-dependent expressions of "a" for CO_2 and H_2O. Values of "b" and expressions of a(T) for CO_2, H_2O, CO, CH_4, H_2, H_2S, SO_2, and N_2 may be retrieved from the listing of a computer program in Holloway (1981). Because of equation (28), fugacity coefficients for species in a pure fluid of i may be derived from an equation of state for i of the form of equation (42).

The procedure involves solving equation (42) for $\bar{V}$ and integrating it with respect to P at constant T (*cf.* equation 28). Holloway (1977) presents an equation from which fugacity coefficients for species i in a fluid of composition pure i may be directly calculated at T and P from the values of "a" and "b" in equation (42). Fugacity coefficients calculated by the Redlich-Kwong equation for H_2, H_2O, and CO_2 are in good agreement with fugacity coefficients derived from experimentally-determined P-V-T relations (Holloway, 1977).

The power of the Redlich-Kwong equation is that it can be modified to calculate the fugacity of components in a *mixture* of molecular species. Equation (42) is used as an equation of state for the fluid mixture. Values of "a" and "b" for the mixture are derived from "a" and "b" values for the individual species according to mixing rules summarized by Holloway (1977) and Flowers (1979). The fugacity of component i in the fluid mixture, f_i, can then be calculated from the modified P-V-T equation of state for the mixture (Holloway, 1977, lists the appropriate equation). Calculation of f_i in turn allows the non-ideality of the fluid solution to be characterized by the activity coefficient, λ_i, because of the following convention:

$$RT\ell n f_i = \int_1^P \bar{V}_i^o dP + RT\ell n a_{i,fluid} = RT\ell n f_i^o + RT\ell n \lambda_i X_{i,fluid} = RT\ell n (\gamma_i^o P \lambda_i X_{i,fluid})$$
(43)

or

$$f_i = \gamma_i^o P \lambda_i X_{i,fluid}$$
(44)

which is the expression, analogous to equation (40), for real fluids which display *non-ideal* mixing. Notice that equation (44) is equivalent to stating

$$(f_i/f_i^o)^{P,T} = a_{i,fluid} = \lambda_i X_{i,fluid}$$
(45)

Thus the modified Redlich-Kwong equation of state for fluid mixtures allows a semi-empirical calculation of λ_i for a wide range of geologically-important C-O-H-S-N fluid species over the entire range of metamorphic P-T conditions. The calculated values of λ_i represent the best available way to account quantitatively for the non-ideal mixing of C-O-H-S-N species in metamorphic fluids. Equation (22) now becomes (for non-ideal mixing of species in a real fluid):

$$(\Delta\bar{G}^o)^{1,T} + \Delta\bar{V}_s^o P + \sum_j \nu_j RT\ell n a_j + \sum_i \nu_i RT\ell n f_i^o + \sum_i \nu_i RT\ell n (\lambda_i X_{i,fluid}) = 0 \quad (46)$$

or

$$(\Delta\bar{G}^o)^{1,T} + \Delta\bar{V}_s^o P + \sum_j \nu_j RT\ell n a_j + \sum_i \nu_i RT\ell n (\gamma_i^o P \lambda_i X_{i,fluid}) = 0 \ . \quad (47)$$

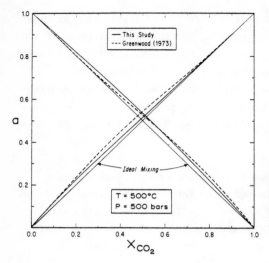

Figure 8. Activity-composition relations for CO_2-H_2O fluids at 500°C, 500 bars. Light solid line: ideal mixing; heavy solid line: calculated from a modified Redlich-Kwong equation of state; dashed line: experimental data of Greenwood (1973). From Kerrick and Jacobs (1981).

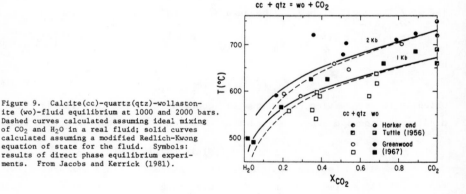

Figure 9. Calcite(cc)-quartz(qtz)-wollastonite (wo)-fluid equilibrium at 1000 and 2000 bars. Dashed curves calculated assuming ideal mixing of CO_2 and H_2O in a real fluid; solid curves calculated assuming a modified Redlich-Kwong equation of state for the fluid. Symbols: results of direct phase equilibrium experiments. From Jacobs and Kerrick (1981).

One common way to display non-ideal mixing of binary solutions is with a-X plots. Figure 8 is an example which illustrates the degree of non-ideal mixing for CO_2-H_2O mixtures at 500°C. Activity-composition relations calculated by the Redlich-Kwong equation agree well with those derived from direct P-V-T experiments on CO_2-H_2O mixtures by Greenwood (1973). In general, non-ideality (departure of λ_i from 1) increases with increasing pressure and decreasing temperature for a particular mixture. Molecules with similar properties (e.g., CO_2 and N_2) mix nearly ideally over a wide range of P-T-X conditions. Molecules with different properties (e.g., H_2O and CH_4) form mixtures that can depart significantly from ideality (Holloway, 1981).

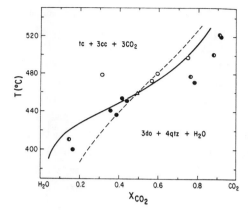

Figure 10. Dolomite(do)-quartz(qtz)-talc(tc)-calcite(cc)-fluid equilibrium at 2000 bars. Dashed curve calculated assuming ideal mixing of CO_2 and H_2O in a real fluid; solid curve calculated assuming a modified Redlich-Kwong equation of state for the fluid. Symbols: results of direct phase equilibrium experiments. From Jacobs and Kerrick (1981).

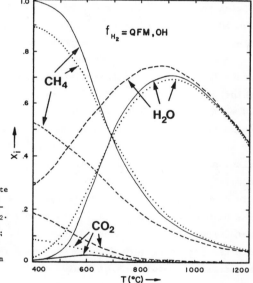

Figure 11. Composition of C-O-H fluid in equilibrium with quartz + fayalite + graphite + magnetite as a function of temperature at 2000 bars. The only species present in significant concentration are H_2O, CH_4, and CO_2. Curves calculated assuming (a) ideal mixing of species in an ideal C-O-H fluid (dashed); (b) ideal mixing of species in a real C-O-H fluid (dotted); and (c) a modified Redlich-Kwong equation of state for the fluid. From Holloway (1977).

The Redlich-Kwong-type equation of state has served as a model for CO_2-H_2O mixtures in recent experimental studies (e.g., Jacobs and Kerrick, 1981). Comparison of fluid composition calculated using an ideal-mixing equation of state versus a Redlich-Kwong non-ideal equation of state may be made by returning to the example of quartz, calcite, and wollastonite coexisting with CO_2-H_2O fluid. The equilibrium is represented on an isobaric T-X_{CO_2} diagram in Figure 9. Figure 9 illustrates the sort of differences in fluid composition which will result depending on whether an assumption of non-ideal mixing or ideal mixing for the fluid is adopted. Differences in inferred fluid composition are $\lesssim \sim 0.05X_{CO_2}$. In general, for mineral equilibria represented by reactions which produce CO_2, H_2O, or $CO_2 + H_2O$, the differences in calculated fluid composition resulting from the choice of mixing model for the fluid are often $\lesssim \sim 0.05X_{CO_2}$ for CO_2-H_2O fluids (Jacobs and Kerrick, 1981). For mineral equilibria represented by reactions which consume H_2O and evolve CO_2 (or vice versa), however, the differences are much greater (Fig. 10). For calculations of fluid composition based on these latter sorts of equilibria, an equation of state for the fluid which accounts for non-ideal mixing of species appears to be much more important. Figure 11 illustrates the composition of C-O-H fluid in equilibrium with quartz, fayalite, magnetite, and graphite at 2000 bars pressure and T = 400°-1200°C assuming three different equations of state for the fluid (ideal mixing in an ideal gas, ideal mixing in a real fluid, modified Redlich-Kwong). Results indicate that while the ideal gas model is not entirely satisfactory, ideal mixing in a real fluid may be a satisfactory approximation for many applications.

The most serious weakness of the modified Redlich-Kwong equation of state is its inability to satisfactorily model fluids which are mixtures of chlorides, fluorides, and C-O-H-S-N species. The presence of up to at least 10 wt % NaCl in H_2O-CO_2 fluid does not apparently significantly affect calculation of the proportions of H_2O and CO_2 in fluid from mineral equilibria by methods described above (Jacobs and Kerrick, 1981). Studies of fluid inclusions in metamorphic rocks, however, suggest that metamorphic fluids may commonly be quite concentrated aqueous brines with over 25 wt % equivalent dissolved NaCl (Rich, 1979; Crawford et al., 1979; Sisson et al., 1982). The brines may be immiscible with nearly pure CO_2 fluid over a wide range of metamorphic P-T conditions (Sisson et al., 1982). An equation of state for fluids containing both C-O-H-S species and halogen complexes must be developed before the compositions of fluids observed by Rich, Crawford et al., and Sisson et al. may be accurately characterized from mineral equilibria.

Aqueous electrolyte solutions. Up to this point equations of state have been described only for fluids composed of C-O-H-S-N species. Although these species comprise most of common metamorphic fluids on a molar basis, studies of fluid inclusions indicate the presence of other species. These species are principally associated and disassociated chlorides (of which the most abundant apparently is NaCl). Mineral equilibria can be used to estimate the concentration of non-C-O-H-S-N species in metamorphic fluid, and it is useful, therefore, to at least briefly explore the thermodynamic treatment of electrolyte solutions. As will be seen later, the calculation of these non-C-O-H-S-N species in solution is hampered by a number of practical problems. In addition, the thermodynamics of electrolyte solutions are more complicated than for C-O-H-S-N fluids. For these reasons, electrolyte solutions are discussed separately from C-O-H-S-N fluids.

The way in which aqueous electrolytes have been treated thermodynamically may be illustrated by a particular mineral-fluid equilibrium:

$$KAl_3Si_3O_{10}(OH)_2 + 6SiO_2 + 2K^+ = 3KAlSi_3O_8 + 2H^+ \ . \qquad (48)$$

Equation (48) indicates that muscovite, K-feldspar, and quartz constrain the concentration of K^+ and H^+ in any coexisting electrolyte solution. Looking at it another way, the muscovite-feldspar-quartz assemblage records information about the composition of any coexisting fluid. The mineral-fluid equilibrium may be treated in a fashion analogous to that for calcite-quartz-wollastonite-fluid (equations 15-20). If the minerals are pure substances (all $a_j = 1$), the expression analogous to equation (20) for relation (48) is:

$$(\Delta\bar{G}^o)^{1,T} + \int_1^P \Delta\bar{v}_s^o dP + 2\int_1^P (\bar{v}_{H^+}^o - \bar{v}_{K^+}^o) dP + 2RT ln(a_{H^+,fluid}/a_{K^+,fluid}) = 0 \ . \qquad (49)$$

The standard state for aqueous species in equation (49) (and others like it which follow) is one of unit activity of the species in a hypothetical one molal solution. The simplest treatment of equilibria involving minerals and aqueous electrolytes is one adopted by Eugster (1970). He assumed

$$\int_1^P (\bar{v}_{H^+}^o - \bar{v}_{K^+}^o) dP = 0 \qquad (50)$$

and used equation (49) with tabulated thermochemical data, $(\Delta\bar{G}^o)^{1,T}$ and $\Delta\bar{v}_s^o$, to calculate the ratio $(a_{H^+,fluid}/a_{K^+,fluid})$ as a function of P and T for fluid coexisting with K-feldspar, muscovite, and quartz. The integral in equation (49) can be several kilocalories (Helgeson and Kirkham, 1974a,b, 1976), and Eugster's model of electrolytes probably should not be used to characterize metamorphic fluid composition except at very low pressures (e.g., P < 500 bars).

A major difficulty in efforts to model electrolyte solution quantitatively at elevated pressures is the evaluation of integrals such as the second in equation (49). A second approach to the problem is that used by Frantz and his co-workers (Frantz et al., 1981; Boctor et al., 1980; Popp and Frantz, 1979; Frantz and Popp, 1979). Their work can be best described by first expanding $(\Delta\bar{G}^o)^{1,T}$ and rearranging equation (49):

$$\{(\Delta \bar{H}^o)^{1,T} + \int_1^P \Delta \bar{V}_s^o dP + 2\int_1^P (\bar{V}_{H^+}^o - \bar{V}_{K^+}^o) dP\} -$$

$$T(\Delta \bar{S}^o)^{1,T} + 2RT\ell n(a_{H^+,fluid}/a_{K^+,fluid}) = 0 \ . \qquad (51)$$

At constant pressure and with the assumption $\Delta c_p^o \sim 0$, the term inside the first set of brackets is a constant. Equation (51), therefore, may be re-written (with the assumption $\Delta c_p^o = 0$) as:

$$y - T\Delta \bar{S}^o + 2RT\ell n(a_{H^+,fluid}/a_{K^+,fluid}) = 0 \ . \qquad (52)$$

Frantz and his co-workers further assumed ideal mixing of species in the electrolyte solution (all $\lambda_i = 1$) so that equation (52) becomes:

$$y - T\Delta \bar{S}^o + 2RT\ell n(m_{H^+,fluid}/m_{K^+,fluid}) = 0 \qquad (53)$$

and rearranging,

$$y/(2.303RT) - \Delta \bar{S}^o/(2.303R) + 2\log(m_{H^+,fluid}/m_{K^+,fluid}) = 0 \ ; \qquad (54)$$

$$A + B/T = 2\log(m_{H^+,fluid}/m_{K^+,fluid}) \ . \qquad (55)$$

For equilibrium in which minerals are solid solutions equation (55) would be modified according to:

$$A + B/T = 2\log(m_{H^+,fluid}/m_{K^+,fluid}) +$$
$$\log\{(a^3_{KAlSi_3O_8,feldspar})/(a_{KAl_3Si_3O_{10}(OH)_2,muscovite})(a^6_{SiO_2,quartz})\} \qquad (56)$$

In the general case of mineral-fluid equilibrium,

$$A + B/T = \sum_i \nu_i \log m_i + \sum_j \nu_j \log a_j \qquad (57)$$

where the electrolyte solution is composed of components i, the minerals of components j, and the ν are stoichiometric coefficients of the components in equations like (48).

Numerical values for A and B may be determined from sets of isobaric high-pressure mineral solubility experiments. Frantz et al. (1981) tabulate values of A and B for 69 mineral-fluid equilibria. Their tabulated constants may be used to calculate ratios of concentrations of species in electrolytes through equation (57). Because the B constant in equation (57) is pressure-specific, calculations can be made only at 1000 and 2000 bars, the only pressures for which Frantz et al. (1981) have tabulated values of A and B.

The limitation of Frantz et al.'s treatment of electrolyte solution is the restriction of calculations to 1000 and 2000 bars pressure and the assumption of ideal mixing. This limitation has been overcome in theoretical treatments of mineral-fluid equilibria by Helgeson and his co-workers (Helgeson et al., 1981; Helgeson and Kirkham, 1974a,b, 1976). Their approach can best be described by the analog to equations (22)-(24) for equilibria involving components i in an electrolyte solution and components j in coexisting minerals:

$$(\Delta \bar{H}^o)^{1,298} + \int_{298}^{T}(\Delta c_p^o)dT - T(\Delta \bar{S}^o)^{1,298} - T\int_{298}^{T}(\Delta c_p^o/T)dT + \Delta \bar{V}_s^o(P-1) +$$

$$\int_{1}^{P}(\sum_i \nu_i \bar{V}_i^o)dP + \sum_j \nu_j RT \ln a_j + \sum_i \nu_i RT \ln(\lambda_i m_{i,fluid}) = 0 . \tag{58}$$

In a remarkable series of papers Helgeson and his co-workers have presented equations and numerical data which allow evaluation of each of the terms in equation (58) for a wide variety of minerals and fluid species and for a wide range of pressure-temperature conditions of interest to metamorphic petrologists. The data include: (a) standard thermochemical parameters for mineral and fluid species $[(\Delta \bar{H}^o)^{1,298}, (\Delta \bar{S}^o)^{1,298}]$; (b) expressions for the heat capacity of mineral and fluid species as a function of temperature (Δc_p^o); (c) molar volumes of mineral species $(\Delta \bar{V}_s^o)$; (d) expressions for the molar volumes of fluid species as a function of pressure $(\bar{V}_i^o)$; and (3) stratagems for calculating the activities of species in electrolyte solutions (λ_i). From the data and equations in Helgeson's works and mineral composition data $(a_j$, equation 58), equation (58) may be used to calculate metamorphic fluid composition (the $m_{i,fluid}$) if estimates of pressure and temperature are known. These calculations are most easily performed with a computer program, SUPCRT, which Helgeson has kindly made available to the petrological and geochemical community.

Because the A and B parameters of Frantz et al. (1981) summarize the measured compositions of fluids in equilibrium with various mineral assemblages, the composition of electrolyte solutions can be quite accurately calculated from equation (57) when metamorphic conditions of interest closely match those of their experiments (P = 1000 or 2000 bars; T = 400°-850°C). For metamorphic P-T conditions significantly removed from those of Frantz's experiments, calculation of the composition of electrolyte solutions with Helgeson's methods (i.e., SUPCRT) are preferable.

Calculation of fluid composition

C-O-H-S fluids. We have developed the basic equations by which the concentration of components in metamorphic fluids may be determined quantitatively from mineral equilibria (equations 38 and 47; equation 35 is *not* recommended). We have summarized how each of the various terms in equations (38) and (47) may be numerically evaluated. The composition of C-O-H-S fluids is simply determined by simultaneously solving sets of equations (38) or (47) along with the constraint of Dalton's Law,

$$P = \sum_i p_i = \sum_i \{f_i / (\gamma_i^o \lambda_i)\} \ . \tag{59}$$

If a metamorphic fluid is principally composed of i species, then i-1 linearly independent equations are required. The i-1 equations are solved simultaneously with equation (59) at metamorphic conditions P and T to yield the concentrations, $X_{i,fluid}$, of each of the species in the fluid. Some or all of the i-1 equations are based on mineral equilibria like that expressed by equation (38) or (47). Others may represent conditions of homogeneous equilibrium among species in the fluid. Conditions of homogeneous equilibrium are quantitatively represented simply by equations like (38) or (47) in which $\Delta \bar{V}_s^o = 0$. If ideal mixing in the fluid is assumed, the calculation of concentrations is direct. Such calculations were pioneered by and described in detail by Eugster and Skippen (1967), Skippen (1971, 1975), and Eugster (1975). The activity coefficients in equation (47), λ_i, are functions of composition. Consequently, if a non-ideal mixing model for the fluid is chosen, calculation of concentrations is iterative. Perhaps the simplest stratagem is to first assume ideal mixing. The concentrations can then be used to estimate activity coefficients which are used as input to calculate a second set of concentrations. Values of λ_i are updated and calculations recycled until concentrations do not change between successive cycles.

The most common C-O-H-S species in fluids at metamorphic P-T conditions are CO_2, H_2O, CO, H_2, CH_4, and H_2S. As an example of how the composition of C-O-H-S fluids may be calculated from mineral equilibria, consider the assemblage biotite + K-feldspar + pyrite + pyrrhotite + graphite (Ferry, 1981). The following three linearly independent stoichiometric relations may be written among components in the minerals and species in the fluid:

$$KFe_3AlSi_3O_{10}(OH)_2 + 3H_2S = KAlSi_3O_8 + 3FeS + 4H_2O \ ; \tag{60}$$

$$2FeS_2 + C + 2H_2O = 2FeS + 2H_2S + CO_2 \ ; \tag{61}$$

$$C + 2H_2O = 2H_2 + CO_2 \ . \tag{62}$$

Two additional linearly independent stoichiometric relations may be written among fluid species alone:

$$CO + H_2O = CO_2 + H_2 \; ; \tag{63}$$

$$CH_4 + 3CO_2 = 2H_2O + 4CO \; . \tag{64}$$

For relations (60)-(64), there exist five corresponding equations (38) or (47). Equations like (38) or (47) for relations (63) and (64) refer to conditions of homogeneous equilibrium in the fluid and are characterized by $\Delta\bar{V}_s = 0$. At specified values of P, T, and mineral composition, the five equations like (38) or (47) may be solved simultaneously with equation (59) for the $X_{i,fluid}$ (six equations and six unknown i's: CO_2, H_2O, CH_4, CO, H_2, H_2S). Results are summarized in Table 3 (see below) for several graphitic sulfide-rich schists from Maine. The example illustrates how assumptions about the major species in metamorphic fluid are related to the amount of information needed to calculate fluid composition from mineral equilibria. If it were assumed, for example, that metamorphic fluids contained only CO_2, H_2O, and H_2S in significant concentrations, then only three equations would be required to estimate fluid composition: the equivalent of equation (38) or (47) for relations (60) and (61) and equation (59). The simpler the model of metamorphic fluid, the less information in the form of equations like (38) or (47) is required for characterization of fluid composition.

Aqueous electrolyte solutions. The concentration of dissolved species in aqueous electrolytes is calculated using a strategy similar to that for C-O-H-S fluids. A set of equations of the form (57) or (58) is solved simultaneously with an equation which ensures charge balance:

$$\sum_i \nu_i z_i m_i = 0 \tag{65}$$

and equations which ensure the conservation of the mass of certain anions in the fluid. In the case of Cl, for example, such an equation would be

$$m_{Cl_T} = \sum_i \nu_{Cl,i} m_{i,fluid} \tag{66}$$

If a metamorphic fluid is treated as an aqueous electrolyte whose sole anion is Cl and composed of i dissolved species, then i-2 linearly independent equations of the form of equation (57) or (58) are required. An example of this exercise is given by Frantz et al. (1981). They calculate the composition of an aqueous fluid with $m_{Cl_T} = 1.0$ in equilibrium with a model spilite (an albite-epidote-chlorite-tremolite-magnetite rock). Results are illustrated in Figure 12. The ability to characterize the composition of aqueous

TABLE 1. Concentrations of dissolved species in fluid coexisting with muscovite, albite, and quartz at 400°C and 2000 bars. Calculated from data of Frantz et al. (1981).

Species, i	m_i	m_i
Cl_T	0.1	1.0
KCl^o	<0.001	0.002
$NaCl^o$	0.0135	0.482
H_4SiO_4	0.042	0.042
K^+	<0.001	0.001
Na^+	0.086	0.515
Cl^-	0.086	0.516
H^+	<0.001	<0.001
$(OH)^-$	0.001	0.001
HCl^o	<0.001	<0.001

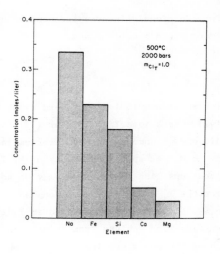

Figure 12. Concentrations of elements in an aqueous solution in equilibrium with a model spilite (chlorite + tremolite + epidote + albite + magnetite) at 500°C, 2000 bar. Total dissolved Cl is 1.0 molal. From Frantz et al. (1981).

electrolytes in equilibrium with common metamorphic rocks is hampered by two problems. The first of these, illustrated by the calculations of Frantz et al. (1981) above, is that some total concentration of chloride (the commonest anion in nature) in the electrolyte must be chosen. Frantz et al. arbitrarily chose m_{Cl_T} = 1.0. Although fluid inclusions may provide some insight, petrologists generally have a poor understanding of the chloride concentration of metamorphic fluids. Table 1 compares the composition of fluid in equilibrium with an albite-muscovite-quartz rock at 400°C, 2000 bars calculated by the method of Frantz et al. (1981). A choice of m_{Cl_T} = 0.1 or 1.0 makes a substantial difference in the concentration of dissolved species and in the ratios of concentration of some species. Better characterization of the chloride contents of metamorphic fluids is necessary before progress can be made characterizing metamorphic electrolyte solutions.

One strategy of characterizing the composition of electrolytes, which avoids a knowledge of m_{Cl_T}, is to calculate only ratios of activities of dissolved species from mineral equilibria. The equilibrium defined by equation (48), for example, permits the calculation of the ratio $a_{H^+, fluid}/a_{K^+, fluid}$ by equation (49). For a wide range of m_{Cl_T} the ratio, as defined by coexisting muscovite, K-feldspar, and quartz, is nearly constant.

An example of the application of this approach is the study of Bird and Norton (1981). Thus, even in the absence of accurate values of m_{Cl_T}, certain aspects of composition of metamorphic electrolytes may be characterized.

A second problem in characterizing natural electrolytes is the question of solvation number (Walther and Helgeson, 1980). Bare ions do not exist in aqueous solutions but are surrounded by clusters of H_2O molecules. The equilibrium among muscovite, K-feldspar, quartz, and fluid is therefore perhaps better represented not by relation (48) but by

$$KAl_3Si_3O_{10}(OH)_2 + 6SiO_2 + 2(K^+ \cdot nH_2O) =$$
$$3KAlSi_3O_8 + 2(H^+ \cdot mH_2O) + 2(n-m)H_2O \qquad (67)$$

where n and m are the avarage numbers of H_2O molecules clustered around K^+ and H^+, respectively (i.e., the solvation numbers). Because in general $n \neq m$, the equilibrium is dependent on $a_{H_2O, fluid}$. Consequently, for electrolyte solutions in which $a_{H_2O, fluid} < 1$, additional information is required in order to characterize the $a_{K^+, fluid}/a_{H^+, fluid}$ from the muscovite-feldspar-quartz-fluid equilibrium: $a_{H_2O, fluid}$ and the solvation numbers n and m. Because solvation numbers are poorly known, existing models for calculating the concentrations of species in electrolyte solutions are ill-suited for application to the many metamorphic fluids for which $a_{H_2O, fluid} < 1$. Progress in characterizing metamorphic electrolyte solutions would be greatly assisted by knowledge of the solvation number of species in CO_2-H_2O fluids at elevated temperatures and pressures (cf. Walther and Helgeson, 1980).

Error analysis. Any calculation of the concentration of a component in metamorphic fluid should be done with the awareness that errors may be quite large. The principal source of error derives from the simple fact that $X_{i, fluid}$ is an exponential function of other measured parameters. For the case of a mineral-fluid equilibrium involving no solid solution and just one fluid component, equation (47), for example, becomes:

$$(\Delta \bar{G}^o)^{1,T} + \Delta \bar{V}_s^o P + RT \ln \lambda_i \gamma_i^o P + RT \ln X_{i, fluid} = 0 \qquad (68)$$

and

$$\ln X_{i, fluid} = -[(\Delta \bar{G}^o)^{1,T}/RT + (\Delta \bar{V}_s^o P)/RT + \ln \lambda_i \gamma_i^o P] . \qquad (69)$$

Small errors in $(\Delta \bar{G})^{1,T}$, γ_i^o, λ_i, P, and/or T are greatly magnified into corresponding errors in $X_{i, fluid}$ when the exponent of both sides to equation (69) is taken. In addition, estimations of $X_{i, fluid}$ may be very sensitive to estimates of temperature for certain combinations of T and X. For example, with reference to Figure 13, large errors in estimated X_{CO_2} will be associated

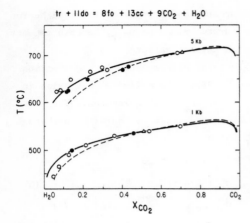

tr + 11do = 8fo + 13cc + 9CO$_2$ + H$_2$O

Figure 13. Tremolite(tr)-dolomite(do)-forsterite(fo)-calcite(cc)-fluid equilibrium at 1000 and 5000 bars. Curves calculated assuming ideal mixing in a real fluid (dashed) or a modified Redlich-Kwong equation of state (solid) for the fluid. Symbols: results of direct phase equilibrium experiments. From Jacobs and Kerrick (1981).

with small errors in temperature when fluid composition is determined from the tremolite-dolomite-forsterite-calcite-CO$_2$/H$_2$O fluid equilibrium in the range $0.2 < X_{CO_2} < 0.9$. As a generalization, the best monitors of fluid composition are equilibria which do *not* exclusively involve the principal constituent of the fluid. With reference to CO$_2$-H$_2$O fluids, dehydration equilibria serve as good indicators of the composition of CO$_2$-rich fluids [because for them, $(\partial X_{CO_2}/\partial T)_P$ is small at high X_{CO_2}] while decarbonation equilibria serve as good indicators of the composition of H$_2$O-rich fluids [because for them, $(\partial X_{CO_2}/\partial T)_P$ is small at low X_{CO_2}]. Knowledge of the quantitative behavior of mineral equilibria as functions of P, T, and $X_{i,fluid}$ makes calculations of fluid composition with large associated errors less likely. Anderson (1976) provides a good discussion of the estimation of errors in calculations of phase equilibria. He presents a Monte Carlo-type strategy which is useful in appraising errors associated with calculations involving many adjustable parameters as is required in the quantitative determination of metamorphic fluid composition.

Calculation of differences in chemical potentials of fluid components between mineral assemblages

An additional way in which the composition of metamorphic fluids may be quantitatively characterized through mineral equilibria is the determination of differences in μ_i between mineral assemblages where i are components in the fluid phase. These methods and results of their application to metamorphic rocks are reviewed by Spear, Ferry, and Rumble in Chapter 4, this volume.

GRAPHICAL REPRESENTATION OF METAMORPHIC FLUID COMPOSITION

It is often useful to graphically display the quantitative relationship between mineral assemblages and the composition of coexisting metamorphic fluid. The graphical representation of the compositions of polycomponent fluids, of necessity, involves some simplifications. Two-dimensional figures can plot only two independent variables while mineral-fluid equilibrium is a function of many variables (equation 38 or 47).

$T-X_i$ and $P-X_i$ diagrams

The most familiar simplification is to neglect mineral solubilities and all but two (i and k) fluid species. Mineral equilibria then may be plotted on an isobaric $T-X_i$ or isothermal $P-X_i$ diagram, where X_i = molar $i/(i + k)$. Because many metamorphic fluids may be approximated as CO_2-H_2O solutions, the most commonly used of these diagrams involves $X_{CO_2}^* = CO_2/(CO_2 + H_2O)$ as the composition variable. Isobaric univariant curves on $T-X_{CO_2}$ diagrams simply represent the graphical solution to equation (38) or (47) at specified values of P and a_j and with i = CO_2 or H_2O or both. Isothermal univariant curves on $P-X_{CO_2}$ diagrams simply represent the graphical solution to equation (38) or (47) at specified values of T and a_j and with i = CO_2 or H_2O or both. In both cases a graphical solution is possible because of the additional constraint that the fluid is binary (i.e., $X_{H_2O} = 1 - X_{CO_2}$).

Experimental phase equilibrium studies relevant to the construction of quantitative $T-X_{CO_2}$ diagrams are summarized by Kerrick (1974), Hewitt and Gilbert (1975) and Chernosky (1979). Articles by Kerrick (1974) and Eugster (1981) review numerous quantitative $T-X_{CO_2}$ diagrams applicable to a variety of rock systems.

Figure 14 is an example of a quantitative isobaric $T-X_{CO_2}$ diagram which refers to the metamorphism of siliceous dolomitic limestones. If the temperature-pressure conditions of metamorphism are known, curves in Figure 14 can be used to quantitatively estimate the composition of CO_2-H_2O metamorphic fluid that was in equilibrium with a variety of mineral assemblages.

Problems in the application of $T-X_i$ and $P-X_i$ diagrams to the characterization of metamorphic fluid composition include: (a) the presence of reduced C-O-H species in metamorphic fluids; (b) the occurrence of dissolved chloride species in metamorphic fluids (Rich, 1974; Crawford et al., 1979; Sisson et al., 1982); and (c) the unmixing of CO_2-H_2O fluid solutions at low temperature (Jacobs and Kerrick, 1981; Sisson et al., 1982).

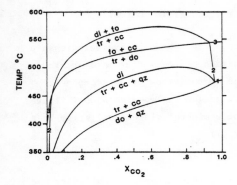

Figure 14. Isobaric T-X_{CO_2} diagram illustrating mineral equilibria in the system CaO-MgO-SiO$_2$-H$_2$O-CO$_2$ at 1000 bars pertinent to metamorphosed siliceous dolomitic limestones.

Abbreviations: do = dolomite; qz = quartz; tr = tremolite; cc = calcite; di = diopside; fo = forsterite. Curve 1: do + qz = di + CO$_2$; curve 2: tr + cc = di + do + CO$_2$ + H$_2$O; curve 3: di + do = fo + cc + CO$_2$. From Skippen (1974).

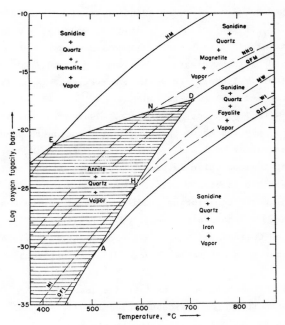

Figure 15. Isobaric T-$log f_{O_2}$ diagram for mineral equilibria in the system KAlSi$_3$O$_8$-SiO$_2$-Fe-O-H. Shaded area is the stability region of annite + quartz. From Eugster and Wones (1962).

Graphical representation of redox equilibria involving C-O-H fluids

Oxidation-reduction equilibria involving O-H fluids have long been illustrated on $log f_{O_2}$-T diagrams popularized by H.P. Eugster (1959). Figure 15 is an example of a $log f_{O_2}$-T diagram pertinent to rocks which contain annite + quartz. Solid curves in Figure 15 may be thought of as graphical solutions to equation (38) or (47) at specified values of P and a$_j$ and with i = O$_2$ or

242

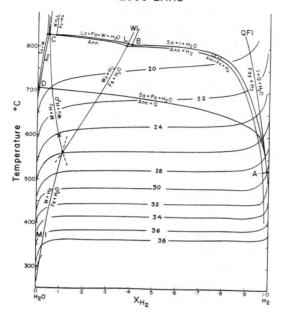

2000 BARS

Figure 16. Isobaric T-X_{H_2} diagram illustrating mineral equilibrium in the system $KAlSi_3O_8$-SiO_2-Fe-O-H at 2000 bars. Abbreviations: M = magnetite; Wu = wüstite; I = iron; Q = quartz; Fay = fayalite; Ann = annite; Sa = sanidine; Lc = leucite; Ks = kalsilite. From Greenwood (1975).

H_2 or both. The solutions are possible because O-H fluids at the conditions of the diagram are essentially binary H_2-H_2O fluids (X_{H_2} + X_{H_2O} = 1). The species O_2 is only an extremely minor constituent of metamorphic fluids, and it is really being used in figures like 15 as an "operator" to perform the reactions, somewhat akin to the "exchange operators" discussed earlier. Because O-H fluids are principally binary H_2-H_2O mixtures at the conditions of metamorphism in the Earth's crust, it is actually more useful to represent redox equilibria on T-X_{H_2} diagrams rather than on the more conventional $log f_{O_2}$-T diagrams (Greenwood, 1975). Figure 16 represents the same mineral equilibria in Figure 15 on a T-X_{H_2} diagram. The composition of O-H fluid in equilibrium with various mineral assemblages at specified T and P may be read directly from Figure 16. The same information could be obtained from Figure 15 but only after the intermediate exercise of calculating the speciation of an O-H fluid at a value of f_{O_2} read from Figure 15.

Graphical treatments of mineral equilibria involving reduced C-O-H species are discussed by Eugster and Skippen (1967), Burt (1972), and

243

Frost (1972). One strategy is to only consider fluids in equilibrium with graphite. A second strategy is to assume either a fixed concentration of a particular fluid species or a fixed concentration of an element in the fluid [e.g., fixed $X_C = C/(C+O+H)$]. A third strategy is to abandon T and P as plotting variables and represent fluid compositions on an isobaric, isothermal triangular C-O-H composition diagram (Fig. 1; Fig. 2 of Chapter 8, this volume; Rumble, 1980). Plotting coordinates refer to the bulk composition of a C-O-H fluid and the diagram may be contoured with the concentration of various species. Isothermal, isobaric univariant mineral equilibria may be represented as curves on the diagram, and they are graphical solutions to equation (38) or (47) at specified values of P, T, and a_j.

Graphical representation of mineral equilibria involving aqueous electrolyte solutions

The simplest method to graphically portray mineral equilibria involving components in aqueous electrolyte solutions is on triangular isobaric, isothermal composition diagrams with the non-C-O-H-S components representing one or more vertices (Petersen, 1965). Figure 17 is an example and illustrates the proportions of Ca, Si, and Mg in aqueous fluid in equilibrium with various combinations of minerals commonly found in metamorphosed siliceous dolomitic limestones. The curves in diagrams like Figure 17 have many interesting similarities with the curves on ternary liquidus diagrams used in igneous petrology. Both congruent and incongruent solution of individual minerals may be depicted in a manner analogous to the portrayal of congruent and incongruent melting in liquidus diagrams.

A second type of diagram represents mineral equilibria involving aqueous electrolyte solutions in terms of activities or activity ratios of dissolved species in the fluid at constant P and T (Hemley, 1959; see reviews by Meyer and Hemley, 1967 and Rose and Burt, 1979). Figure 18 is an example of an isobaric, isothermal $log a_1$-$log a_2$ diagram which can be used to characterize the composition of metamorphic fluid in equilibrium with combinations of feldspars, micas, and Al-silicates. Curves in Figure 18 represent the graphical solution to equations like (58) at specified values of P, T, and a_j and with i = K^+, Na^+, and/or H^+. For anions such as chloride, which do not commonly enter minerals, activity ratios such as a_{K^+}/a_{H^+} or a_{KCl}/a_{HCl} may be used as variables. For anions such as fluoride, borate, and phosphate, which do enter minerals, acidity and salinity may be considered as separate variables (as on $log a_{HF}$-$log a_{KF}$ diagrams, Burt, 1981).

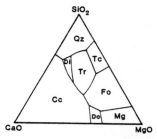

Figure 17. Isobaric, isothermal diagram showing schematic solution compositions in equilibrium with quartz (Q), talc (Ta), tremolite (Tr), forsterite (Fo), magnesite (M), dolomite (Do), calcite (Cc), and diopside (Di) in the system $CaO-MgO-SiO_2-H_2O-CO_2$. Solution compositions are projected through H_2O and CO_2. From Frantz and Mao (1979).

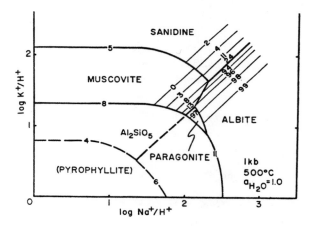

Figure 18. Mineral equilibria in the system $K_2O-Na_2O-Al_2O_3-SiO_2-H_2O-HCl$ at 500°C; 1000 bars as a function of $log(a_{K^+}/a_{H^+})$ (symbolized as $logK^+/H^+$) and of $log(a_{Na^+}/a_{H^+})$ (symbolized as $log(Na^+/H^+)$). Light straight lines show how isobaric, isothermal univariant curves shift as a function of solid solution in mica and feldspar. From Wintsch (1975).

We have already mentioned the problem that solvation numbers present in characterizing the composition of aqueous electrolytes for which $a_{H_2O,fluid} < 1$. In the graphical portrayal of the composition of electrolytes, Walther and Helgeson (1980) present a partial solution to the problem. They first define a parameter Δn_i:

$$\Delta n_i = n_i - Z_i n_{H^+} \tag{70}$$

where n_i is the solvation number of charged species i in solution, Z_i is the charge on i, and n_{H^+} is the solvation number of H^+ in solution. Next they define a second parameter σ_i:

$$\sigma_i = a_{H_2O,fluid}^{\Delta n_i} \cdot \tag{71}$$

245

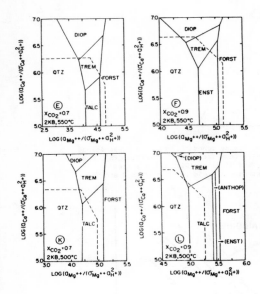

Figure 19. Mineral equilibria in the system
CaO-MgO-SiO$_2$-H$_2$O-CO$_2$-HCl at 2000 bars and 500°C
or 550°C. Fluid composition variables, $(a_i)/$
$(a_{H^+}^{Zi} \sigma_i)$, are discussed in the text. From
Walther and Helgeson (1980).

For example, with reference to reaction (67), $\Delta n_{K^+} = n - m$ and $\sigma_{K^+} = a_{H_2O}^{n-m}$.
For equilibria between minerals and an aqueous fluid containing uncharged
species k and charged species i, equation (58) may be rewritten involving σ_i:

$$\{(\Delta \bar{G}^o)^{1,298} + \int_{298}^{T} \Delta c_p^o dT - T\int_{298}^{T} (\Delta c_p^o/T) dT + \Delta \bar{V}_s^o P + \sum_j \nu_j RT\ln a_j$$

$$\sum_k \int_1^P (\nu_k \bar{V}_k^o) dP + \sum_k \nu_k RT\ln a_k + \sum_i \int_1^P (\nu_i \bar{V}_i^o) dP\} + \sum_i \nu_i RT\ln(a_i/\sigma_i a_{H^+}^{z_i}) = 0 \quad (72)$$

(k = H$_2$O, CO$_2$, etc.; i = K$^+$, Na$^+$, Ca^{++}, Mg^{++}, etc.). At specified values of
P, T, a_j and a_k the sum of the terms in brackets is a constant. Representation of equilibria involving minerals and fluid with $a_{H_2O} < 1$ therefore may
be graphically portrayed under these circumstances on diagrams like Figure 19
in which the variables $(a_i)/(a_{H^+})^{z_i}$ are replaced by the variables $(a_i)/$
$[(a_{H^+})^{z_i}\sigma_i]$. For example, mineral-fluid equilibria in Figure 18 at conditions of $a_{H_2O} < 1$ would be illustrated as a function of $\log(a_{K^+}/a_{H^+}\sigma_{K^+})$
versus $\log(a_{Na^+}/a_{H^+}\sigma_{Na^+})$.

Figure 19 is an example of a $\log[a_i/a_{H^+}^{z_i}\sigma_i]$ versus $\log[a_j/a_{H^+}^{z_j}\sigma_j]$ where
i = Ca^{++} and j = Mg^{++}. The diagram illustrates aspects of the composition of
an aqueous electrolyte in equilibrium with minerals commonly observed in
siliceous dolomitic limestones at two temperatures and at two values of
$a_{H_2O} < 1$. Lines in the diagram represent solutions to equation (72) at
specified values of P, T, a_{H_2O}, a_{CO_2}, and a_j with i = Ca^{++} or Mg^{++} or both
and k = H$_2$O or CO$_2$ or both. Regardless of whether σ is constant throughout

the diagram, Walther and Helgeson claim that phase relations are accurately represented by diagrams such as Figure 19 which involve terms of the form $log[a_i/a_{H^+}^{z_i}_i]$. Values of Δn_i (and hence σ_i) are not now available for aqueous species at the elevated pressures and temperatures of metamorphism. When the Δn_i do become available, fluid composition may readily be determined from mineral equilibria and diagrams such as Figure 19.

Applications of the graphical representation of metamorphic fluid composition

For certain mineral assemblages, fluid composition obviously may be directly read from the curves in figures like 14-19 (if the pressure temperature conditions of metamorphism have been estimated). An important advantage of the graphical representation of mineral equilibria is that metamorphic fluid composition may often be quite accurately *bracketed* by figures such as 14-19 even if the assemblage in question does not correspond to a curve or an intersection of curves. The assemblage diopside + dolomite, for example, is stable only in the portion of Figure 14 bounded by curves 1, 2, and 3. Even though the composition of CO_2-H_2O fluid coexisting with diopside + dolomite cannot be directly calculated, Figure 14 narrowly brackets the composition of fluid in equilibrium with the two minerals between $X_{CO_2} = 0.9$-1.0. Similarly, the f_{O_2} of fluid coexisting with annite + quartz is variable and a unique value cannot be calculated. At 650°C, 2070 bars, however, fluid coexisting with annite + quartz must be characterized by $f_{O_2} = 10^{-18}$-10^{-21} (Fig. 15). Graphical representation of mineral equilibria and fluid composition thus is often useful in quantitatively characterizing the composition of fluid in equilibrium with a mineral assemblage whose variance is too high to permit an exact calculation.

THE COMPOSITION OF METAMORPHIC FLUIDS: A REVIEW

C-O-H-S fluids

Many studies have characterized the concentration of C-O-H-S species in metamorphic fluids through the study of mineral equilibria. We summarize selected results according to common rock types found in metamorphic terrains.

Metamorphosed carbonate rocks.   It is commonly assumed that carbonate rocks are metamorphosed in equilibrium with CO_2-H_2O fluids. More detailed evaluation of the composition of fluid in equilibrium with metacarbonates suggests that this is not a bad approximation (Ferry, 1976b, 1979). Table 2 summarizes the composition of CO_2-H_2O fluid that attended metamorphism of

Table 2A. Summary of metamorphic fluid compositions calculated from mineral equilibria in metamorphosed carbonate rocks.

Reference	Environment	Mineral Assemblage	X_{CO_2}
Trommsdorf (1972)	regional	(a) dolomite–talc–quartz–tremolite–diopside–forsterite–calcite*	0.5–1.0
		(b) dolomite–quartz–diopside–calcite	0.95–1.0
		(c) dolomite–forsterite–diopside–calcite	0.95–1.0
Joesten (1974)	contact	wollastonite–calcite–tilleyite–rankinite–spurrite–merwinite–melilite*	0.6–1.0
Suzuki (1977)	contact	(a) dolomite–quartz–talc–calcite–tremolite–diopside–forsterite*	0.25–1.0
		(b) tremolite–calcite–talc–dolomite	0.25
		(c) dolomite–quartz–diopside	0.95–1.0
		(d) diopside–dolomite–forsterite–calcite	0.95–1.0
Rice (1977a)	contact	calcite–dolomite–quartz–phlogopite–K-feldspar–tremolite–diopside*	≈0.5–0.95
Rice (1977b)	contact	calcite–dolomite–amphibole–chlorite–quartz–spinel–olivine*	≈0.5–0.95
Ghent et al. (1979)	regional	dolomite–calcite–quartz–tremolite	0.76–0.82
Hoersch (1981)	contact	dolomite–calcite–quartz–talc–tremolite–forsterite–diopside*	≈0.6–0.95

* *Individual rocks contain subsets of these minerals.*

248

Table 2B. Summary of metamorphic fluid compositions calculated from mineral equilibria in metamorphosed carbonate rocks.

Reference	Environment	Mineral Assemblage	X_{CO_2}
Greenwood (1967)	contact	calcite-quartz-wollastonite	≈0.1
Carmichael (1970); Carmichael and Skippen (1977)	regional	zoisite-calcite-plagioclase	0.06-0.25**
Hewitt (1973a)	regional	(a) calcite-dolomite-muscovite-biotite-K-feldspar-amphibole-zoisite-plagioclase* (b) zoisite-calcite-plagioclase	<0.1-? ≈0.05-0.10
Kerrick et al. (1973)	contact	garnet-calcite-plagioclase-wollastonite-quartz-zoisite*	≈0.05-0.15
Moore and Kerrick (1976)	contact	dolomite-calcite-quartz-talc-tremolite-forsterite-clinohumite-periclase-wollastonite-diopside*	0.2-0.7
Ghent et al. (1979)	regional	zoisite-plagioclase-calcite-quartz-diopside-amphibole	0.23-0.27
Crawford et al. (1979)	regional	(a) zoisite-calcite-plagioclase-quartz-biotite-amphibole-diopside-microcline (b) zoisite-calcite-plagioclase-quartz-muscovite-microcline	0.02-0.05 0.02-0.05
Ferry (1980b)	regional	(a) zoisite-calcite-plagioclase-quartz-garnet (b) zoisite-calcite-plagioclase-quartz-amphibole-diopside (c) zoisite-calcite-plagioclase-quartz-biotite-amphibole-microcline	0.05-0.3
Valley and Essene (1980)	regional	calcite-quartz-tremolite-phlogopite-K-feldspar-diopside-dolomite-forsterite-rutile-sphene*	0.1-?
Bowman and Essene (1982)	contact	dolomite-calcite-quartz-talc-tremolite-chlorite-forsterite-spinel-periclase-phlogopite*	0.05-0.6
Rumble et al. (1982)	regional	(a) calcite-wollastonite-quartz (b) calcite-quartz-zoisite-grossularite	0.09-0.14

* _Individual rocks contain subsets of these minerals; **Calculated by JMF._

249

carbonate rocks expressed as $X_{CO_2} = CO_2/(CO_2 + H_2O)$. No attempt has been
made to identify the grade of metamorphism because most studies have characterized fluid composition over a range of prograde metamorphic conditions.
Studies of contact metamorphosed rocks, however, are differentiated from
studies of regionally metamorphosed rocks. The difference in environment
primarily is one of pressure (depth). Mineral barometers indicate that the
contact environments correspond to metamorphism at depths of ~1-7 km. The
regional metamorphic environments, in turn, correspond to metamorphism at
depths of ~13-28 km. Studies of terrains metamorphosed at depths of 7-13 km
have not been found.

Fluids in equilibrium with carbonate rocks during metamorphism apparently
may be nearly pure CO_2, nearly pure H_2O, or any mixture in between. The
studies have been segregated into two groups on the basis of fluid composition. In Table 2A most fluid compositions are CO_2-rich and in all cases at
least some fluids are nearly pure CO_2 (X_{CO_2} = 0.8-1.0). In Table 2B most
fluid compositions are H_2O-rich and in all cases at least some fluids are
nearly pure H_2O (X_{CO_2} = 0.02-0.25). Both sorts of fluids are found in both
contact and regional metamorphic environments. Furthermore, the studies of
Suzuki (1977) and of Ghent et al. (1979) demonstrate that H_2O-rich fluids and
CO_2-rich fluids may exist in different parts of the same terrain during the
same metamorphic event.

Devolitilization reactions which occur during progressive metamorphism
of carbonate rocks release CO_2-rich fluids. Essentially all the commonly-
observed reactions generate a CO_2-H_2O fluid in the composition range X_{CO_2}
≥ 0.5. The studies in Table 2A are therefore interpreted as cases in which
metamorphic rocks equilibrated with the same CO_2-rich fluids that they generated internally by metamorphic decarbonation/dehydration reactions. As is
discussed in detail by Rice and Ferry (this volume), the composition of these
CO_2-rich fluids was controlled internally by the buffer capacity of the prograde mineral reactions.

The studies in Table 2B are ones in which the equilibrium metamorphic
fluid composition is significantly more H_2O-rich than the composition of
fluid which was generated internally by mineral reactions in the metacarbonate
rocks. In these cases the carbonate rocks did not equilibrate simply with
fluid which they produced through devolitilization. The fluid compositions
in Table 2B are interpreted in terms of a mixing process. The fluid compositions are a result of the mixing of a CO_2-rich component (derived internally
from metamorphic mineral reactions) with an H_2O-rich component (derived from

Table 3. Summary of metamorphic fluid compositions
calculated from mineral equilibria in pelitic schists.

Reference	Mineral Assemblage	X_i, fluid				
		H_2O	CO_2	CH_4	H_2	H_2S
Guidotti (1970)	muscovite-biotite-sillimanite-quartz-pyrrhotite-graphite	0.853	0.063	–	–	0.084
Jones (1972)	paragonite-plagioclase-quartz-pyrite-pyrrhotite-carbonate-graphite	0.81	0.00	0.15	0.00	0.04
Ghent (1975)*	muscovite-kyanite-plagioclase-quartz-graphite	0.898	0.049	0.051	0.002	–
Ghent et al. (1979)	muscovite-kyanite-plagioclase-quartz	0.43–0.64	–	–	–	–
Ferry (1980a)	muscovite-andalusite-plagioclase-quartz	0.31–1.00	–	–	–	–
Ferry (1981)	biotite-microcline-pyrite-pyrrhotite-graphite	0.24–0.71	0.00–0.23	0.01–0.74	0.00–0.01	0.02–0.05
Labotka (1981)	muscovite-andalusite-plagioclase-quartz	0.32–0.83	–	–	–	–

* *Recalculated by JMF.*

some unknown source external to the carbonates). The two components mixed
within the pore space of the carbonate rocks and the minerals in the carbonates equilibrated with the mixture. This process is described in more
detail by Rice and Ferry (Chapter 7, this volume); they show that equilibrium
fluid compositions can be used with the measured progress of the prograde
mineral reactions to estimate how much H_2O-rich, externally-derived fluid
interacted with the carbonate rocks in Table 2B.

The studies in Table 2A thus may be interpreted as examples of rocks
which were metamorphic systems closed to interaction with external fluids.
Such "closed" systems occur both in contact and regional settings. Studies
in Table 2B may be interpreted as metamorphic systems open to interaction
with external fluids. These "open" systems also occur both in contact and
regional environments.

Pelitic schists. The composition of metamorphic fluids recorded by
mineral equilibria in pelitic schists is listed in Table 3. All samples
studied are from regional metamorphic terrains. Inferred depths of metamorphism, based on mineral barometers, range from ∿13 to ∿28 km. Most fluids
are H_2O-rich as would be expected from the observation that dehydration reactions predominate during metamorphism of shales and their equivalents.

251

Even though fluids in Table 3 are H_2O-rich, the tabulation suggests that pure H_2O may be a poor model for the fluid attending metamorphism of pelitic schists. The results collected in Table 3 are in agreement with the theoretical study by Ohmoto and Kerrick (1977) of the composition of fluids in equilibrium with graphite-bearing model pelitic schists at P-T conditions of metamorphism. They found that graphitic schists in general were in equilibrium with fluids characterized by $X_{H_2O} < 0.9$. Regrettably, only a few studies have attempted to characterize what composes the non-H_2O portion of fluids in equilibrium with pelitic schists. Limited data suggest that the balance is dominated by either CO_2 or CH_4 or both. The CO_2 and CH_4 are likely often produced by the reaction of H_2O with graphite during metamorphism:

$$2C + 2H_2O = CH_4 + CO_2 . \qquad (73)$$

The equal concentration of CO_2 and CH_4 in the fluid whose composition we calculated from the data of Ghent (1975) almost certainly may be taken as evidence that the carbon-bearing species formed by reaction (73).

In two studies, however, the non-H_2O fraction of the fluid is dominated by CH_4 (Jones, 1972; Ferry, 1981). Some calculated fluid compositions are over 70 mol % CH_4. This concentration of methane and CH_4/CO_2 ratios > 1 cannot be explained by graphite-H_2O reactions such as (73). Perhaps these compositions are best explained by infiltration of rock by CH_4-rich fluids in analogy with our explanation of the fluid compositions listed in Table 2B. The origin of CH_4-rich fluids is uncertain; they may be generated by the decomposition of hydrocarbons or organic matter during prograde metamorphism.

Metamorphosed mafic igneous rocks.   We found only one study which characterized the composition of fluid in equilibrium with a metamorphosed mafic igneous rock (Ghent and DeVries, 1972). They used the assemblage garnet-plagioclase-quartz-epidote±calcite in regionally metamorphosed amphibolites to estimate that $X_{H_2O, fluid} = 0.9-1.0$ during metamorphism. The presence of calcite in some samples indicated that the remaining portion of the fluid was dominantly composed of CO_2.

Granulites.   Granulite terrains have widely attracted attention both as samples of lower continental crust and as samples of very ancient crustal material. It has long been known or suspected that f_{H_2O} is very low during granulite facies metamorphism. Low f_{H_2O} is, for the most part, supported by estimations of $X_{H_2O, fluid}$ made from mineral equilibria in rocks from granulite terrains (Table 4). The low f_{H_2O} may either indicate that granulite facies

Table 4. Summary of metamorphic fluid composition calculated
from mineral equilibria in rocks from the granulite facies.

Reference	Mineral Assemblage	$X_{H_2O, fluid}$
Wones and Eugster (1965); Wones (1972)	biotite-K-feldspar-magnetite-quartz	<0.01
Wells (1976)	augite-hypersthene-amphibole-plagioclase-quartz-magnetite±garnet	~0-0.3
Boone (1978)	quartz-K-feldspar-sillimanite-plagioclase-biotite-garnet-spinel	0.02-0.25
Bohlen et al. (1980)	biotite-K-feldspar-magnetite-ilmenite	<0.25
Valley and Essene (1980)	phlogopite-calcite-quartz-K-feldspar-tremolite-diopside-forsterite-sphene-rutile-dolomite*	>0.90

* Individual rocks contain subsets of these minerals.

metamorphism occurs in the absence of fluids or that fluids in granulite
terrains are dominated by species other than H_2O. The latter explanation is
supported by the frequent observation of CO_2-rich fluid inclusions in minerals from granulites (Touret, 1971, 1977). The presence of the CO_2-rich
fluid inclusions, in fact, has helped prompt the development of models which
consider granulite facies metamorphism as large-scale CO_2 metasomatism (Newton
et al., 1980). The study of Valley and Essene (1980) presents a stimulating
alternative to conventional views of granulite facies metamorphism. Their
investigation of mineral equilibria in granulite facies marbles indicates
not only that $P_{fluid} \sim P_{lithostatic}$ during metamorphism, but that at least
some fluids were very H_2O-rich. Further work obviously is needed to determine to what extent the rocks studied by Valley and Essene are representative
of the products of granulite facies metamorphism.

Halogens in metamorphic fluids

   Fluorine.   The concentration of F-bearing species in metamorphic fluids
may be calculated through equations like (38), (47), (57), or (58) based on
mineral equilibria involving fluorides and fluorine-bearing silicate minerals.
Experimental studies (Munoz and Ludington, 1974; Duffy and Greenwood, 1979)
and analyses of natural assemblages (e.g., Rice, 1980a,b) demonstrate that
fluorine is systematically partitioned among coexisting (OH-F) silicate
minerals such as talc, tremolite, phlogopite, brucite, and the humite group.

Furthermore, fluorine is strongly partitioned into hydroxyl sites relative to the fluid phase. Experimental results and calculations based on tabulated thermodynamic data indicate that HF is the most abundant F-bearing species in the fluid and that $X_{HF,fluid}$ << 0.01. Fluorine-bearing species therefore are a negligible contribution to the composition of metamorphic fluids and are unlikely to seriously affect the thermodynamic properties of the dominant species. Rice (1980a,b) has shown how variations in the fugacity of HF in C-O-H-F fluids can be calculated by solving a set of simultaneous equations and graphically portrayed on isobaric $T-X(CO_2-H_2O)$ diagrams contoured for the $f_{HF,fluid}$.

Chlorine. In contrast to fluorine, chlorine is strongly partitioned into the fluid relative to minerals. The presence of chlorine in silicate minerals therefore suggests that the coexisting fluid contained significant amounts of chlorine, with NaCl being the dominant associated species. The concentration of NaCl in metamorphic fluid may be calculated through equations like (57) or (58) based on mineral equilibria involving scapolite (Ellis, 1978). Results indicate that on a molar basis NaCl may constitute zero to approximately 50 percent of metamorphic fluid coexisting with scapolite.

Because of the relative scarcity of Cl-bearing minerals in common metamorphic rocks, studies of fluid inclusions, rather than studies of mineral-fluid equilibria, hold more promise for the characterization of the concentration of Cl-bearing species in metamorphic fluid.

Nitrogen in metamorphic fluids

The characterization of the concentration of N-bearing species in metamorphic fluids from mineral equilibria represents an instructive example of the limitations of this technique. Petrologists are often limited in the degree to which they can characterize metamorphic fluid composition not by theory or thermochemical data but by the mineral assemblages under consideration. An extreme example is the case of N-bearing species. Recent fluid inclusion studies report the presence of significant concentrations of N (Kreulin and Schuiling, 1982). The concentration of N-bearing species in metamorphic fluids cannot be accurately characterized from mineral equilibria, however, because metamorphic rocks do not contain nitrates, nitrides, or ammonia-bearing minerals.

Aqueous metamorphic electrolyte solutions

We noted above two practical problems which hamper the characterization of the composition of aqueous metamorphic electrolyte solutions: (a) determination of total dissolved chloride; and (b) determination of solvation numbers. Problem (a), at least, can be circumvented by calculating the ratio of the concentrations of dissolved species, rather than actual concentrations, as has been advocated in numerous studies (e.g., Helgeson, 1967; Eugster, 1970; Wintsch, 1975). While this approach is commonly applied to ore deposits (e.g., review by Rose and Burt, 1979), it has not often been utilized by metamorphic petrologists. An example of a successful application of the method to metamorphic rocks is the study of mineral-fluid equilibria in the Salton Sea geothermal field by Bird and Norton (1981). They determined, for numerous samples, the ratios a_{K^+}/a_{H^+}, a_{Na^+}/a_{H^+}, $a_{Ca^{++}}/a_{H^+}^2$, and $a_{Mg^{++}}/a_{H^+}^2$ of coexisting metamorphic fluid. Problem (b) can be partially circumvented, as discussed above, by calculating ratios of the form $(a_i/a_{H^+}^{z_i}\sigma_i)$ rather than of the form $(a_i/a_{H^+}^{z_i})$.

Although few studies have attempted to apply equations (58) or (72) to actual metamorphic rocks, theoretical investigations of mineral-fluid equilibria provide some insight into the probable elemental composition of metamorphic electrolyte solutions. Figures 12, 18, and 19, as well as results in Table 1, summarize the general relationships. Na-bearing species are apparently the most abundant of the non-C-O-H-S-N fluid species (Fig. 12; Table 1). The ratio Na/K can vary substantially as a function of temperature and coexisting mineral assemblage (Fig. 18; Table 1). Among the divalent cations, Fe-bearing species probably occur in the greatest concentration (Fig. 18). In contrast, Mg-bearing species occur in concentrations 2-1000 times less than those for Fe-bearing or Ca-bearing species (Figs. 12 and 19). The concentration of Si (as H_4SiO_4) is apparently intermediate between that for Na- and Fe-bearing species (which are higher) and that for Ca- and Mg-bearing species (Fig. 12; Table 1). The compositions of aqueous electrolytes vary greatly, of course, as a function of coexisting mineral assemblage (Figs. 18 and 19). Nevertheless, we may look for Na-, Fe-, Ca-, and Si-bearing species as composing the majority of any non-C-O-H-S fraction of metamorphic fluids.

ACKNOWLEDGMENTS

We thank J.M. Rice for a careful review of an earlier version of the chapter. J.M.F. acknowledges the support of a Cottrell Grant from Research Corporation and of NSF grant EAR 80-20567 (Earth Sciences Section) in preparing the manuscript.

APPENDIX 1

NOTATION

P, P_s total or lithostatic pressure (bars)

T temperature ($°K$)

R universal gas constant (cal/degree-mole)

$\mu_{i,j}$ chemical potential of component i in phase j (cal)

$\bar{G}_{i,j}^{P,T}$ partial molar Gibbs energy of component i in phase j at pressure P and temperature T (cal)

$(\bar{G}_i^o)^{P,T}$ molar Gibbs energy of a phase of pure i at pressure P and temperature T (cal)

$(\bar{G}_i^o)^{1,T}$ molar Gibbs energy of a phase of pure i at 1 bar and temperature T (cal)

$(\bar{G}_i^o)^{1,298}$ molar Gibbs energy of a phase of pure i at 1 bar and 298°K (cal)

$a_{i,j}$ activity of component i in phase j

$\bar{V}_i^o$ molar volume of a phase of pure i

γ_i^o fugacity coefficient of component i in a fluid of pure i

λ_i activity coefficient of component i in solution

$X_{i,j}$ mole fraction of component i in phase j

f_i fugacity of component i in fluid phase

ν_i stoichiometric coefficient of i in a stoichiometric or reaction relationship

Δ operator which performs the following function on variable Y: $\Delta Y = \sum \nu_i Y_i$

$(\bar{H}_i^o)^{1,T}$ molar enthalpy of a phase of pure i at 1 bar and temperature T (cal)

$(\bar{H}_i^o)^{1,298}$ molar enthalpy of a phase of pure i at 1 bar and 298°K (cal)

$(\bar{S}_i^o)^{1,T}$ molar entropy of a phase of pure i at 1 bar and temperature T (cal/degree)

$(\bar{S}_i^o)^{1,298}$ molar entropy of a phase of pure i at 1 bar and 298°K (cal/degree)

c_p^o heat capacity of a phase of pure i

$\bar{V}$ volume of a fluid

$\bar{V}_i$ partial molar volume of component i in a fluid solution

$m_{i,fluid}$ molality of component i in a fluid solution

p_i partial pressure of component i in a fluid solution

z_i charge on a species in a fluid solution

n_i solvation number of charged species in solution

Δn_i $n_i - z_i n_{H^+}$

σ_i $\dfrac{\Delta n_i}{a_{H_2O,fluid}}$

257

Anderson, G.M. (1976) Error propagation by the Monte Carlo method in geochemical calculations. Geochim. Cosmochim. Acta 40, 1533-1538.

Barnes, I. (1970) Metamorphic waters from the Pacific Tectonic Belt of the west coast of the United States. Science 168, 973-975.

_____ and J.R. O'Neil (1969) The relationship between fluids in some fresh alpine-type ultra-mafics and possible modern serpentinization, western United States. Bull. Geol. Soc. Am. 80, 1947-1960.

_____, J.B. Rapp, and J.R. O'Neil (1972) Metamorphic assemblages and the direction of flow of metamorphic fluids in four instances of serpentinization. Contrib. Mineral. Petrol. 35, 263-276.

Batzle, M.L. and G. Simmons (1976) Microfractures in rocks from two geothermal areas. Earth Planet. Sci. Lett. 30, 71-93.

Bird, D.K. and D.L. Norton (1981) Theoretical prediction of phase relations among aqueous solutions and minerals: Salton Sea geothermal system. Geochim. Cosmochim. Acta 45, 1479-1493.

Boctor, N.Z., R.K. Popp, and J.D. Frantz (1980) Mineral-solution equilibria - IV. Solubilities and the thermodynamic properties of $FeCl_2^0$ in the system Fe_2O_3-H_2-H_2O-HCl. Geochim. Cosmochim. Acta 44, 1509-1518.

Bohlen, S.R., D.R. Peacor, and E.J. Essene (1980) Crystal chemistry of a metamorphic biotite and its significance in water barometry. Am. Mineral. 65, 55-62.

Boone, G.M. (1978) Kyanite in Adirondack Highlands sillimanite-rich gneiss, and P-T estimates of metamorphism. Geol. Soc. Am. Abstr. Prog. 10, 34.

Bowman, J.R. and E.J. Essene (1982) P-T-X(CO_2) conditions of contact metamorphism in the Black Butte aureole, Elkhorn, Montana. Am. J. Sci. 282, 311-340.

Burnham, C.W., J.R. Holloway, and N.F. Davis (1969) Thermodynamic properties of water to 1000°C and 10,000 bars. Geol. Soc. Am. Spec. Paper 132.

Burt, D.M. (1971a) Some phase equilibria in the system Ca-Fe-Si-C-O. Carnegie Inst. Wash. Year Book 70, 178-184.

_____ (1971b) Multisystems analysis of the relative stabilities of babingtonite and ilvaite. Carnegie Inst. Wash. Year Book 70, 189-197.

_____ (1972a) The influence of fluorine on the facies of Ca-Fe-Si skarns. Carnegie Inst. Wash. Year Book 71, 443-450.

_____ (1972b) The system Fe-Si-C-O-H, a model for metamorphosed iron formations. Carnegie Inst. Wash. Year Book 71, 435-443.

_____ (1974) Concepts of acidity and basicity in petrology - the exchange operator approach. Geol. Soc. Am. Abstr. Prog. 6, 674-676.

_____ (1976) Hydrolysis equilibria in the system K_2O-Al_2O_3-SiO_2-H_2O-Cl_2O_{-1}: Comments on topology. Econ. Geol. 71, 665-671.

_____ (1978) Multisystems analysis of beryllium mineral stabilities: The system BeO-Al_2O_3-SiO_2-H_2O. Am. Mineral. 63, 664-676.

_____ (1979) Exchange operators, acids, and bases (in Russian). In Zharikov, V.A., V.I. Fonarev, and D.S. Korikovskii, Eds., Problemy fiziko-khimicheskoi Petrologii, 'Nauka' Press, Moscow, 2, 3-15.

_____ (1981) Acidity-salinity diagrams - application to greisen and porphyry deposits. Econ. Geol. 76, 832-843.

Carmichael, D.M. (1970) Intersecting isograds in the Whetstone Lake area, Ontario. J. Petrol. 11, 147-181.

Chatterjee, N.D. and W. Johannes (1974) Thermal stability and standard thermodynamic properties of synthetic $2M_1$-muscovite, $KAl_2[AlSi_3O_{10}(OH)_2]$. Contrib. Mineral. Petrol. 48, 89-114.

Chernosky, J.V., Jr. (1979) Experimental metamorphic petrology. Rev. Geophys. Space Phys. 17, 860-872.

Churayev, N.V., V.D. Sobolev, and Z.M. Zorin (1970) Measurement of viscosity of liquids in quartz capillaries. Spec. Discuss. Faraday Soc. 1, 213.

Crawford, M.L. (1981) Fluid inclusions in metamorphic rocks - low and medium grade. In Hollister, L.S. and M.L. Crawford, Eds., Fluid Inclusions: Applications to Petrology. Mineral. Assoc. Canada, Calgary, 157-181.

_____, D.W. Kraus, and L.S. Hollister (1979) Petrologic and fluid inclusion study of calc-silicate rocks, Prince Rupert, British Columbia. Am. J. Sci. 279, 1135-1159.

Day, H.W. (1972) Geometrical analysis of phase equilibria in ternary systems of six phases. Am. J. Sci. 272, 711-734.

Duffy, C.J. and H.J. Greenwood (1979) Phase equilibria in the system $MgO-MgF_2-SiO_2-H_2O$. Am. Mineral. 64, 1156-1173.

Ellis, D.E. (1978) Mineralogy and petrology of chloride and carbonate bearing scapolite synthesized at 750°C and 4000 bars. Geochim. Cosmochim. Acta 42, 1271-1283.

Eugster, H.P. (1959) Reduction and oxidation in metamorphism. *In* Abelson, P.H., Ed., *Researches in Geochemistry*, I, John Wiley and Sons, N.Y., 397-426.

_____ (1977) Compositions and thermodynamics of metamorphic solutions. *In* Fraser, D.G., Ed., *Thermodynamics in Geology*, Reidel, Boston, 183-203.

_____ (1981) Metamorphic solutions and reactions. Phys. Chem. Earth 13-14, 461-507.

_____ and W.D. Gunter (1981) The compositions of supercritical metamorphic solutions. Bull. Mineral. 104, 817-826.

_____ and G.B. Skippen (1967) Igneous and metamorphic reactions involving gas equilibria. *In* Abelson, P.H., Ed., *Researches in Geochemistry*, II, John Wiley and Sons, N.Y., 492-521.

_____ and D.R. Wones (1962) Stability relations of the ferruginous biotite, annite. J. Petrol. 3, 82-125.

Ferry, J.M. (1976a) Metamorphism of calcareous sediments in the Waterville-Vassalboro area, south-central Maine: Mineral reactions and graphical analysis. Am. J. Sci. 276, 841-882.

_____ (1976b) P, T, f_{CO_2}, and f_{H_2O} during metamorphism of calcareous sediments in the Waterville-Vassalboro area, south-central Maine. Contrib. Mineral. Petrol. 57, 119-143.

_____ (1979) A map of chemical potential differences within an outcrop. Am. Mineral. 64, 966-985.

_____ (1980a) A case study of the amount and distribution of heat and fluid during metamorphism. Contrib. Mineral. Petrol. 71, 373-385.

_____ (1980b) A comparative study of geothermometers and geobarometers in pelitic schists from south-central Maine. Am. Mineral. 65, 720-732.

_____ (1981) Petrology of graphitic sulfide-rich schists from south-central Maine: An example of desulfidation during prograde regional metamorphism. Am. Mineral. 66, 908-930.

Flowers, G.C. (1979) Correction of Holloway's (1977) adaption of the modified Redlich-Kwong equation of state for calculation of the fugacities of molecular species in supercritical fluids of geologic interest. Contrib. Mineral. Petrol. 69, 315-318.

Frantz, J.D. and R.K. Popp (1979) Mineral-solution equilibria - I. An experimental study of complexing and the thermodynamic properties of aqueous $MgCl_2$ in the system $MgO-SiO_2-H_2O-HCl$. Geochim. Cosmochim. Acta 43, 1223-1239.

_____, _____, and N.Z. Boctor (1981) Mineral-solution equilibria - V. Solubilities of rock-forming minerals in supercritical fluids. Geochim. Cosmochim. Acta 45, 69-77.

_____ and H.K. Mao (1979) Bimetasomatism resulting from intergranular diffusion: II. Prediction of multimineralic zone sequences. Am. J. Sci. 279, 302-323.

French, B.M. (1966) Some geological implications of equilibrium between graphite and a C-H-O gas phase at high temperatures and pressures. Rev. Geophys. 4, 223-253.

Frost, B.R. (1979) Mineral equilibria involving mixed volatiles in a C-O-H fluid phase: The stabilities of graphite and siderite. Am. J. Sci. 279, 1033-1059.

Fyfe, W.S., N.J. Price, and A.B. Thompson (1978) *Fluids in the Earth's Crust*. Elsevier, N.Y.

Ghent, E.D. (1975) Temperature, pressure, and mixed-volatile equilibria attending metamorphism of staurolite-kyanite-bearing assemblages, Esplanade Range, British Columbia. Bull. Geol. Soc. Am. 86, 1654-1660.

_____ and C.D.S. DeVries (1972) Plagioclase-garnet-epidote equilibria in hornblende - plagioclase bearing rocks from the Esplanade Range, British Columbia. Canadian J. Earth Sci. 9, 618-635.

_____, D.B. Robbins, and M.Z. Stout (1979) Geothermometry, geobarometry, and fluid compositions of metamorphosed calc-silicates and pelites, Mica Creek, British Columbia. Am. Mineral. 64, 874-885.

Greenwood, H.J. (1962) Metamorphic reactions involving two volatile components. Carnegie Inst. Wash. Year Book 61, 82-85.

_____ (1967) Wollastonite: Stability in H_2O-CO_2 mixtures and occurrence in a contact metamorphic aureole near Salmo, British Columbia, Canada. Am. Mineral. 62, 1669-1680.

_____ (1973) Thermodynamic properties of gaseous mixtures of H_2O and CO_2 between 450° and 800°C and 0 to 500 bars. Am. J. Sci. 273, 561-571.

_____ (1975) Buffering of pore fluids by metamorphic reactions. Am. J. Sci. 275, 573-593.

Guidotti, C.V. (1970) The mineralogy and petrology of the transition from the lower to upper sillimanite zone in the Oquossoc area, Maine. J. Petrol. 11, 277-336.

Guo, Qiti (1980) Complete systems of closed nets for unary five-phase (n + 4) multisystems and their application to concrete configurations of phase diagrams. Scientia Sinica 23, 1039-1045.

Helgeson, H.C. and D.H. Kirkham (1974a) Theoretical prediction of the thermodynamic behavior of aqueous electrolytes: I. Summary of the thermodynamic/electrostatic properties of the solvent. Am. J. Sci. 274, 1089-1198.

_____ and _____ (1974b) Theoretical prediction of the thermodynamic behavior of aqueous electrolytes at high pressures and temperatures: II. Debye-Huckel parameters for activity coefficients and relative partial molal properties. Am. J. Sci. 274, 1199-1261.

_____ and _____ (1976) Theoretical prediction of the thermodynamic behavior of aqueous electrolytes: III. Equation of state for aqueous species at infinite dilution. Am. J. Sci. 276, 97-240.

_____, _____, and G.C. Flowers (1981) Theoretical prediction of the thermodynamic behavior of aqueous electrolytes at high pressures and temperatures: IV. Calculation of activity coefficients, osmotic coefficients, and apparent molal and standard and relative partial molal properties to 600°C and 5kb. Am. J. Sci. 281, 1249-1516.

_____, J.M. Delany, H.W. Nesbitt, and D.K. Bird (1978) Summary and critique of the thermodynamic properties of rock-forming minerals. Am. J. Sci. 278-A, 1-229.

Hemley, J.J. (1959) Some mineralogical equilibria in the system $K_2O-Al_2O_3-SiO_2-H_2O$. Am. J. Sci. 257, 241-270.

Hewitt, D.A. (1973a) The metamorphism of micaceous limestones from south-central Connecticut. Am. J. Sci. 273-A, 444-469.

_____ (1973b) Stability of the assemblage muscovite-calcite-quartz. Am. Mineral. 58, 785-791.

_____ (1975) Stability of the assemblage phlogopite-calcite-quartz. Am. Mineral. 60, 391-397.

_____ and M.C. Gilbert (1975) Experimental metamorphic petrology. Rev. Geophys. Space Phys. 13, 79-128.

Hoersch, A.L. (1981) Progressive metamorphism of the chert-bearing Durness limestone in the Beinn an Dubhaich aureole, Isle of Skye, Scotland: A reexamination. Am. Mineral. 66, 491-506.

Hollister, L.S. and M.L. Crawford (1981) *Fluid Inclusions: Applications to Petrology*. Mineral. Assoc. Canada, Calgary.

Holloway, J.R. (1977) Fugacity and activity of molecular species in supercritical fluids. *In* Fraser, D.G., Ed., *Thermodynamics in Geology*, Reidel, Boston, 161-181.

_____ (1981) Compositions and volumes of supercritical fluids in the Earth's crust. *In* Hollister, L.S. and M.L. Crawford, Eds., *Fluid Inclusions: Applications to Petrology*. Mineral. Assoc. Canada, Calgary, 13-38.

Hoy, T. (1976) Calc-silicate isograds in the Riondel area, southeastern British Columbia. Canadian J. Earth Sci. 13, 1093-1104.

Jacobs, G.K. and D.M. Kerrick (1981) Devolitilization equilibria in H_2O-CO_2 and H_2O-CO_2-NaCl fluids: An experimental and thermodynamic evaluation at elevated pressures and temperatures. Am. Mineral. 66, 1135-1153.

Joesten, R. (1974) Local equilibrium and metasomatic growth of zoned calc-silicate modules from a contact aureole, Christmas Mountains, Big Bend region, Texas. Am. J. Sci. 274, 876-901.

Jones, J.W. (1972) An almandine garnet isograd in the Rogers Pass area, British Columbia: The nature of the reaction and an estimation of the physical conditions during its formation. Contrib. Mineral. Petrol. 37, 291-306.

Kerrick, D.M. (1974) Review of metamorphic mixed-volatile (H_2O-CO_2) equilibria. Am. Mineral. 59, 729-762.

_____ and G.K. Jacobs (1981) A modified Redlich-Kwong equation for H_2O, CO_2, and H_2O-CO_2 mixtures at elevated pressures and temperatures. Am. J. Sci. 281, 735-767.

_____ and J. Slaughter (1976) Comparison of methods for calculating and extrapolating equilibria in $P-T-X_{CO_2}$ space. Am. J. Sci. 274, 883-916.

_____, K.E. Crawford, and A.F. Randazzo (1973) Metamorphism of calcareous rocks in three roof pendants in the Sierra Nevada, California. J. Petrol. 14, 303-325.

Korzhinskii, D.S. (1959) *Physicochemical Basis of the Analysis of the Paragenesis of Minerals*. Consultants Bureau, New York (in translation).

Kreulen, R. and R.D. Schuiling (1982) $N_2-CH_4-CO_2$ fluids during formation of the Dome de l'Agout, France. Geochim. Cosmochim. Acta 46, 193-203.

Labotka, T.C. (1981) Petrology of an andalusite-type regional metamorphic terrain, Panamint Mountains, California. J. Petrol. 22, 261-296.

Lewis, G.N. and M. Randall (1961) *Thermodynamics*. McGraw Hill, N.Y.

Malbrunot, P. and B. Vodar (1973) Experimental PVT data and thermodynamic properties of nitrogen up to 1000°C and 5000 bar. Physica 66, 351-363.

Marakushev, A.A. (1973) *Petrologia Metamorficheskikh Gornykh Porod*. Moscow State University Publishing House, Moscow.

McDowell, D. and W.A. Elders (1980) Mineralogical variations in borehole Elmore No. 1, Salton Sea geothermal field, California, U.S.A. Contrib. Mineral. Petrol. 74, 293-310.

_____ and M. McCurry (1977) Active metamorphism in the Salton Sea geothermal field, California: Mineralogical and mineral chemical changes with depth and temperature in sandstone. Geol. Soc. Am. Abstr. Prog. 9, 1088.

Meyer, C. and J.J. Hemley (1967) Wall rock alteration. *In* Barnes, H.L., Ed., *Geochemistry of Hydrothermal Ore Deposits*, Holt, Rinehart and Winston, Inc., N.Y., 166-235.

Moore, J.N. and D.M. Kerrick (1976) Equilibria in siliceous dolomites of the Alta aureole, Utah. Am. J. Sci. 276, 502-524.

Muffler, L.F.P. and D.E. White (1969) Active metamorphism of Upper Cenozoic sediments in the Salton Sea geothermal field and the Salton Trough, southeastern California. Geol. Soc. Am. Abstr. Prog. 80, 157-182.

Munoz, J.L. and S.D. Ludington (1974) Fluorine-hydroxl exchange in biotite. Am. J. Sci. 274, 396-413.

Newton, R.C., J.V. Smith, and B.F. Windley (1980) Carbonic metamorphism, granulites and crustal growth. Nature 288, 45-50.

Ohmoto, H. and D.M. Kerrick (1970) Devolitization equilibria in graphitic systems. Am. J. Sci. 277, 1013-1044.

Olejnik, S. and J.W. White (1972) Thin layers of water in vermiculites and montmorillonites - modification of water diffusion. Nature Phys. Sci. 236, 15-17.

Orville, P.M. and H.J. Greenwood (1965) Determination of ΔH of reaction from experimental P-T curves. Am. J. Sci. 263, 678-683.

Padovani, E.R., M.L. Batzle, and G. Simmons (1978) Characteristics of microcracks in samples from the drill hole Nordlingen 1973 in the Ries crater, Germany. Proc. 9th Lunar Planet. Sci. Conf. 2731-2748.

Petersen, U. (1965) Application of saturation (solubility) diagrams to problems in ore deposits. Econ. Geol. 60, 853-893.

Popp, R.K. and J.D. Frantz (1979) Mineral-solution equilibria - II. An experimental study of mineral solubilities and the thermodynamic properties of aqueous $CaCl_2$ in the system $CaO-SiO_2-H_2O-HCl$. Geochim. Cosmochim. Acta 43, 1777-1790.

Presnall, D.C. (1969) Pressure-volume-temperature measurements on hydrogen from 200°C to 600°C and up to 1800 atmospheres. J. Geophys. Res. 74, 6026-6033.

Rice, J.M. (1977a) Contact metamorphism of impure dolomitic limestone in the Boulder aureole, Montana. Contrib. Mineral. Petrol. 59, 237-259.

_____ (1977b) Progressive metamorphism of impure dolomitic limestone in the Marysville aureole, Montana. Am. J. Sci. 277, 1-24.

_____ (1980a) Phase equilibria involving humite minerals in impure dolomitic limestones: Part I. Calculated stability of clinohumite. Contrib. Mineral. Petrol. 71, 219-235.

_____ (1980b) Phase equilibria involving humite minerals in impure dolomitic limestones: Part II. Calculated stability of chondrodite and norbergite. Contrib. Mineral. Petrol. 75, 205-223.

Rich, R.A. (1979) Fluid inclusion evidence of Silurian evaporites in southeastern Vermont. Bull. Geol. Soc. Am. 90, part II, 1628-1643.

Robie, R.A., P.M. Bethke, and K.M. Beardsley (1967) Selected X-ray Crystallographic Data, Molar Volumes, and Densities of Minerals and Related Substances. U.S. Geol. Surv. Bull. 1248.

_____, B.S. Hemingway, and J.R. Fisher (1978) Thermodynamic properties of minerals and related substances at 298.15K and 1 bar (10^5 pascals) pressure and at higher temperatures. U.S. Geol. Surv. Bull. 1452.

Roedder, E. (1972) *Data of Geochemistry*, 6th edition. Chapter J.J: Composition of fluid inclusions. U.S. Geol. Surv. Prof. Paper 440-JJ.

Rose, A.W. and D.M. Burt (1979) Hydrothermal alteration. *In* Barnes, H.L., Ed., *Geochemistry of Hydrothermal Ore Deposits*, 2nd edition, Wiley-Interscience, N.Y., 173-235.

Roseboom, E.H., Jr. and E-an Zen (1982) Unary and binary multisystems, topologic classification of phase diagrams and relation to Euler's theorem on polyhedra. Am. J. Sci. 282, 286-310.

Rumble, D., III (1980) Oxygen isotope fractionation during regional metamorphism. Carnegie Inst. Wash. Year Book 79, 328-332.

_____, J.M. Ferry, T.C. Hoering, and A.J. Boucot (1982) Fluid flow during metamorphism at the Beaver Brook fossil locality, New Hampshire. Am. J. Sci. 282, 886-919.

Ryzhenko, B.N. and S.D. Malinin (1971) The fugacity rule for the systems CO_2-H_2O, CO_2-CH_4, CO_2-N_2, and CO_2-H_2. Geochem. Int'l. 8, 562-574.

261

_____ and V.P. Volkov (1971) Fugacity coefficients of some gases in a broad range of temperatures and pressures. Geochem. Int'l. 8, 468-481.

Shaw, D.M. (1956) Geochemistry of pelitic rocks. Part III. Major elements and general geochemistry. Bull. Geol. Soc. Am. 67, 919-934.

Shaw, H.R. and D.R. Wones (1964) Fugacity coefficients for hydrogen gas between 0° and 1000°C, for pressures to 3000 atm. Am. J. Sci. 262, 918-929.

Shmonov, V.M. and K.I. Shmulovich (1974) Molar volumes and equation of state of CO_2 at temperatures from 100 to 1000°C and pressures from 2000 to 10,000 bars. Dokl. Akad. Nauk SSSR 217, 206-209.

Sisson, V.B., M.L. Crawford, and P.H. Thompson (1982) Carbon dioxide-brine immiscibility at high temperatures, evidence from calcareous metasedimentary rocks. Contrib. Mineral. Petrol. 78, 371-378.

Skippen, G.B. (1971) Experimental data for reactions in siliceous marbles. J. Geol. 79, 457-481.

_____ (1974) An experimental model for low pressure metamorphism of siliceous dolomitic marble. Am. J. Sci. 274, 487-509.

_____ (1975) Thermodynamics of experimental sub-solidus silicate systems including mixed volatiles. Fortschr. Mineral. 52, 75-99.

_____ and D.M. Carmichael (1977) Mixed-volatile equilibria. In Greenwood, H.J., Ed., Short Course in Application of Thermodynamics to Petrology and Ore Deposits. Mineral. Assoc. Canada, Vancouver, 109-125.

Slaughter, J., D.M. Kerrick, and V.J. Wall (1975) Experimental and thermodynamic study of equilibria in the system CaO-MgO-SiO$_2$-H$_2$O-CO$_2$. Am. J. Sci. 275, 143-162.

Suzuki, K. (1977) Local equilbrium during contact metamorphism of siliceous dolomites in Kasuga-mura, Gifu-ken, Japan. Contrib. Mineral. Petrol. 61, 79-89.

Thompson, J.B., Jr., J. Laird, and A.B. Thompson (1982) Reactions in amphibolite, greenschist and blueschist. J. Petrol. 23, 1-27.

Touret, J. (1971) Le facies granulite en Norvege meridionale, II. Les inclusions fluides. Lithos 4, 423-436.

_____ (1977) The significance of fluid inclusions in metamorphic rocks. In Fraser, D.G., Ed., Thermodynamics in Geology. Reidel, Boston, 203-228.

_____ (1981) Fluid inclusions in high grade metamorphic rocks. In Hollister, L.S. and M.L. Crawford, Eds., Fluid Inclusions: Applications to Petrology. Mineral. Assoc. Canada, Calgary, 182-208.

Trommsdorf, V. (1972) Change in T-X during metamorphism of siliceous dolomite rocks of the central Alps. Schweiz. Mineral. Petrogr. Mitt. 52, 567-571.

Valley, J.W. and E.J. Essene (1980) Calc-silicate reactions in Adirondack marbles: The role of fluids and solid solution: Summary. Bull. Beol. Soc. Am. 91, 114-117.

_____, E.U. Petersen, E.J. Essene, and J.R. Bowman (1982) Fluorphlogopite and fluortremolite in Adirondack marbles and calculated C-O-H-F fluid compositions. Am. Mineral. 67, 545-557.

Walther, J.V. and H.C. Helgeson (1980) Description and interpretation of metasomatic phase relations at high pressures and temperatures: I. Equilibrium activities of ionic species in nonideal mixtures of CO$_2$ and H$_2$O. Am. J. Sci. 280, 575-606.

Wells, P.R.A. (1976) Late Archean metamorphism in the Buksefjord region, S.W. Greenland. Contrib. Mineral. Petrol. 56, 229-242.

White, D.E., I. Barnes, and J.R. O'Neil (1973) Thermal and mineral waters of nonmeteoric origin, California Coast Ranges. Bull. Geol. Soc. Am. 84, 547-560.

Wintsch, R.P. (1975) Solid-fluid equilibria in the system KAlSi$_3$O$_8$-NaAlSi$_3$O$_8$-Al$_2$SiO$_5$-SiO$_2$-H$_2$O-HCl. J. Petrol. 16, 57-79.

Wones, D.R. (1972) Stability of biotite: A reply. Am. Mineral. 57, 316-317.

_____ and H.P. Eugster (1965) Stability of biotite: Experiment, theory and application. Am. Mineral. 50, 1228-1272.

Zen, E-an (1966a) Construction of Pressure-Temperature Diagrams for Multicomponent Systems after the Method of Schreinemakers - a Geometric Approach. U.S. Geol. Surv. Bull. 1225.

_____ (1966b) Some topological relationships in multisystems of n + 3 phases - I. General theory, unary and binary systems. Am. J. Sci. 264, 401-427.

_____ (1967) Some topological relationships in multisystems of n + 3 phases - II. Unary and binary metastable sequences. Am. J. Sci. 265, 871-897.

_____ (1974) Prehnite- and pumpellyite-bearing mineral assemblages, west side of the Appalachian metamorphic belt, Pennsylvania to Newfoundland. J. Petrol. 15, 197-242.

_____ and E.H. Roseboom, Jr. (1972) Some topological relationships in multisystems of n + 3 phases. III. Ternary systems. Am. J. Sci. 272, 677-710.

Chapter 7

BUFFERING, INFILTRATION, and the CONTROL of INTENSIVE VARIABLES during METAMORPHISM

J.M. Rice and J.M. Ferry

INTRODUCTION

A fundamental goal of metamorphic petrology is to understand the in-terrelationships and controls of intensive variables during metamorphism. Identification of the processes that control pressure, temperature, mineral composition, and fluid composition should lead to a better understanding of the overall process of metamorphism, and how metamorphism integrates with other geochemical and geophysical phenomena that occur within the earth's crust. Mineral assemblages and the compositions of coexisting minerals and fluids in metamorphic rocks constitute a very powerful tool for the identification and characterization of the controls of intensive variables during metamorphism.

Significant advances in our understanding of the controls of metamor-phism were made possible when Greenwood (1962) recognized that sensitive relationships exist between the stable mineral paragenesis in a rock and the composition of the coexisting fluid phase. For example, over a small range of fluid composition calc-silicate rocks may contain a number of dif-ferent mineral assemblages. Because sensitive relationships exist between mineralogy and fluid composition, it is natural to question which is the controlling factor. Do mineral assemblages in a rock control the composi-tion of the coexisting fluid, do external reservoirs of fluid control the mineralogy of rocks with which they are in communication, or does some intermediate situation occur? These questions have been the basis of a pro-longed debate (e.g., Korzhinskii, 1959, 1967; Zen, 1963; Weill and Fyfe, 1967; J.B. Thompson, 1970; Gordon and Greenwood, 1971; Kerrick, 1974; Frost, 1982; Labotka et al., 1982). A major contribution to the resolution of these questions was a conceptual framework developed by Greenwood (1975) which permitted an evaluation of the degree to which mineral assemblages are capable of controlling or buffering the composition of the coexisting fluid. Greenwood's results suggest that if rocks do not interact with fluids derived from sources external to the rock under consideration, mineral assemblages in metamorphic rocks should have a large capacity to buffer the concentra-tion of volatile species in metamorphic fluids. The situation originates when metamorphic rocks interact only with initial pore fluid modified by prograde devolatilization reactions. We refer to this circumstance as control of fluid composition by internal buffering. There is good evidence both from the regional study of mineral reactions (as represented by isograds) and from the investigation of mineral parageneses at single outcrops, that this sort of control of metamorphic fluid composition occurs in some terrains. Internal buffering is now regarded as one of the most fundamental mechanisms by which fluid composition and other intensive variables are controlled during metamorphism.

Control of fluid composition by internal buffering represents one extreme sort of behavior. At the other extreme, rocks can be continuously infiltrated by large amounts of fluid derived from, and with composition controlled by, sources external to the rocks in question. Under these conditions, the composition of fluid coexisting with a particular rock is simply the composition of the reservoir fluid. Mineral equilibria effectively exert no control over the composition of the fluid in contact with rocks during metamorphism. The mineralogy and composition of minerals in rocks adjust until they are in equilibrium with the external reservoir of fluid. We refer to this circumstance as control of fluid composition by infiltration. There is also good evidence, both from the regional study of mineral reactions (isograds) and from the investigation of single or closely-spaced outcrops that this sort of control occurs in some terrains.

As with so many natural phenomena, there exists a complete, continuous spectrum from internal buffering of fluid composition without infiltration, through a combination of buffering and infiltration, to infiltration alone without buffering. In this chapter we will review the principles of the buffering and infiltration processes, and show how detailed studies of metamorphic rocks can be used to recognize buffering and infiltration, acting either alone or in concert. Although the concepts of buffering and infiltration in petrologic studies have been applied most successfully to discussion of the control of fluid composition by mineral reactions, we will demonstrate that the thermodynamic basis for chemical equilibrium requires that these processes be capable of controlling not only compositional variables in the fluid phase, but also temperature, pressure, and compositional variables of the condensed mineral phases. Our chapter begins with a discussion of mineral equilibria as buffers of fluid composition and illustrates the concept with a number of case studies. Next we take a more general view of buffering and infiltration, and consider mineral equilibria as buffers of mineral composition, temperature and pressure. Specific examples illustrate that natural mineral assemblages have acted as effective buffers of various intensive variables during metamorphism. Although the buffering of fluid composition, mineral composition, temperature, and pressure are treated separately, we emphasize from the beginning that mineral equilibria in a rock may simultaneously buffer two or more of these intensive variables during a metamorphic event.

The buffering process

The concept of buffering in the chemical literature is commonly restricted to homogeneous equilibria involving an aqueous solution containing either a mixture of a weak acid and its salt (conjugate base) or a weak base and its salt (conjugate acid). For example, the following reaction defines an equilibrium relation among species in a solution containing a mixture of acetic acid molecules and acetate ions: $HC_2H_3O_2 + H_2O = H_3O^+ + C_2H_3O_2^-$. If additional acid is added to this solution, much of the added hydronium ion will be consumed by reaction with the acetate ion present in solution to produce undissociated acetic acid. On the other hand, if a strong base is added, much of the added hydroxide ions of the base will react with the hydronium ions present to form water. More acetic acid then dissociates to replace the hydronium ion consumed by reaction. As long as the concentrations of the weak acid and salt are not greatly altered, the pH of the solution will remain nearly constant. Solutions of weak acids and their salts are thus referred to as buffer solutions because they "resist attempts to change" the concentrations of the dissociated species (e.g., through addition of strong acids or bases). The operation of such a buffer solution is an example of LeChatelier's principle, which states that if a system at equilibrium is subjected to a disturbance that changes any of the factors that determine the state of equilibrium, the system will react in such a way as to minimize the effect of the disturbance.

Although buffering conventionally refers to homogeneous equilibria which hold the pH of aqueous solutions nearly constant, the same basic principles apply to many heterogeneous equilibria which proceed within the earth during metamorphism. For example, the equilibrium among ferrosilite, magnetite, and quartz buffers the concentration of oxygen in any coexisting fluid according to the reaction: $2Fe_3O_4 + 6SiO_2 = 3Fe_2Si_2O_6 + O_2$. At constant temperature and pressure, this assemblage is invariant and uniquely defines the f_{O_2} of coexisting fluid. If external processes act either to oxidize or to reduce the assemblage, f_{O_2} remains fixed and the reaction simply progresses forward (reduction) or backwards (oxidation). Only after one or more of the minerals is consumed by reaction may f_{O_2} change in response to the external processes of reduction or oxidation. In this fashion, the assemblage of reactants and products controls or buffers the concentration of oxygen in the fluid phase.

The qualitative notion of buffering, and the idea that mineral equilibria hold intensive variables constant, can be understood in a more rigorous fashion by examining the thermodynamic requirements for heterogeneous phase equilibrium. In general, a mass balance equation can be written among a group of phases when the number of phases exceeds the number of thermodynamic components by one. The requirement for chemical equilibrium among the components is given by the following energy balance equation:

$$\Delta \bar{G}_r = \Delta \bar{H}_r - T\Delta \bar{S}_r + \Delta \bar{V}_s (P-1) + RT(\Sigma_i \nu_i \ln f_i + \Sigma_j \nu_j \ln a_j) = 0 \qquad (1)$$

where $\Delta \bar{G}_r$, $\Delta \bar{H}_r$, $\Delta \bar{S}_r$ refer to the molar reaction free energy, enthalpy and entropy at the temperature and pressure of equilibrium, $\Delta \bar{V}_s$ corresponds to the reaction volume change for condensed phases, ν_i is the stoichiometric coefficient of the ith phase in the equilibrium, and f_i and a_j refer to the fugacity and activities of the ith and j^{th} components, respectively (i and j refer to volatile and condensed components, respectively). If the standard states for the components are appropriately defined, equation (1) may also be expressed in the form:

$$\Delta \mu_{r_{(T,P)}} = \Delta \mu^o_{r_{(T,P)}} + RT(\Sigma_i \nu_i \ln a_i + \Sigma_j \nu_j \ln a_j) = 0 \; . \qquad (2)$$

For complex heterogeneous equilibria the total number of thermodynamic variables, both extensive and intensive, may be large. In many experimental and/or theoretical studies, variables such as f_{O_2}, f_{H_2}, f_{H_2O}, f_{CO_2}, mole fractions such as X_{CO_2}, activities of condensed components, and of course temperature and pressure are separated from one another whenever possible. This separation of variables is commonly employed in order to facilitate graphical representation; it is thermodynamically justified because the variation of free energy is an exact differential. However, it is fundamental to recognize that equation (1) requires that equilibrium among the components constrains possible variations in the intensive variables so that they are not all independent, but rather related to one another by the equilibrium itself. The phase-rule variance of a given reaction corresponds to the number of intensive parameters that can be independently varied; once this number of variables is fixed, then all other intensive variables must behave in a dependent fashion. As long as equilibrium is maintained among products and reactants, then $i + j + 2$ variables are constrained, and may thus not vary in an independent fashion. Equation (1) predicts that in nature, many of the possible intensive variables may not always be separable, but dependent upon one another through buffering by various mineral reactions.

Returning to the example of coexisting orthopyroxene, magnetite, and quartz, it should be recognized that, in contrast to the pure substances generally dealt with in the laboratory or in theoretical discussions, mineral equilibria in nature usually involve solid solution. Natural pyroxenes are, in the simplest case, $(Fe,Mg)_2Si_2O_6$ solid solutions. The coexistence of the three minerals nevertheless constitutes an oxygen buffer because the assemblage defines a specific f_{O_2} as illustrated by the following relationship:

$$\Delta\bar{H}_r - T\Delta\bar{S}_r + \Delta\bar{V}_s (P-1) + 3RTln(a^{opx}_{Fe_2Si_2O_6}) + RTlnf_{O_2} = 0 \qquad (3)$$

where $\Delta\bar{H}_r$, $\Delta\bar{S}_r$, and $\Delta\bar{V}_s$ refer to the end-member equilibrium. For a particular pyroxene composition, at a particular pressure and temperature, pyroxene + magnetite + quartz defines a particular concentration of oxygen in any coexisting fluid. Thus natural mineral assemblages are nearly as effective as ones involving pure substances in controlling fluid composition.

There is, however, an important difference between natural assemblages and assemblages of pure substances which involves their response to changes in intensive variables. For example, if coexisting orthopyroxene, magnetite and quartz are subjected to oxidation, the reaction will proceed from right to left. Because the reaction does not involve the $Mg_2Si_2O_6$ component of pyroxene, the pyroxene composition must change with reaction progress. As pyroxene changes composition, the f_{O_2} of coexisting fluid also changes in a continuous fashion (see equation 3). During oxidation, the assemblage pyroxene + magnetite + quartz does not fix f_{O_2} at a unique value (as would the assemblage ferrosilite + magnetite + quartz) but the equilibrium among pyroxene, magnetite, and quartz nevertheless retards changes in f_{O_2} that would otherwise occur due to the oxidation process. Petrologists commonly think of buffers only as mineral assemblages that fix intensive variables rather than as assemblages that retard changes in intensive variables. According to the definition of buffering given in standard chemistry texts, however, mineral assemblages that retard changes in intensive variables

constitute perfectly respectable buffers. It is important to understand
that buffers involving pure substances behave differently in detail from
those natural buffers that involve solid solution (or more exactly those
that have variance greater than zero at constant temperature and pressure).
It is also important, though, to recognize that many natural assemblages are
buffers; they serve as constraints which inhibit (but not prevent) changes
in intensive variables that might otherwise occur during metamorphism.
Following the treatment of buffers in the chemical literature, we adopt the
more general concept of buffering phenomena. Buffers are considered as those
equilibria that *resist changes* in certain intensive variables that are imposed on the system under consideration. The conventional petrologic definition of buffers -- equilibria that *fix* intensive variables -- is then simply
a special case of this more general definition of buffers.

BUFFERING, INFILTRATION, AND THE CONTROL OF FLUID COMPOSITION

Sources of fluids

Many, if not all, metamorphic rocks attest to the presence of volatiles
at the time of recrystallization. Evidence may be indirect, as in the
presence of hydrous and/or carbonate minerals which require a finite fugacity
of volatile species, or direct, as in the presence of fluid inclusions which
can be inferred to have been trapped in minerals at the time of their recrystallization. Although several investigators have suggested "dry" conditions attending metamorphism in the granulite and eclogite facies, the
commonly observed, anhydrous assemblages may in fact simply reflect equilibrium between the anhydrous paragenesis and a volatile phase at the appropriate
physical conditions (P, T, a_{H_2O}, etc.). The fluids attending metamorphism
may have in part been (a) present as initial pore fluid (meteoric or seawater) in the unmetamorphosed sediment, (b) generated internally as a result
of progressive devolatilization reactions, (c) introduced into the volume of
reacting rock from an external source, or (d) some mixture of fluid from two
or all three of these sources.

When sediment is first deposited, it may have a high porosity of up to
50% or more, occupied by pore fluid of meteoric or seawater origin. Numerous
studies have shown, however, that this initial porosity decreases rapidly
with depth of burial. Mechanical squeezing, compaction, and cementation
during lithification effectively reduce the original porosity to vanishingly
small values at the depths of burial commonly associated with the beginning

of regional metamorphism. It is therefore unlikely that original pore
fluids make a significant volumetric contribution to the fluid phase
attending subsequent metamorphism (Fyfe et al., 1958).

On the other hand, the amount of fluid that is contained in sediments
in the form of hydrous minerals (e.g., clays, sericite, chlorite, etc.) and
carbonates is of considerable importance during metamorphism. According to
the tabulation of Shaw (1956), average low-grade pelites (slates) contain
approximately 4.34 wt % H_2O and 2.31 wt % CO_2, whereas high-grade schists
contain 2.42 and 0.22 wt %, respectively. This loss of "bound" CO_2 and H_2O
released by devolatilization reactions taking place during metamorphism corresponds to approximately −1.55 moles of fluid per kilogram of rock. If
this fluid was entirely released at 500°C and 5 kbar, it would occupy approximately 12 volume percent of the rock (Walther and Orville, 1980).
Because devolatilization reactions occur over a range of temperature, the
1.55 moles of CO_2 and H_2O are unlikely to exist together as fluid at any one
point in time. The final porosity of high-grade rocks is extremely low
(e.g., <<0.1%); internally-generated fluid therefore must eventually escape
from the volume of rock in which it was generated. The actual amount of
internally generated fluid present at any given time is determined by the
relative rates of volatile production and escape from the reacting volume
of rock. As long as no other fluids are introduced, the composition of this
internally-generated fluid will be controlled (buffered) by the local devolatilization reactions that produce it.

The fluids attending metamorphism may also be introduced into a given
volume of rock from an external source by infiltration or diffusion. Such
external fluids may be meteoric, magmatic, metamorphic or juvenile in origin.
For example, stable isotope studies leave little doubt that meteoric and/or
magmatic fluids have interacted with some metamorphic terrains. In general,
these tend to be in shallow-level plutonic environments associated with
hydrothermal circulation of relatively large quantities of fluid. In addition, metamorphic fluids, generated by devolatilization reactions, may rise
upward into lower-grade rocks higher in the metamorphic pile. This fluid may
be transported either by grain boundary diffusion or by flow along fractures
and vein systems. Although conclusive studies are lacking, it is also possible that juvenile fluid may be introduced into deep-seated metamorphic
terrains from below the crust (Newton et al., 1980).

Buffering and the internal control of fluid composition

The progressive metamorphism of rocks which do not interact with external fluids involves an interplay between pore fluid present in the rocks before devolatilization reactions proceed, and the fluid evolved by those reactions. The view that local mineral assemblages can internally control the fluid composition by buffering was predicted by Greenwood (1975), based on calculations that demonstrate the conditions under which such buffering can operate. Greenwood derived equations which show the extent of reaction required for mineral assemblages to buffer the composition of coexisting fluid. For equilibria involving two volatile components, 1 and 2 (e.g., H_2O and CO_2, respectively), the variation in fluid composition with extent of reaction is given by the following expression (Greenwood, 1975):

$$\frac{dX_2}{dn_2} = \frac{1 - AX_2}{n_1 + n_2} \quad , \quad \text{where } A = \frac{\nu_1 + \nu_2}{\nu_2} \qquad (4)$$

with n_1 and n_2 being the number of moles of components 1 and 2, respectively. $X_2 = n_2/(n_1+n_2)$, and ν_1 and ν_2 are the stoichiometric coefficients of components 1 and 2 in the reaction (positive for products and negative for reactants). The variable dX_2 corresponds to the change in fluid composition associated with a given amount of reaction; dX_2/dn_2, therefore, defines what can be called the buffering capacity. Equilibria with high values of dX_2/dn_2 are efficient buffers, or have a large buffering capacity. In a binary fluid, n_1 and n_2 are coupled, and equation (4) can therefore be integrated to give:

$$\frac{1 - AX_2}{1 - AX_2^o} = \frac{n_1^o + n_2^o}{n_1^o + n_2^o + A(n_2 - n_2^o)} \qquad (5)$$

where null superscripts refer to initial values. Greenwood further simplified equation (5) by considering the buffering of one mole of initial fluid ($n_1^o + n_2^o = 1$). By making this assumption, the following relationships are readily derived:

$$X_2 = \frac{n_2}{1 + A(n_2 - n_2^o)} \qquad (6)$$

and

$$n_2 = \frac{X_2(1 - An_2^o)}{1 - AX_2} \quad . \qquad (7)$$

Plots of X_2 versus n_2 for different types of mixed-volatile equilibria

(Greenwood, 1975, Figs. 1-4) illustrate the influence of reaction stoichiometry on buffering capacity. Furthermore, they show that the buffering
capacity of mineral reactions is greatest for fluids initially far removed
in composition from the composition of the evolved fluid, and for small extents of reaction.

Greenwood used equations (6) and (7) to examine some specific examples
in order to determine the magnitude of X_2 and n_2 necessary to buffer one
mole of fluid. It is, however, not necessary, or in some cases even desirable, to restrict the calculations to one mole of initial fluid. Retaining
$(n_1^o + n_2^o)$ as a variable, equation (5) can be rearranged to give the following
more general expression:

$$X_2 = \frac{n_2}{n_1^o + n_2^o + A(n_2 - n_2^o)} \quad . \tag{8}$$

Note that in this expression, $(n_2 - n_2^o)$ gives the amount of component 2 which
is produced or consumed by the reaction and is related to what can be
defined as a reaction progress variable ξ:

$$\xi \equiv \frac{1}{\nu_2} (n_2 - n_2^o) \quad .$$

Written in terms of this progress variable, equation (8) takes the form:

$$X_2 = \frac{X_2^o(n_1^o + n_2^o) + \nu_2 \xi}{(n_1^o + n_2^o) + (\nu_1 + \nu_2)\xi} \quad . \tag{9}$$

The extent of reaction, or the progress variable itself, is given by:

$$\xi = \frac{(n_1^o + n_2^o)(X_2^o - X_2)}{X_2(\nu_1 + \nu_2) - \nu_2} \quad . \tag{10}$$

Equations (9) or (10) may be used to calculate the evolution of fluid
composition with extent of mineral reaction (or vice versa) for any desired
amount of initial fluid.

In order to examine the amount of reaction necessary to buffer fluid
composition for typical mixed-volatile equilibria, consider the hypothetical
metamorphism of 100 cm^3 of an impure dolomitic limestone with a very high
porosity of 5% consisting of quartz, calcite, dolomite and K-feldspar in the
molecular ratio 10:4:4:1, respectively. In terms of volume this rock would
contain 29.1% quartz, 18.9% calcite, 33.0% dolomite and 13.9% K-feldspar.
The equilibria governing the metamorphism of such bulk compositions are

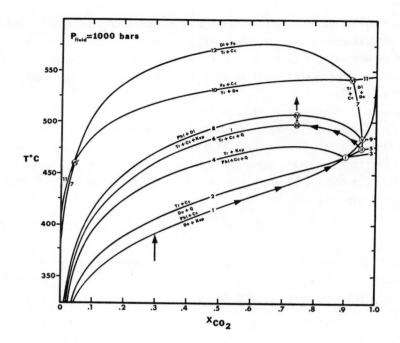

Figure 1. Temperature - fluid composition phase diagram for reactions within the phase volume calcite - dolomite - quartz - K-feldspar - CO_2 - H_2O at a total pressure of 1000 bars. From Rice (1977b).

illustrated in the isobaric temperature–fluid composition diagram shown in Figure 1.

Starting with an initial fluid composition of $X_{CO_2} = 0.3$, addition of heat to this rock increases the temperature until the following reaction boundary is reached at 392°C:

$$3 \text{ dolomite} + \text{K-feldspar} + H_2O = \text{phlogopite} + 3 \text{ calcite} + 3CO_2 \ . \quad (R1)$$

Further addition of heat causes the reaction to proceed, consuming H_2O and producing CO_2, thereby increasing the mole fraction of CO_2 in the fluid phase. Based on data given by Burnham et al. (1969) and Bottinga and Richet (1981), at 392°C and 1000 bars, a rock with 5% porosity which is filled with fluid of composition $X_{CO_2} = 0.3$, contains 0.087 moles H_2O and 0.037 moles CO_2. Defining the reaction progress variable ξ_1, as

$$\xi_1 = \text{moles phlogopite produced} = 1/3 \text{ moles } CO_2 \text{ produced} \ , \quad (11)$$

equation (10) can be used to describe the variation in moles of volatiles and condensed phases as a function of the reaction progress. Continuous

272

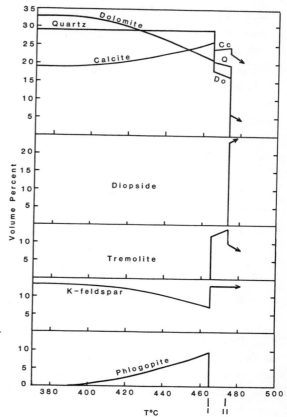

Figure 2. Variation in abundance of calcite, dolomite, quartz, K-feldspar, phlogopite, tremolite, and diopside with temperature in a system with the fluid composition internally buffered. Abrupt appearance or disappearance of phases takes place at isobaric invariant points labeled I and II on the phase diagram in Figure 1.

reaction from X_{CO_2} = 0.3 (392°C) to invariant point I (T = 465°C, X_{CO_2} = 0.90) corresponds to a ξ_1 value of 0.0624, a net increase of 0.1872 moles of CO_2, and a net decrease of 0.0624 moles of H_2O. The proportions of condensed phases developed during the interval T = 392°–465°C are illustrated in Figure 2, showing the volume percent of each mineral versus temperature.[1] Upon arrival at invariant point I, 0.0624 moles (9.3 cm^3) of phlogopite and 0.1872 moles (6.9 cm^3) of calcite have been produced and 0.1872 moles

----

[1]The calculations presented here assume that all fluid generated by a mineral reaction enters a reservoir where it remains in contact with and in equilibrium with the rock under consideration. As more fluid is evolved, the reservoir continuously expands. Amounts of reaction required to buffer the fluid composition would be less than the calculated amount if the fluid were allowed to escape from the reacting rock.

273

(12.0 cm^3) of dolomite and 0.0624 moles (6.8 cm^3) of K-feldspar have been consumed. The growth of phlogopite is non-linear with increasing temperature because of the stoichiometry of the reaction as expressed in equation (10). Even though the composition of the coexisting CO_2-H_2O fluid was buffered completely over the temperature range of 392°-465°C, only 9.3 modal % phlogopite was produced. Lower initial amounts of fluid would have resulted in even less reaction. If, for example, the rock initially contained 1% porosity, complete buffering of fluid composition over the interval T = 392°-465°C would produce only 0.0125 moles (1.9 cm^3) of phlogopite. This small amount of phlogopite might be difficult to observe petrographically.

At invariant point I, the following reactions occur simultaneously at constant temperature:

$$3 \text{ dolomite} + \text{K-feldspar} + H_2O = \text{phlogopite} + 3 \text{ calcite} + 3CO_2 \qquad \text{(R1)}$$

$$5 \text{ dolomite} + 8 \text{ quartz} + H_2O = \text{tremolite} + 3 \text{ calcite} + 7CO_2 \qquad \text{(R2)}$$

$$\text{phlogopite} + 2 \text{ dolomite} + 8 \text{ quartz} = \text{tremolite} + \text{K-feldspar} + 4CO_2 \qquad \text{(R3)}$$

$$5 \text{ phlogopite} + 6 \text{ calcite} + 24 \text{ quartz} = 3 \text{ tremolite} + 5 \text{ K-feldspar} + 6CO_2 + 2H_2O \ . \qquad \text{(R4)}$$

In a non-equilibrium or near-equilibrium situation, the relative rates of reaction will be determined by the kinetics of the individual reactions coupled with the requirement that the evolved fluid from all four reactions must be of the composition of the invariant point. All reactions will proceed in this fashion until one or more of the mineral reactants is totally consumed. Because kinetic considerations are not a part of classical thermodynamics, it is not possible to determine the relative amounts of heat consumed by reactions occurring simultaneously at the invariant point in a perfect equilibrium situation. Greenwood (1975), however, proposed that the linear dependence among the reactions affords a solution to the kinetic problem. In the example under consideration, note that reaction (R2) is equal to the sum of reactions (R1) and (R3). Under conditions of equilibrium, progress of reactions (R1) and (R3) together is exactly equivalent to the progress of (R2) alone. There are, in fact, only two linearly independent reactions associated with a given invariant point; all others can be derived by the appropriate linear combination of the two independent reactions. The overall combination of reactions at any invariant point, in any chemical system, regardless of size, can therefore be described in terms of two arbitrarily selected univariant reactions. Let us, for example, choose to model the overall reaction at invariant point I with reactions (R2) and (R4). With this selection, the process at the invariant point can be visualized as one in which phlogopite is destroyed by reaction (R4) and tremolite is produced by (R2) and (R4). The combination of these two equilibria must produce a fluid of the invariant point's composition. Let progress variables for (R2) and (R4) be defined as:

$$\xi_2 = \text{moles tremolite produced by (R2)} \qquad (12)$$

$$\xi_4 = 1/5 \text{ moles phlogopite consumed by (R4)} . \qquad (13)$$

The relative amounts of reaction can be calculated from the following general equation:

$$X_{CO_2} = \frac{n_{CO_2}}{n_{CO_2} + n_{H_2O}} = \frac{\Sigma_i (\nu_{CO_2,i}) \xi_i}{\Sigma_i (\nu_{CO_2,i} + \nu_{H_2O,i}) \xi_i} . \qquad (14)$$

For reactions (R2) and (R4) at invariant point I, equation (14) becomes:

$$X_{CO_2} = 0.90 = \frac{7\xi_2 + 6\xi_4}{7\xi_2 + 6\xi_4 - \xi_2 + 2\xi_4} = \frac{7\xi_2 + 6\xi_4}{6\xi_2 + 8\xi_4} . \qquad (15)$$

From this expression it is apparent that $\xi_4 = 1.33\ \xi_2$. The destruction of the 0.0624 moles of phlogopite formed by reaction (R1) enroute to the invariant point requires:

$$\xi_4 = 0.0125 \qquad (16)$$

$$\xi_2 = 0.0094 .$$

The net process at the invariant point is the production of 0.0469 moles (12.8 cm^3) of tremolite and 0.0624 moles (6.8 cm^3) of K-feldspar while consuming 0.375 moles (8.5 cm^3) of quartz, 0.0461 moles (1.70 cm^3) of calcite and 0.0470 moles (3.0 cm^3) of dolomite.

Further addition of heat results in buffering of the fluid composition along equilibrium curve (R2), the phlogopite-absent reaction, to invariant point II or until either dolomite or quartz is exhausted. The close proximity of points I and II virtually assures that point II will be reached. In fact equation (10) predicts (R2) will produce only 0.006 moles (1.61 cm^3) of tremolite while consuming 0.030 (1.91 cm^3) of dolomite and 0.047 moles (1.08 cm^3) of quartz. Upon arrival at invariant point II, four reactions proceed simultaneously. Again, because only two of these reactions are linearly independent, the net process at the invariant point can be described in terms of progress variables for two arbitrarily selected, linearly independent reactions such as the following:

$$\text{dolomite} + 2 \text{ quartz} = \text{diopside} + 2CO_2 \qquad \text{(R5)}$$

$$\xi_5 = \text{moles diopside produced by (R5)} = \text{moles dolomite consumed} \qquad \text{(18)}$$

$$\text{tremolite} + 3 \text{ calcite} + 2 \text{ quartz} = 5 \text{ diopside} + 3CO_2 + H_2O \qquad \text{(R6)}$$

$$\xi_6 = 1/5 \text{ moles diopside produced by (R6)} . \qquad \text{(19)}$$

Calculations of the relative amounts of the two reactions with equation (14) show that $\xi_5 = 10.5 \, \xi_6$. Owing to the silica-saturated bulk composition being considered, dolomite will be consumed by reaction (R5) at invariant point II. The previous calculations show that by the time point II is reached, 0.249 moles of dolomite remain; consumption of this amount of dolomite requires that $\xi_5 = 0.249$ and $\xi_6 = 0.024$. The net reaction at the invariant point produces 0.369 moles (24.4 cm^3) of diopside while consuming all dolomite, 0.072 moles (2.6 cm^3) calcite, 0.546 moles (12.4 cm^3) quartz and 0.024 moles (6.5 cm^3) tremolite. Once dolomite has been consumed, further addition of heat results in buffering of fluid composition along equilibrium curve (R6), the dolomite-absent reaction.

The important features of the buffering process, as illustrated in Figure 2, are the very gradual changes in the mode accompanying reaction along the isobaric univariant reaction curves, and the abrupt changes corresponding to reaction at the invariant points. Changes taking place at the invariant points are characterized by the sudden and volumetrically significant appearance of a new phase and the simultaneous disappearance of a previously abundant phase. The importance of these abrupt changes is that they are more likely to be observed in the field than the gradual changes taking place along the univariant buffer curves. The discontinuous changes occurring at the isobaric invariant points would make good isograds, in contrast to the individual isobaric univariant equilibria, which would not, both because they result in gradual changes in the mineral assemblage and because they do not represent a unique temperature at fixed pressure. Metamorphic terrains dominated by internal control of fluid composition should therefore exhibit divariant zones (or univariant zones in isobaric contact aureoles) composed of fluid-buffering mineral parageneses separated by abrupt isograds marking the polybaric traces of univariant (or isobaric invariant) discontinuous reactions. It is also apparent that the buffering process results in a more simple sequence of reaction steps than one might initially predict from examination of the multitude of reactions present on a given phase diagram. Under conditions of complete internal control, the path of progressive metamorphism is determined by the initial bulk composition, and many of the possible reactions are effectively bypassed during the buffering process. This feature suggests that interlayered rocks of contrasting bulk composition may develop independently, thereby producing local variations in chemical potentials of volatile species from layer to layer. Under these conditions, individual layers of a given outcrop may contain different fluid-buffering assemblages. Gradients in the chemical potentials of volatile species may be reflected in mineralogic zoning at the margins of individual layers (see, for example, Hewitt, 1973; Vidale and Hewitt, 1973). Finally, it is important to note from equation (10) that the mineral proportions developed during progressive metamorphism can be used as indicators of the amount of fluid that interacted with a given volume of rock.

The process by which fluid composition may be internally controlled by mineral reactions during progressive metamorphism has been described in detail, and evidence by which it might be recognized in the field has been presented. The question now becomes, do petrologists find such evidence in nature? Perhaps the best example of metamorphic rocks that equilibrated with an internally-generated, and compositionally-buffered, fluid is found in the contact aureole surrounding the Marysville stock in west-central Montana (Rice, 1977b). On the eastern side of the intrusion, impure dolomitic limestones have been progressively metamorphosed across a zone 1 to 3 km wide. The bulk composition of the majority of the limestones is not unlike the hypothetical example discussed above. Reactions affecting such bulk compositions are illustrated on Figure 1, and the observed distribution of mineral assemblages is shown on Figure 3. The most striking feature of the Marysville aureole is the fact that the spatial distribution of mineral paragenesis in the low- and medium-grade portions corresponds exactly to what one would predict from a buffering path with increasing temperature (shown by arrows on Fig. 1) on the $T-X_{CO_2}$ diagram. The general features predicted in the hypothetical example discussed above are almost exactly matched in the field where the progressive sequence is characterized by zones of isobarically univariant buffering assemblages which are separated by distinctive isograds resulting from reaction at isobaric invariant points or singular points ($T-X_{CO_2}$ maxima). For example, zone A contains the diagnostic assemblage of calcite + dolomite + quartz + K-feldspar + phlogopite corresponding to reaction (R1). At isograd I, tremolite first appears and phlogopite completely disappears in all dolomite-bearing bulk compositions. This

277

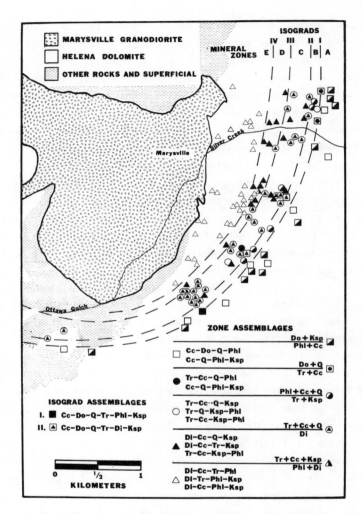

Figure 3. Distribution of mineral assemblages in the Marysville contact aureole. From Rice (1977b).

distribution is consistent with reaction at invariant point I, and confirmed by observation of the six-mineral paragenesis defining point I. Inward from isograd I there exists a very narrow zone (B) defined by the assemblage calcite + dolomite + quartz + tremolite + K-feldspar corresponding to reaction (R2). In the field, the most distinctive isograd is II, which is based on the abrupt appearance of abundant diopside and the simultaneous disappearance of dolomite. Although the modal amounts of minerals in zones A and B are highly variable owing to fine-scale compositional layering, the

278

general features match those of Figure 2. There is little doubt that reaction at invariant point II (Fig. 1) was responsible for isograd II. Inward from isograd II, there exists a zone (C) of uniform width defined by
the isobaric univariant assemblage corresponding to reaction (R6) which
apparently buffered the fluid composition near the $T-X_{CO_2}$ maximum located
at $X_{CO_2} = 0.75$.

It should be apparent that in the Marysville aureole the spatial distribution of low-variance assemblages requires efficient buffering of the
fluid composition. The overall similarity between modes of metacarbonate
rocks from the Marysville aureole and the predicted modes in Figure 2 suggests that little or no externally-derived fluid interacted with the limestones in the outer and central portions of the aureole. There is ample
evidence for introduction of H_2O-rich fluids near the contact with the
granodiorite. These fluids were, however, apparently localized in discrete vein systems, and transported away from reacting rocks in the central
and outer zones of the aureole. In this fashion, the chemical potentials
of CO_2 and H_2O in samples undergoing progressive increase in temperature
were controlled internally by the local mineral assemblages.

Internal control of metamorphic fluid composition by local mineral
reactions can be identified not only by examining the spatial distribution
of mineral assemblages over a range of grades (e.g., Fig. 3), but also by
the investigation of mineral equilibria within individual outcrops which
can be inferred to have equilibrated at the same pressure and temperature.
If fluid composition is controlled during metamorphism by some agent or
process external to an outcrop under consideration, then minerals in each
part of the outcrop should record equilibrium with a common fluid composition. Conversely, if mineral equilibria in samples from different parts of
an outcrop (or different compositional layers) record differences in fluid
composition, the chemical potentials of components in the fluid could not
have been controlled externally. Differences in fluid composition recorded
by mineral equilibria from different parts of the outcrop can be explained
by internal control of fluid composition by local mineral reactions. Different mineral reactions in different samples buffer the composition of
coexisting fluid at different values.

Using this approach, Rumble (1978) demonstrated the internal control
of fluid composition during metamorphism of a single large outcrop of
quartzites and pelitic schists on Black Mountain, New Hampshire. He studied
numerous samples containing either quartz + muscovite + staurolite +

279

chloritoid + chlorite + ilmenite + magnetite or quartz + muscovite + kyanite
+ staurolite + chloritoid + ilmenite +
magnetite (as well as other assemblages),
and found significant variation in mineral composition in different samples
containing the same assemblage (Fig. 4).
The two assemblages are invariant at
constant pressure, temperature, μ_{H_2O},
and μ_{O_2} in a model $K_2O-Al_2O_3-FeO-MgO-$
$MnO-TiO_2-SiO_2-H_2O-O_2$ system. The variation in mineral composition therefore
records a variation in one or more of
the intensive variables within the outcrop during metamorphism. Rumble presented chemical, isotopic, and geologic
arguments that pressure and temperature
were uniform over the area sampled.

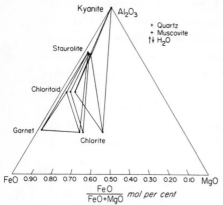

Figure 4. Kyanite, staurolite, chloritoid, garnet, and chlorite compositions from Black Mountain, New Hampshire, plotted on an AFM projection. The symbol next to H_2O signifies that
projection is made through H_2O. From Rumble
(1978).

Differences in μ_{H_2O} and/or μ_{O_2} must have existed during metamorphism between
samples spaced 15–650 m apart. Rumble also found significant differences in
$\delta^{18}O$ of quartz and magnetite from different samples. Because relationships
may be written among μ_{18_O}, μ_{16_O}, μ_{H_2O}, μ_{O_2}, and the chemical potentials of
components in the coexisting minerals found at Black Mountain, the mineral
assemblages were capable of buffering these chemical potentials in a coexisting fluid during metamorphism. The existence of the buffer assemblages, combined with the inferred differences in the chemical potential
of volatile species between samples, suggests that the composition of fluid
within samples was controlled internally, and independently of nearby samples, by mineral-fluid equilibria.

Infiltration and the external control of fluid composition

Internal control of the composition of a metamorphic fluid generated
and buffered by local devolatilization reactions, obviously comprises an
end-member situation in nature. At the other extreme exists the situation
where the chemical potentials of volatile components are controlled, or
fixed at constant values, by an external fluid reservoir. If the external
reservoir is large relative to the buffering capacity of local mineral assemblages, then reactions among the minerals will have little ability to
modify the externally-controlled chemical potentials. Under these conditions, mineral reactions will proceed in such a manner as to adjust the
mineralogy of the rock to be in equilibrium with the external fluid.

Many, if not most, devolatilization reactions have the capacity to buffer significant quantities of fluid. This, coupled with the relative rates of mineral reactions, suggests that small volumes of fluid transported by slow grain boundary diffusion will, in all likelihood, be buffered. We therefore believe that the most important mechanism for the external control of fluid composition is infiltration of rocks by large quantities of fluid. The lower limit of what constitutes "large" is arbitrary, and can only be evaluated for specific rocks undergoing specific reactions. With reference to equation (9), "large" is $(n_1^o + n_2^o)$ of sufficient magnitude that any amount of reaction in terms of ξ results in a negligible difference between X_2 and X_2^o.

In metamorphic terrains where fluid composition is internally controlled (buffered) by local mineral reactions, $T-X_{CO_2}$ conditions during metamorphism should closely follow the appropriate isobaric univariant reaction curves on the phase diagram (Fig. 1). Consequently, when fluid composition is internally controlled, reactants and products for the univariant reactions should coexist in individual hand specimens and be widely distributed across the terrain (Fig. 3). In contrast, if metamorphic fluid composition is externally controlled by an infiltration process, the $T-X_{CO_2}$ conditions during metamorphism should correspond to vertical or near vertical paths on the phase diagram with the position of the prograde path representing the fluid composition imposed on all rocks of the region (*cf.* the dashed path on Fig. 6). Consequently, isobaric divariant assemblages will be very common, whereas the coexistence of reactants and products for isobaric univariant reactions will be rare. Because the isograd boundaries are isothermal at constant fluid composition, the odds of finding mineral assemblages corresponding to the univariant reaction would be very low. In fact, they will be found only in those samples metamorphosed to a peak temperature precisely equal to the temperature of the univariant reaction for the fixed fluid composition (T_{rxn} in Fig. 6). The distribution of reactants and products for isobaric univariant devolatilization reactions serves as an important probe which monitors the mechanism by which fluid composition is controlled during metamorphism.

Another significant feature to be expected in terrains in which fluid composition is externally controlled by infiltration is the lack of evidence for local gradients in chemical potentials of volatile constituents. Layered rocks of contrasting bulk composition will not be able to independently control the composition of coexisting pore fluid, but rather will

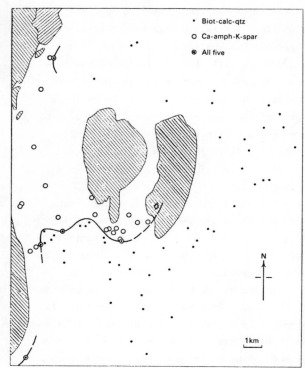

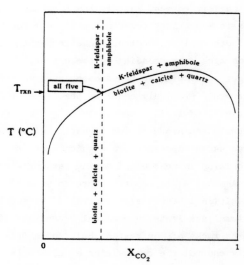

Figure 5. Isograd in the Whetstone Lake area, Ontario based on reaction (R4) (biotite + calcite + quartz = amphibole + K-feldspar). Patterned areas are synmetamorphic granitic intrusions. From Carmichael (1970).

Figure 6. Schematic representation of reaction (R4) on an isobaric $T-X_{CO_2}$ diagram. Dashed $T-X_{CO_2}$ path explained in the text.

282

adjust their mineralogies to be in equilibrium with the infiltrating fluid. The resulting mineral assemblages in different bulk compositions will all be compatible with the common fluid composition. In addition, the modal proportions of minerals will not continuously change in any systematic fashion related to temperature, because all mineral reaction occurs discontinuously at those unique temperatures at which the T-X_{CO_2} path of metamorphism crosses the isobaric univariant reaction curves.

An example in which the spatial distribution of prograde reactants and products indicates external control of fluid composition is Carmichael's (1970) study of the Whetstone Lake area, Ontario. Carmichael identified reaction (R4) in calcareous rocks and mapped the distribution of reactants (biotite + calcite + quartz), products (calcic amphibole + K-feldspar), and reactants + products as shown in Figure 5. The distribution of reactants and products can be separated by a line (isograd) along which only a very few occurrences (four out of more than 60 samples) of both reactants and products coexist. The end-member reaction (R4) corresponds to a univariant curve on an isobaric T-X_{CO_2} diagram (Fig. 6). Comparison of Figures 5 and 6 with Figures 3 and 1, respectively, indicates a lack of internal buffering of fluid composition in the Whetstone Lake area. In particular, the paucity of rocks containing both reactants and products suggests that metamorphism occurred with increasing temperature at constant or nearly constant fluid composition (dashed path, Fig. 6). Such a path implies that X_{CO_2} was externally controlled at a particular value in the region during the metamorphic event.

As noted above, studies of mineral equilibria in individual outcrops may also demonstrate external control of fluid composition if samples representing different bulk compositions in different parts of the outcrop are consistent with equilibrium with fluid of a common composition. Albee (1965), for example, studied numerous samples containing either muscovite + quartz + kyanite + chloritoid + chlorite, muscovite + quartz + biotite + garnet + chlorite, or muscovite + quartz + garnet + chloritoid + chlorite, collected from a small area of exposures on Mount Grant, Vermont. Within error of measurement, he detected no variation in mineral composition for different samples containing the same assemblage (Fig. 7). On the basis of his analyses, Albee was able to construct a topologically consistent Thompson AFM diagram to illustrate chemographic relations among minerals in all samples that he analyzed (Fig. 7). The three assemblages are univariant at constant pressure, temperature, and μ_{H_2O} in the model AFM system,

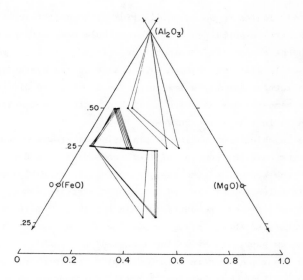

Figure 7. Compositions of kyanite, chloritoid, garnet, chlorite and biotite from Mount Grant, Vermont, plotted on an AFM projection. From Albee (1965).

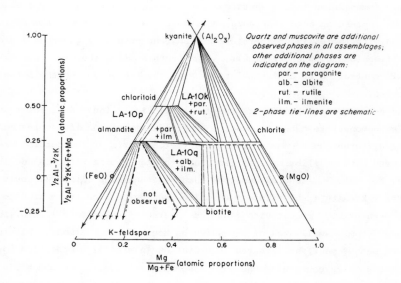

Figure 8. Observed mineral assemblages and compositions of coexisting minerals, shown on a Thompson AFM projection, for pelitic schists from Mount Grant, Vermont. Additional accessory minerals include hematite, magnetite, apatite, tourmaline, graphite, clinozoisite-allanite, and pyrite. From Albee (1965).

$K_2O-Al_2O_3-FeO-MgO-SiO_2-H_2O$. The uniform mineral compositions within each assemblage and the consistent chemographic relations in Figure 8 are consistent with uniform pressure, temperature, and μ_{H_2O} during metamorphism over the area sampled. Albee also reports that there are no measurable differences in $\delta^{18}O$ of any particular mineral between samples. The mineralogical and isotopic data led Albee to conclude that concentrations of H_2O, ^{18}O, and ^{16}O were the same in the metamorphic fluid in equilibrium with each sample. A logical mechanism by which this could be achieved is through continuous infiltration of metamorphic rock by large volumes of fluid with constant concentration of H_2O, ^{18}O, ^{16}O, etc. The rocks from Mount Grant evidently adjusted their mineralogy, mineral chemistry, and isotopic composition so as to be in equilibrium with the infiltrating fluid.

An interesting aspect of Albee's study is that although data are consistent with external control of the concentrations of H_2O, ^{18}O, and ^{16}O in fluid, they are inconsistent with external control of the concentration of O_2. Albee found various combinations of magnetite, ilmenite, hematite, and rutile in the samples he studied. At constant pressure and temperature, the different oxide assemblages record different μ_{O_2} in any coexisting fluid. Albee's studies emphasize the importance, in any discussion of internal versus external control of fluid composition, of which particular species is in question. For any particular metamorphic fluid-rock system, the concentration of some fluid species may be controlled externally while others are controlled internally by mineral-fluid equilibria.

In our discussion of external control of metamorphic fluid composition, we have not specified the scale on which this phenomenon might occur. Over what distance can a process such as infiltration control fluid composition during metamorphism? The study of Carmichael (1970) provides insight into the magnitude of the process. In addition to the calc-silicate isograd defined by reaction (R4), Carmichael mapped four other isograds based on mineral reactions in pelitic schists from the Whetstone Lake area (Fig. 9). The mineral reactions may be represented as univariant curves on an isobaric $T-X_{CO_2}$ diagram (Fig. 10) with reaction curve 5 corresponding to the calc-silicate isograd shown in Figure 5 and the equilibrium curve in Figure 6. The isograds mapped from mineral assemblages in the pelitic schists (1-4) intersect the isograd defined by reaction (R4) in carbonate rocks. Carmichael interpreted the intersecting isograds in terms of a regional gradient in metamorphic fluid composition. For example, the sequence of isograds along traverse B-B' in Figure 9 corresponds to an externally

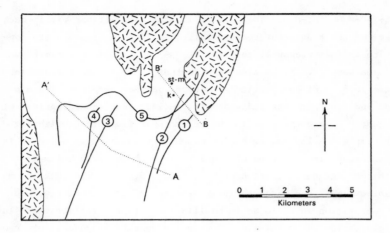

Figure 9. Isograd map for the Whetstone Lake area, Ontario. Isograds are based on the following mineral reactions (from Carmichael, 1970):

(1) chlorite + muscovite + garnet = staurolite + biotite + quartz + H_2O;

(2) chlorite + muscovite + staurolite + quartz = biotite + kyanite + H_2O;

(3) kyanite = sillimanite;

(4) staurolite + muscovite + quartz = sillimanite + garnet + biotite + H_2O;

(5) biotite + calcite + quartz = amphibole + K-feldspar + CO_2 + H_2O.

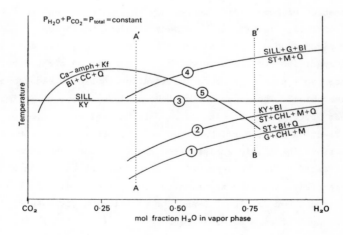

Figure 10. The $T-X_{CO_2}$ diagram which schematically represents the five mineral reactions mapped in Figure 9 under conditions of a constant pressure greater that the Al_2SiO_5 triple point. From Carmichael (1970).

286

controlled, H_2O-rich fluid composition (B-B', Fig. 10). The sequence of isograds along A-A' several kilometers away (Fig. 9) corresponds to an externally-controlled, but more CO_2-rich fluid composition (A-A', Fig. 10). These observations suggest that fluid composition may be regionally controlled on a scale of 1 to 5 kilometers. In the case of the Whetstone Lake area, the more H_2O-rich fluids along traverse B-B' are thought to have resulted from infiltration of aqueous magmatic fluids expelled from crystallizing synmetamorphic plutons (patterned areas, Figs. 5 and 9).

Fluid composition controlled by combined infiltration and buffering

In this section we consider a control of fluid composition that combines the phenomena of infiltration and buffering. Rocks are considered to be in communication with some external fluid reservoir. Either because of the limited size of the reservoir and/or because of small permeability, the rock's buffer capacity with respect to major fluid species is retained throughout metamoprhism. Devolatilization reactions occur during metamorphism within each rock under consideration, and these mineral-fluid reactions control the composition of the fluid occupying the pore space of the rock at any particular instant. Such a situation would occur, for example, if the fluid from an external reservoir flowed very slowly through the metamorphic terrain and if the rate of introduction of fluid were slower than the rate at which the devolatilization reactions occurred. Although the devolatilization reactions buffer the composition of the pore fluid, the fluid is actually a mixture. Part of the fluid is composed of molecules introduced into the rock from the external fluid reservoir. The remainder of the fluid is composed of molecules generated internally within the rock by the devolatilization reactions. Thus the molecular constituency of the fluid (as opposed to its chemical composition) results from, and is controlled by, an interplay between infiltration and buffering.

We now turn to how combined infiltration and buffering may be recognized from the study of metamorphic rocks and how the combined processes may be differentiated from the end-member cases acting alone. Because mineral equilibria control fluid compositions during combined buffering and infiltration, many observed field relations will be consistent with either buffering alone or combined buffering and infiltration. For example, in both cases (a) mineral equilibria are likely to record differences in the chemical potentials of volatile components over small exposures of metamorphic rock, and (b) the coexisting products and reactants of isobaric univariant devolatilization reactions will be widely distributed over a range of metamorphic grades. In order to differentiate buffering alone from combined buffering and infiltration, additional observations are necessary. These observations include modal analysis of rocks and the determination of $\Delta \bar{H}$ of reaction for isobaric univariant devolatilization reactions that occur spread out over a range of grade.

Exothermic mineral-fluid reactions. Prograde metamorphism, in part, is the result of the addition of heat to rocks. The heat raises the temperature of the rocks and drives endothermic devolatilization reactions. If the mineral reactions buffer metamorphic fluid composition, coexisting reactants and products will be distributed over a range of metamorphic grades. In some rocks, however, observed prograde mineral-fluid reactions may be exothermic ($\Delta \bar{H}$ of reaction < 0), for example:

$$3 \text{ anorthite} + \text{calcite} + H_2O = 2 \text{ zoisite} + CO_2 . \qquad (R7)$$

Reactants and products of reaction (R7) are commonly found to coexist and be widely distributed in metamorphic terrains (e.g., Hewitt, 1973; Ferry, 1976a) and such observations indicate that reaction (R7) buffered metamorphic fluid composition in rocks containing zoisite + calcite + anorthite. Exothermic reactions such as (R7) cannot be driven from left to right by increasing temperature. It is unlikely that pressure changes in a metamorphic terrain are sufficiently large to drive a reaction like (R7). From the stoichiometry of the reaction, however, it is evident that it may proceed from left to right through the interaction of zoisite + anorthite + calcite in the rock with an H_2O-rich fluid. Both Hewitt (1973) and Ferry (1976a) interpreted the prograde production of zoisite by reaction (R7) as a result of infiltration of H_2O-rich fluid into zoisite + anorthite + calcite-bearing carbonate rocks during metamorphism. Identification of exothermic mineral-fluid buffer reactions thus is one line of evidence for combined buffering and infiltration during a metamorphic event.

Mineral proportions in isobaric univariant mineral assemblages. The widespread occurrence of isobaric univariant devolatilization reactions in a metamorphic terrain is evidence that the reaction buffered metamorphic fluid composition. Examination of the proportions of reactants to products in the rocks often reveals whether infiltration occurred along with the buffering or not (*cf.* Fig. 2). Consider a rock composed of calcite, quartz,

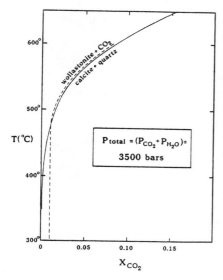

Figure 11. Equilibrium among calcite, quartz, wollastonite, and CO_2-H_2O fluid as a function of temperature and fluid composition (solid line). Calculated from data in Ferry (1976b). The dashed line follows the metamorphism of a hypothetical rock described in the text.

and 1% pore space filled with CO_2-H_2O fluid of composition arbitrarily taken as $X_{CO_2} = 0.01$. If metamorphism occurs at 3500 bars pressure and to a maximum temperature of 600°C, Figure 11 describes the mineralogical changes that will occur in the rock. Below 470°C calcite + quartz will be stable. At 470°C wollastonite will form according to the reaction

$$calcite + quartz = wollastonite + CO_2 \ . \tag{R8}$$

Between 470°C and 600°C reaction (R8) will continue to produce wollastonite from calcite + quartz and buffer X_{CO_2} of the pore fluid along the univariant curve from $X_{CO_2} = 0.01$ at 470°C to $X_{CO_2} = 0.086$ at 600°C. The amount of wollastonite produced in the interval between 470°C and 600°C can be readily calculated, if no infiltration occurs. Let a progress variable for reaction (R8) be defined as:

$$\xi_8 = moles\ wollastonite\ per\ 1000\ cm^3\ metamorphic\ rock \ . \tag{20}$$

If $X_2^o = 0.01$ is the initial pore fluid composition and $X_2 = 0.086$ is the final pore fluid composition at 600°C, then equation (10) can be used to calculate the amount of wollastonite produced by reaction (R8) in the interval 470°–600°C. At 470°C, 3500 bars, a rock with 1% porosity which is filled with fluid of composition $X_{CO_2} = 0.01$, contains 0.447 moles H_2O and 0.005 moles CO_2. Using relation (10), a $\xi_8 = 0.0375$ moles wollastonite/ 1000 cm^3 rock is obtained. This amount of wollastonite corresponds to 1.50 cm^3/1000 cm^3 rock, or a modal amount of 0.15%.

Because the amount of wollastonite produced by reaction (R8) can be predicted for the case of pure buffering, the modal amount of wollastonite in a rock composed of wollastonite + calcite + quartz can be used to test whether buffering was perhaps accompanied by infiltration. An example of this exercise is the study by Rumble et al. (1982) of a bed of metamorphosed silicified brachiopods (Fig. 12). The rock is composed of calcite + quartz

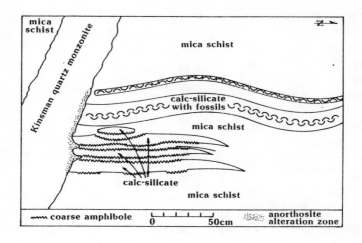

Figure 12. Geologic sketch map of the outcrop at Beaver Brook, west-central New Hampshire, studied by Rumble et al. (1982).

+ wollastonite and peak metamorphic conditions, inferred from mineral equilibria in nearby rocks, were P = 3500 bars; T = 600°C. Reaction (R8) and no other reaction occurred during metamorphism. The rock is composed of 70 modal percent wollastonite. This amount of wollastonite is far more than would be produced by buffering alone, as illustrated by calculations previously described. Even if pore space were an unreasonably high 10% and initial pore fluid were pure H_2O, the amount of wollastonite now observed in the rock is over 40 times what could be produced by buffering without infiltration.

Rumble et al. (1982) interpreted the 70% wollastonite as a result of combined buffering and infiltration during metamorphism. Reaction (R8) buffered the composition of metamorphic fluid in contact with the rock at the peak of metamorphism. The "excess" wollastonite was produced by infiltration of the carbonate rock by H_2O-rich fluids as reaction (R8) buffered fluid composition (during and after infiltration) to the equilibrium value represented by the curve in Figure 11.

An important reason for determining the amount of wollastonite in a rock composed of wollastonite + calcite + quartz is that a minimum estimate of the amount of fluid involved can readily be determined when infiltration is indicated. Modal analysis of the rock studied by Rumble et al. (1982) indicates that ξ_8 = 17.531. If all infiltration occurred at 600°C and involved pure H_2O, then

$$X_{CO_2} = 0.086 = \frac{\xi_8}{\xi_8 + n_{H_2O}} \qquad (21)$$

where n_{H_2O} = number of moles H_2O that infiltrated 1000 cm^3 rock. From
equation (21) a value of n_{H_2O} = 186.32 moles is obtained and this corresponds to a volumetric fluid:rock ratio of 4.6:1. If infiltration also
occurred at T < 600°C, or if infiltrating fluid had composition $0 < X_{CO_2}$
< 0.086, the calculated fluid-rock ratio would be larger.

This example demonstrates that modal amounts of minerals which constitute an isobaric univariant buffer assemblage can be used to determine
whether buffering occurred with or without infiltration during metamorphism.
If infiltration is indicated, the mode of such a rock can be used to obtain
a minimum estimate of the amount of fluid that infiltrated a particular
specimen during the metamorphic event. For a rock in which i linearly
independent dehydration/decarbonation reactions have occurred, the general
expression that relates its mode to the amount of infiltration of CO_2-H_2O
fluid that occurred during metamorphism is:

$$X^e_{CO_2} = \frac{\Sigma_i (\nu_{CO_2,i})\xi_i + n_{CO_2}}{\Sigma_i (\nu_{CO_2,i} + \nu_{H_2O,i})\xi_i + n_{CO_2} + n_{H_2O}} \qquad (22)$$

where $X^e_{CO_2}$ is the composition of CO_2-H_2O fluid in equilibrium with the rock
while each mineral reaction i progressed by an amount ξ_i. The ν_i are
stoichiometric coefficients for CO_2 and H_2O in the ith reaction. The n_{CO_2}
and n_{H_2O} are the number of moles CO_2 and H_2O that infiltrated the rock.
The composition of the infiltrating fluid, X_{CO_2} is equal to $n_{CO_2}/(n_{CO_2} +$
$n_{H_2O})$. Expressions analogous to (22) may be readily derived for mineral
reactions that involve fluid species other than CO_2 or H_2O (e.g., Ferry,
1981).

Mineral proportions in isobaric invariant mineral assemblages. One
consequence of buffering is the common occurrence of iosbaric invariant
mineral assemblages in metamorphic terrains. Such assemblages may be mapped
as isograds because they crystallize isobarically at a unique temperature
(Greenwood, 1975; Rice, 1977a,b), and define a univariant line in P-T space.
The composition of fluid in equilibrium with an isobaric invariant assemblage is fixed through the buffering of fluid composition by mineral reactions. As with isobaric univariant assemblages, modal proportions of minerals in the invariant assemblages constitute evidence that can be used to
determine whether or not infiltration accompanied the buffering process.

Consider a rock composed of the pure phases anorthite, calcite, quartz, zoisite, tremolite, and diopside coexisting with a CO_2-H_2O fluid at 3500 bars. The equilibrium temperature and fluid composition for this isobaric invariant assemblage may be calculated using data from Ferry (1976b) and Allen and Fawcett (1982). Results are T = 481°C; X_{CO_2} = 0.094. At constant temperature and fluid composition, the proportions of minerals will change in the rock either if heat is added or extracted, or if the rock is allowed to interact with CO_2-H_2O fluid whose composition is $X_{CO_2} \neq 0.094$. The composition of fluid in equilibrium with the rock is held constant at X_{CO_2} = 0.094 during fluid-rock and/or heat-rock interaction by mineral reactions that buffer fluid composition. Reaction among minerals at the invariant point can be described in terms of any two linearly independent isobaric univariant reactions that can be written among the end-member minerals, CO_2 and H_2O. If reaction at the invariant point produces diopside and anorthite, two logical choices are:

$$\text{tremolite + 3 calcite + 2 quartz = 5 diopside + 3CO}_2 + \text{H}_2\text{O} \qquad \text{(R6)}$$

and $\qquad$ 2 zoisite + CO_2 = 3 anorthite + calcite + H_2O . $\qquad$ (R9)

Let progress variables for reactions (R6) and (R9), ξ_6 and ξ_9, respectively, be defined as:

$$\xi_6 = 1/5(\text{moles diopside per 1000 cm}^3 \text{ metamorphic rock}) \qquad \text{(23)}$$

$$\xi_9 = 1/3(\text{moles anorthite per 1000 cm}^3 \text{ metamorphic rock}) . \qquad \text{(24)}$$

If reaction at the invariant point involves buffering alone, then reactions (R6) and (R9) will proceed in such a way that evolved fluid has the equilibrium composition at the invariant point (X_{CO_2} = 0.094). In other words, from equation (14):

$$X_{CO_2} = 0.094 = \frac{3\xi_6 - \xi_9}{4\xi_6} . \qquad \text{(25)}$$

On a plot of ξ_6 versus ξ_9 (Fig. 13), equation (25) defines the solid straight line in quadrant I that is labelled "0.0." If anorthite and diopside are produced at the invariant point without infiltration, amounts of diopside and anorthite must be produced in such a ratio that the mode of the rock, expressed in values of ξ, plots on this unique line in the reaction progress diagram. If infiltration accompanies reaction, minerals will be produced in such a manner that the mode will plot off the "0.0" line in the $\xi_6-\xi_9$ diagram. For example, if a rock composed of zoisite + calcite + tremolite

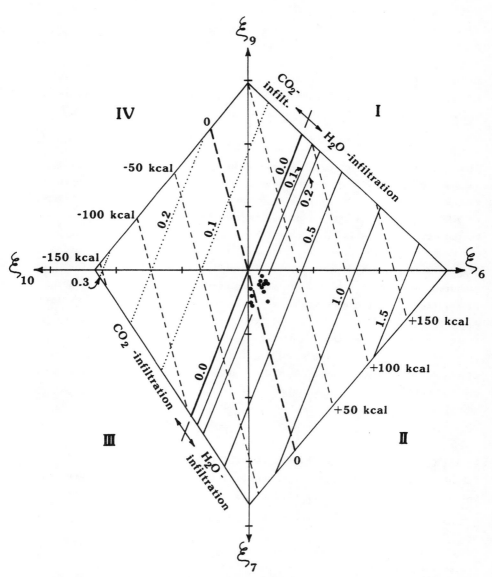

Figure 13. Relation among heat-rock interaction, fluid-rock interaction, and reaction progress in the equilibrium assemblage zoisite-anorthite-calcite-tremolite-quartz-diopside-fluid at P = 3500 bars, T = 481°C. The ξ_i are measures of the progress of reactions 6, 7, 9 and 10 in the text. Dashed contours refer to heat-rock interaction during reaction progress (positive values indicate heat added to rock; negative values indicate heat extracted from rock). Solid and dotted contours refer to fluid-rock interaction during reaction progress and are labelled in values of volumetric fluid-rock ratio. Solid lines are for H_2O-rock interaction; dotted lines are for CO_2 − rock interaction.

+ quartz interacts with n_{H_2O} moles of pure H_2O, buffer reactions (R6) and
(R9) will proceed until fluid coexisting with the rock is of equilibrium
composition, i.e., until

$$X_{CO_2} = 0.094 = \frac{3\xi_6 - \xi_9}{4\xi_6 + n_{H_2O}} . \tag{26}$$

At 481°C, 3500 bars the molar volume of H_2O is 22.11 cm^3 (Burnham et al.,
1969). Equation (26) therefore may be rearranged and written substituting
a certain volume of H_2O, V_{H_2O}, for a certain number of moles, n, of H_2O:

$$V_{H_2O}(cm^3) = 617.20 \, \xi_6 - 235.21 \, \xi_9 . \tag{27}$$

Equation (27) defines straight lines on the $\xi_6 - \xi_9$ diagram which refer to
volumes of H_2O fluid that have resulted, through infiltration, in observed
values of ξ_6 and ξ_9 for any rock. Contours in Figure 13 actually correspond
to $0.001 V_{H_2O}$ which is equivalent to the volumetric fluid-rock ratio associated with infiltration. Rocks which plot in the region of quadrant I
clockwise from "0.0" line must have been infiltrated by H_2O-rich fluid
($X_{CO_2} < 0.094$) during metamorphism. The solid contours indicate, for
measured values of ξ_6 and ξ_9, minimum amounts of H_2O that infiltrated the
rock while diopside and anorthite crystallized. If the infiltrating fluid
was not pure H_2O ($0 < X_{CO_2} < 0.094$), then total amounts of infiltrating
fluid were larger than those indicated by the contours.

If the original zoisite + calcite + tremolite + quartz rock were infiltrated by n_{CO_2} moles pure CO_2 fluid, buffer reactions (R6) and (R9) would
proceed until

$$X_{CO_2} = 0.094 = \frac{3\xi_6 - \xi_9 + n_{CO_2}}{4\xi_6 + n_{CO_2}} . \tag{28}$$

At 481°C, 3500 bars the molar volume of CO_2 is 48.98 cm^3 (Shmonov and
Schmulovich, 1974). Equation (28) can be written substituting a certain
volume of CO_2, V_{CO_2}, for a certain number of moles CO_2 and rearranged:

$$V_{CO_2} = 51.85 \, \xi_9 - 136.07 \, \xi_6 . \tag{29}$$

Equation (29) defines the dotted contours in quadrant I of Figure 13. The
dotted contours refer to volumes of CO_2 fluid which have resulted in observed values of ξ_6 and ξ_9 for a particular rock through infiltration.
Contours are labelled with values of $0.001 V_{CO_2}$ which is equivalent to a
volumetric fluid-rock ratio associated with the infiltration. Rocks that
plot in the region of quadrant I counterclockwise from the "0.0" line must

have been infiltrated by CO_2-rich fluid ($X_{CO_2} > 0.094$) during metamorphism. The dotted contours indicate, for measured values of ξ_6 and ξ_9, minimum amounts of CO_2 that infiltrated the rock while diopside and anorthite crystallized. If infiltrating fluid was not pure CO_2 ($0.094 < X_{CO_2} < 1$), then total amounts of infiltrating fluid were larger than those indicated by the contours.

Quadrant I refers to rocks in which reaction produces anorthite and diopside by reactions (R6) and (R9). The remaining three quadrants cover the other three possible outcomes of reaction at the invariant point. Quadrant II refers to rocks that initially contained combinations of anorthite, calcite, quartz, and tremolite, and in which net reaction has produced zoisite and diopside according to reactions (R6) and (R7). A progress variable for reaction (R7) can be defined as:

$$\xi_7 = 1/2(\text{moles zoisite per 1000 cm}^3 \text{ metamorphic rock}) . \qquad (30)$$

Quadrant II has been contoured with solid lines of constant volumetric fluid-rock ratio in a manner analogous to quadrant I. The contours indicate that all rocks with measured values of ξ_6 and ξ_7 that plot in quadrant II must have interacted with H_2O fluids while zoisite and diopside crystallized. Rocks cannot plot in quadrant II if they were either infiltrated by CO_2-rich fluids or were not infiltrated by any fluid at all.

Quadrant III refers to rocks that initially contained combinations of anorthite, calcite, and diopside and in which net reaction at the invariant point produced zoisite by reaction (R7) and tremolite + calcite + quartz by the reaction:

$$5 \text{ diopside} + 3CO_2 + H_2O = \text{tremolite} + 3 \text{ calcite} + 2 \text{ quartz} . \quad (R10)$$

The variable ξ_{10} is defined by

$$\xi_{10} = (\text{moles tremolite per 1000 cm}^3 \text{ metamorphic rock}) . \qquad (31)$$

Quadrant III has been contoured with solid and dotted lines of constant volumetric fluid-rock ratio in the manner analogous to quadrant I. Quadrant III is accessible when infiltration of rocks involves either CO_2- or H_2O-rich fluids as well as when reaction occurs without infiltration (i.e., buffering alone). Rocks with a metamorphic history of reactions (R7) and (R10) without infiltration must plot along the solid contour marked "0.0" in quadrant III. Rocks that plot in quadrant III away from the "0.0" contour must have had a metamorphic history that involved both buffering and infiltration.

Quadrant IV refers to rocks that initially contained combinations of zoisite and diopside and in which net reaction produced anorthite, calcite, quartz, and tremolite by reactions (R9) and (R10). Quadrant IV has been contoured with dotted lines of constant volumetric fluid-rock ratio in a manner analogous to quadrant I. The contours indicate that all rocks which plot in quadrant IV must have interacted with CO_2-rich fluid while anorthite, calcite, quartz, and tremolite crystallized. Quadrant IV is not accessible either through infiltration of rocks by H_2O-rich fluid (X_{CO_2} < 0.094) or through metamorphism without infiltration.

The four quadrants are bounded by lines along which contours of fluid-rock ratio terminate. The boundaries are a result of the simple fact that 1000 cm^3 rock can contain no more than 1000 cm^3 diopside (ξ_6 = 3.025); no more than 1000 cm^3 zoisite (ξ_7 = 3.662); no more than 1000 cm^3 anorthite + calcite in the molar ratio 3:1 (ξ_9 = 2.2947); or no more than 1000 cm^3 tremolite + calcite + quartz in the molar ratio 1:3:2 (ξ_{10} = 2.330).

With a knowledge of the mode and reaction history of any rock containing tremolite + calcite + quartz + diopside + zoisite + anorthite, the rock may be plotted as a point in Figure 13. All rocks that plot off of the "0.0" contour record a history of buffering combined with infiltration. Rocks that plot northwest of the "0.0" contour were infiltrated by CO_2-rich fluids (0.094 < X_{CO_2} $\leq$ 1); rocks that plot southeast of the "0.0" contour were infiltrated by H_2O-rich fluids (0 $\leq$ X_{CO_2} < 0.094). Solid and dotted contours provide minimum estimates of the amount of fluid that interacted with any particular specimen during the metamorphic event. Diagrams analogous to Figure 13 may be easily constructed for many isobaric invariant mineral assemblages.

An example of the application of a diagram such as Figure 13 is the study of Ferry (1982a). He collected, from a single outcrop, 14 samples of metamorphosed carbonate rocks containing the assemblage calcic plagioclase-zoisite-calcite-calcic amphibole-quartz-diopside. Pressure during metamorphism, inferred from mineral equilibria in nearby outcrops, was 3500 bars. Although the natural minerals have compositions that deviate somewhat from ideal anorthite, calcite, zoisite, tremolite, and diopside, Figure 13 may be applied to these rocks in at least a semi-quantitative fashion. Regional study of the metacarbonate rocks indicates that zoisite formed from calcite + calcic plagioclase and that reaction (R7) serves as a model reaction. Diopside formed from calcite + calcic amphibole + quartz, and reaction (R6) serves as a model reaction. Values of ξ_6 and ξ_7 were calculated from

296

measured amounts of zoisite and diopside in each sample by equations (23) and (30). Results are plotted in Figure 13 as filled circles. All samples fall in quadrant II: Each sample must therefore have been infiltrated by H_2O-rich fluid as zoisite and diopside crystallized. Fluid-rock ratios varied from place to place in the outcrop from 0.1 to 0.3 by volume. In another related study (Ferry, 1980a), similar analysis indicated that all metamorphosed carbonate rocks collected over a 400 km^2 area in south-central Maine had been infiltrated by up to 0.4 rock volumes of H_2O fluid during metamorphism. Both studies point to the importance of modal analysis in unravelling aspects of the control of fluid composition and fluid-rock interaction during metamorphism.

Review of the literature

We have presented two extreme mechanisms for the control of fluid composition during metamorphism: infiltration and buffering. In this section the literature is briefly reviewed with the aim of assessing the relative importance of infiltration versus buffering in the control of metamorphic fluid composition. Table 1 summarizes the results. If the list of studies can be taken as representative of the spectrum of metamorphic environments, it is clear that buffering is a far more important control on fluid composition than is infiltration. We take our tabulation as a tribute to the remarkable insight of Hugh Greenwood who first systematically developed the concept of buffering of fluid composition by mineral reactions (Greenwood, 1962, 1967a, 1975).

The tabulation of studies (Table 1) in which mineral equilibria buffer fluid composition does not differentiate between cases in which buffering was accompanied by infiltration and cases in which it was not. As noted previously, a rigorous assessment of whether buffering occurred with or without infiltration requires modal analysis. Regrettably, in recent years modal analysis has fallen out of fashion, and the required data are simply not reported (if ever collected). We have devised a somewhat less rigorous test to detect infiltration in those studies for which there are no reported modes. Figure 2 emphasizes that when buffering occurs without infiltration, almost all mineral reaction occurs at isobaric invariant points. At the invariant points, mineral reaction will produce a fluid with composition appropriate to equilibrium with minerals at the invariant point. Consequently for those studies of buffering processes in which infiltration did not occur, reported mineral reactions should be capable of producing a fluid

TABLE 1.

Literature review of studies related to the buffering of the composition of metamorphic fluid by mineral equilibria.

I. Studies Consistent with External Control of the Concentration of Species an the Metamorphic Fluid

Reference	Evidence	Species
1. Albee (1965)	a.,c.	H_2O, CO_2
2. Allen (1978)	b.	H_2O, CO_2
3. Bucher-Nurminen (1981)	b.	H_2O, CO_2
4. Carmichael (1970)	b.	H_2O, CO_2
5. Greenwood (1967a)	b.	CO_2
6. Gordon and Greenwood (1971)	b.	H_2O, CO_2
7. Misch (1964)	b.	H_2O, CO_2
8. Rumble et al. (1982)	c.	^{18}O, ^{16}O

— — — — —

II. Studies Consistent with Internal Control of the Concentration of Species in Metamorphic Fluid

Reference	Evidence	Species
1. Albee (1965)	f.	O_2
2. Bowman et al. (1980)	m.	H_2O, CO_2
3. Butler (1969)	g.	H_2O, CO_2, O_2
4. Cermignani and Anderson (1973)	k.	H_2O, CO_2
5. Chatterjee (1971)	h.	H_2O, CO_2
6. Chinner (1960)	f.,g.	O_2
7. Crawford et al. (1979)	l.	H_2O, CO_2, CH_4
8. Evans and Trommsdorff (1970)	h.	H_2O
9. Evans and Trommsdorff (1974)	h.	H_2O, CO_2
10. Erdmer (1981)	d.,h.	H_2O, CO_2
11. Frost (1982)	a.,g.	O_2
12. Ferry (1976a)	d.,h.	H_2O, CO_2
13. Ferry (1979)	f.,g.,h., j.	H_2O*, CO_2*, O_2* S_2, H_2S, CH_4* $CO*$, H_2*
14. Ferry (1981)	d.,g.	H_2O, H_2S
15. Ghent et al. (1979)	i.	H_2O, CO_2
16. Glassley (1975)	h.	H_2O, CO_2, HF
17. Grambling (1981)	e.,g., j.	H_2O*
18. Grew (1978)	g.,j.	H_2O*, O_2*
19. Guidotti (1970)	e.,g.	H_2O
20. Hewitt (1973)	d.,h.	H_2O, CO_2

298

21.	Hoy (1976)	d.	H_2O
22.	James and Howland (1955)	d., f.	O_2
23.	Jansen et al. (1978)	d., 1.	H_2O, CO_2
24.	Kerrick et al. (1973)	d.,f., h.	H_2O, CO_2
25.	Kranck (1961)	f.,g.	H_2O, CO_2
26.	Melson (1966)	h.	H_2O, CO_2
27.	Moore and Kerrick (1976)	d., h.	H_2O, CO_2
28.	Morgan (1970)	h.	H_2O
29.	Osberg (1974)	f.	H_2O
30.	Rice (1977a)	d.,h.	CO_2, H_2O
31.	Rice (1977b)	d.,h.	H_2O, CO_2
32.	Rumble (1978)	e.,f.,g.,h.,j.	H_2O*,O_2*
33.	Rumble et al. (1982)	f.,h.,i.	H_2O, CO_2
34.	Spear (1977)	g.	H_2O*
35.	Spear (1981)	g.	H_2O*, CO_2
36.	Suzuki (1977)	d.,h.	H_2O, CO_2
37.	Trommsdorff (1972)	f.	H_2O, CO_2
38.	Trommsdorff and Evans (1969)	h.	H_2O
39.	Trommsdorff and Evans (1974)	h.	CO_2
39.	Valley and Essene (1980)	d.,h.	H_2O, CO_2
40.	Williams-Jones (1981)	d.,h.	H_2O, CO_2
41.	Zen (1961)	f.	H_2O

a. topology of phase diagrams

b. distribution of isobaric univariant and divariant mineral assemblages (P = P_{fluid} = constant)

c. $\delta^{18}O$ of minerals

d. widespread occurrence of isobaric univariant mineral assemblages.

e. crossing tie lines on a phase diagram constructed from mineral assemblages collected from a small area.

f. occurrence of different isobaric univariant mineral assemblages within a small area.

g. variation in mineral composition in isobaric univariant mineral assemblages collected from a small area.

h. mineral assemblages from a small area plot at different points on a $\mu-\mu$ or $T-X_{CO_2}$ diagram.

i. calculated differences in equilibrium X_{CO_2} between mineral assemblages collected from a small area.

j. calculated differences in the chemical potentials of components in fluid in equilibrium with different mineral assemblages collected from a small area.

k. textural evidence in a sample for a change from one isobaric univariant mineral assemblage to another with time.

l. fluid inclusions.

m. unspecified.

* local differences in μ_i quantitatively determined by method of Rumble (1974).

TABLE 2. Relation of the composition of equilibrium metamorphic fluid ($X_{CO_2}^{eq}$) to the composition of fluid produced internally by mineral reactions ($X_{CO_2}^{rxn}$) in metamorphosed carbonate rocks.

Locality	Age of Metamorphism	Depth of Metamorphism	$X_{CO_2}^{rxn}$	$X_{CO_2}^{eq}$	Reference
Mount Royal, Quebec	Cretaceous	≈2 km	≥0.75	>0.75	Williams-Jones (1981)
Elkhorn, Montana	Tertiary	≈4 km	>0.75	0.03-0.8	Bowman and Essene (1982)
Helena area, Montana	Cretaceous	≈4 km	≥0.75	≥0.75	Rice (1977a,b)
Alta, Utah	Tertiary	4-7 km	>0.5	0.2-0.7	Moore and Kerrick (1976)
Salmo, British Columbia	Cretaceous	4-9 km	1.0	≈0.1	Greenwood (1967a)
Sierra Nevada, California	Mesozoic	≈7 km	1.0	<0.1	Kerrick et al. (1973)
Stanhope area, Quebec	Devonian	6-10 km	>0.75	0.05-0.15	Erdmer (1981)
Waterville area, Maine	Devonian	13 km	>0.75	0.04-0.31	Ferry (1980)
Beaver Brook, New Hampshire	Devonian	13 km	>0.83	0.09-0.14	Rumble et al. (1982)
Whetstone Lake, Ontario	Precambrian	15-19 km	>0.75	0.06-0.25	Carmichael (1970); Skippen and Carmichael (1977)
Tudor, Ontario	Precambrian	≈19 km	>0.75	<0.5	Allen (1978)
Lepontine Alps	Tertiary	≈20 km	>0.50	>0.50	Trommsdorff (1972)
South-central Connecticut	Devonian	≈22 km	>0.75	0.1-0.4	Hewitt (1973)
Mica Creek, British Columbia	Mesozoic	20-28 km	>0.75	0.23-0.27	Ghent et al. (1979)
Adirondacks, New York	Precambrian	22-28 km	>0.63	0.1-?	Valley and Essene (1980)

whose composition matches the composition of fluid with which the rocks were in equilibrium as the reactions proceeded. Conversely, if reported mineral reactions are incapable of producing a fluid whose composition matches that of the fluid with which the rocks were in equilibrium, infiltration must have accompanied mineral reaction.

Table 2 lists those studies from Table 1 for which authors reported both the prograde mineral reactions and the estimated compositions of fluid in equilibrium with the rocks when the reactions occurred. In most of the cases at least some equilibrium fluids were H_2O-rich ($X_{CO_2} < 0.2$) while in all cases the mineral reactions evolved CO_2-rich fluids ($X_{CO_2} > 0.5$). In most cases, therefore, the mineral reactions were incapable of producing a fluid of the composition with which the minerals were in equilibrium, and metamorphism must have been associated with some infiltration. Furthermore, because the equilibrium fluid was more H_2O-rich than the fluid produced internally by the buffer reactions, the infiltrating fluids must have been predominantly H_2O-rich. We know of no case studies, utilizing the criteria outlined here, that document infiltration of large quantities of CO_2-rich fluid (however, see Newton et al., 1980). The metamorphic events listed in

Table 2 range from Tertiary to Precambrian in age. Depth of metamorphism
ranges from 4 to 30 km. We conclude that infiltration of carbonate rock by
H_2O fluid during metamorphism is a common phenomenon in space and time at
least as far back as one billion years ago and as deep as 30 km in the
crust. Our conclusion, however, should not detract from the fact that many
instances of metamorphism involve buffering alone without evidence for infiltration.

Discussion

There is abundant geologic evidence that fluid composition during metamorphism may be controlled by the process either of infiltration or of buffering. Buffering appears to be by far the most important control mechanism.
Although it is commonly thought that buffering and infiltration are somehow
incompatible processes, infiltration may accompany buffering. Limited data,
in fact, indicate that combined buffering and infiltration is probably a
very common phenomenon during metamorphism. Because infiltration accompanies buffering but buffering ultimately controls fluid composition, many
rocks must have a large capacity to buffer the concentration of major
species in the metamorphic fluid.

Buffering and infiltration were presented as extremes. Furthermore,
buffering with infiltration was presented as a variation on the theme of
buffering. In fact, there is a complete, continuous spectrum from buffering alone through buffering with infiltration to infiltration alone. At
one end of the continuum, buffering of fluid composition occurs without
infiltration. If some infiltration occurs, buffering still controls fluid
composition. Beyond a certain critical amount of fluid, however, fluid-rock
reactions exhaust the buffer capacity of the rock (i.e., one or more mineral
reactants in the buffer assemblage is consumed). After the buffer capacity
of the rock is exhausted, the external fluid exerts exclusive control on
the composition of fluid in contact with the rock. This is the point at
which fluid composition is controlled by infiltration alone.

In some cases, the relative importance of buffering and infiltration
changes with time throughout a metamorphic episode. Because buffering of
the fluid composition can only occur while a reaction is actually taking
place, the pore fluid composition and the prograde buffering sequence may
be altered by infiltration of external fluid in the time periods between
different buffer reactions. An example of this phenomenon is the metamorphism of micaceous limestones in south-central Connecticut discussed by

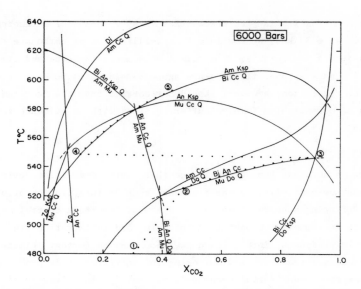

Figure 14. Temperature - fluid composition diagram representing observed reactions in micaceous limestones from central Connecticut. The dotted line represents the proposed fluid composition path for the interior portions of limestone pods and lenses. From Hewitt (1973).

Hewitt (1973) and Hewitt and Vidale (1973). This terrain consists of thin individual layers and pods of limestone surrounded by pelitic schists. The observations made by Hewitt (1973) suggest that the mineral assemblages in the cores of the limestone bodies were nearly always undergoing reaction which buffered the local pore fluid toward CO_2-rich compositions (dotted path along reaction curves, Fig. 14). While reactions were taking place, a steady state transfer of fluid occurred between the schists and limestones (combined infiltration and buffering). Gradients in the chemical potential of mixed-volatile components resulted in mineral-assemblage zoning at the margins of the limestone layers. Whenever reaction ceased in a limestone, owing to the disappearance of one or more reacting minerals, the buffering capacity was lost, and the steady state fluid transfer no longer operated. Under these conditions, the pore fluids became rapidly more H_2O-rich, until the mineral assemblage became unstable and began to react according to a new mineral-fluid equilibrium (subhorizontal dotted path, Fig. 14). In the Connecticut rocks, this process resulted in a nearly continuous sequence of reactions with little time in which the limestones did not contain a fluid-buffering assemblage. However, in the time periods when no reactions were proceeding, the pore fluid composition changed significantly. Because the buffering assemblages occur over a range of metamorphic grade, the rates of reaction were apparently fast enough to effectively buffer the fluid composition. This might imply that transfer of H_2O-rich fluid into the limestone layers occurred predominantly by slow grain-boundary diffusion, rather than by rapid infiltration.

Similar phenomena, but on a scale larger than individual limestone layers, are commonly observed in contact metamorphic aureoles. Mineral assemblages in carbonate rocks, and the spatial distribution of isograds in the outer (low-grade) portions of many aureoles, suggest buffering of the fluid composition to relatively high mole fractions of CO_2. The loss of buffering capacity by consumption of one or more reactants in the central portions of the aureoles subsequently allows the pore fluid to be controlled externally by infiltration of H_2O-rich fluid possibly released from the crystallizing igneous body. This infiltration, however, typically dominates only until additional reactions (commonly found in the inner portions of the aureole) take place which are capable of buffering the local pore fluid as infiltration continues. An example of this type of behavior is found in the contact aureole described by Bowman and Essene (1982). In the outer portion of the aureole fluid composition was apparently controlled by buffering without infiltration; in the inner part of the aureole, fluid composition was controlled by combined infiltration and buffering.

A survey of Table 1 reveals that in any discussion of buffering versus infiltration, the exact species must be specified. For example, at Mount Grant, Vermont, mineral equilibria buffered the concentration of O_2 in metamorphic fluid while the concentrations of H_2O, ^{18}O, and ^{16}O were controlled by infiltration (Albee, 1965). Similarly at Beaver Brook, New Hampshire, mineral equilibria buffered the concentrations of CO_2 and H_2O in fluid, but the concentrations of ^{18}O and ^{16}O were controlled by infiltration (Rumble et al., 1982). Clearly rocks have a variable buffer capacity with respect to different fluid species. At a given point in the evolution of a metamorphic rock, its buffer capacity may be exhausted with respect to some fluid species but not with respect to others. The control of the concentration of each species in metamorphic fluid must be individually considered.

As noted earlier, the modes of metamorphic rocks are the key to recognizing combined buffering and infiltration during metamorphism. If some proselytizing may be briefly tolerated, we wish to encourage the determination of rock modes in all studies of buffering phenomena. With present knowledge, no study can be considered complete without a modal analysis of at least some specimens under consideration. Only from modes can buffering with infiltration be rigorously distinguished from buffering without infiltration. If infiltration did occur, the modes offer the added dividend of a determination of how much fluid flowed through particular rock specimens. Systematic studies that quantify fluid-rock ratios should result in a better understanding of metamorphic fluid-rock interactions both on a regional and on an outcrop scale.

BUFFERING, INFILTRATION, AND THE CONTROL OF MINERAL COMPOSITION

Buffering and the internal control of mineral composition

In light of the previous discussion, it is instructive to examine the capacity of mineral reactions to buffer the composition of mineral solutions as well as fluid solutions. In this section we will discuss some of the mechanisms by which the chemical compositions of minerals may be controlled, or buffered, and examine some natural examples of internal and external control of mineral compositions.

The principles of mineral buffering are the same as those for fluid buffering, and can be understood in terms of the fundamental energy balance expression given in equation (1). As with volatile components, the chemical potentials of thermodynamic components dissolved in mineral solutions in an assemblage undergoing reaction are not entirely independent but rather constrained by equation (1). The control of mineral composition by mineral reactions can most easily be visualized by an example of a mineral equilibrium for which $\Sigma_i \nu_i \ln f_i = 0$ (e.g., for an equilibrium among volatile-free minerals). Consider the reaction among grossularite ($Ca_3Al_2Si_3O_{12}$), quartz (SiO_2), wollastonite ($CaSiO_3$), and plagioclase solid solution ($CaAl_2Si_2O_8$-$NaAlSi_3O_8$). The following equilibrium relationship may be written among the components of the coexisting minerals:

$$Ca_3Al_2Si_3O_{12} + SiO_2 = CaAl_2Si_2O_8 + 2CaSiO_3 \tag{R12}$$

which in turn defines a specific case of equation (1):

$$\Delta\bar{H}_r - T\Delta\bar{S}_r + \Delta\bar{V}_r(P-1) + RT\ln(\gamma_{CaAl_2Si_2O_8}^{plagioclase} \ x_{CaAl_2Si_2O_8}^{plagioclase}) = 0 \ . \tag{32}$$

At a constant pressure of 1000 bars, equation (32) defines the T-X curve shown in Figure 15 ($\Delta\bar{H}_r$, $\Delta\bar{S}_r$, $\Delta\bar{V}_r$ from Ferry, 1976b; activity coefficients for $CaAl_2Si_2O_8$ from Orville, 1972). The similarity between the phase diagrams shown in Figures 11 and 15 is a graphical representation of the similarity between the concepts of buffering of fluid composition by mineral

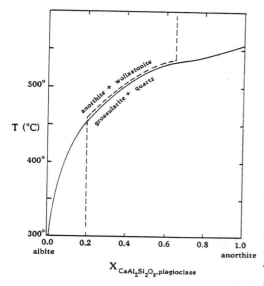

Figure 15. Representation of reaction (R12) on an isobaric T-X diagram at a pressure of 1000 bars. The dashed T-X path is explained in the text.

reaction, and of buffering of mineral composition by mineral reaction. The buffering of plagioclase composition by equilibrium among coexisting plagioclase, grossularite, quartz and wollastonite by reaction (R12) can be illustrated in Figure 15 by considering the progressive metamorphism of a rock at 1000 bars initially containing grossularite + quartz + plagioclase of composition An_{20}. As heat is added to the rock, temperature increases below 453°C at constant plagioclase composition (dashed line, Fig. 15). At 453°C wollastonite + anorthite component forms by reaction (R12), and as temperature increases further, the anorthite content of plagioclase increases as prescribed by the T-X curve for reaction (R12). At any temperature along the equilibrium boundary, plagioclase composition is controlled or buffered by the mineral reaction. The reaction buffers plagioclase composition until one or both of the mineral reactants is consumed (in this hypothetical case at 532°C). The exact temperature and final plagioclase composition depend on the relative proportions and compositions of the minerals initially present in the rock. At temperatures above that at which a reactant is consumed, no reaction is possible and plagioclase composition remains constant at the value attained when reaction ceased (dashed line above 532°C, Fig. 15). During a significant portion of the metamorphism of this hypothetical rock (T = 453°-532°C), the concentration of the anorthite component in plagioclase was controlled by reaction (R12).

In general, heterogeneous phase equilibria affecting solid solutions include not only net-transfer reactions (like R12), but also exchange reactions (see Thompson, Chapter 2). Exchange reactions govern the distribution, or partitioning, of exchangable elements or molecules between coexisting phases such that for a specific bulk composition at fixed T and P, the compositions of the coexisting phases are fixed. Coupled with net-transfer

305

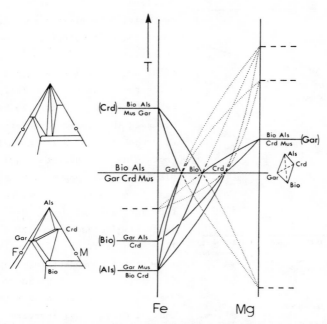

Figure 16. Schematic, isobaric T-X (Fe-Mg) diagram showing phase relations among the minerals biotite (Bio), garnet (Gar) cordierite (Crd), aluminosilicate (Als), and muscovite (Mus). Fe-Mg continuous reactions are represented as isobaric univariant loops which intersect at the discontinuous reaction Gar + Crd + Mus = Bio + Als. AFM diagrams are projected from quartz, Mus, and H2O. From A.B. Thompson (1976).

reactions, exchange reactions therefore can be considered as very effective buffers of mineral composition. A familiar example involving Fe-Mg solid solution in several minerals found in pelites is shown in Figure 16. Under isobaric conditions, the "continuous" reactions are shown as univariant loops on a $T-X_{Fe-Mg}$ diagram. It should be apparent that the "continuous" reactions buffer the composition of participating Fe-Mg solid solutions. For example, the equilibrium among garnet, anhydrous cordierite, Al_2SiO_5, and quartz defines two specific cases of equation (1) based on the reactions:

$$2Fe_3Al_2Si_3O_{12} + 4Al_2SiO_5 + 5SiO_2 = 3Fe_2Al_4Si_5O_{18} \qquad (R13)$$

and $$2Fe_3Al_2Si_3O_{12} + 3Mg_2Al_4Si_5O_{18} = 2Mg_3Al_2Si_3O_{12} + 3Fe_2Al_4Si_5O_{18}. \text{(R14)}$$

They are

$$\Delta\bar{H}_r - T\Delta\bar{S}_r + \Delta\bar{V}_r(P-1) + 3RT\ln(\gamma_{Fe_2Al_4Si_5O_{18}}^{cordierite} \ x_{Fe_2Al_4Si_5O_{18}}^{cordierite})$$

$$- 2RT\ln(\gamma_{Fe_3Al_2Si_3O_{12}}^{garnet} \ x_{Fe_3Al_2Si_3O_{12}}^{garnet}) = 0 \qquad (33)$$

306

for reaction (R13), and

$$\Delta \bar{H}_r - T\Delta \bar{S}_r + \Delta \bar{V}_r (P-1)$$

$$+ 2RT\ln(\gamma_{Mg_3Al_2Si_3O_{12}}^{garnet} \quad x_{Mg_3Al_2Si_3O_{12}}^{garnet} / \gamma_{Fe_3Al_2Si_3O_{12}}^{garnet} \quad x_{Fe_3Al_2Si_3O_{12}}^{garnet})$$

$$+ 3RT\ln(\gamma_{Fe_2Al_4Si_5O_{18}}^{cordierite} \quad x_{Fe_2Al_4Si_5O_{18}}^{cordierite} / \gamma_{Mg_2Al_4Si_5O_{18}}^{cordierite} \quad x_{Mg_2Al_4Si_5O_{18}}^{cordierite}) = 0$$

(34)

for reaction (R14). Because

$$X_{Mg_3Al_2Si_3O_{12}} + X_{Fe_3Al_2Si_3O_{12}} = 1 \quad \text{and} \quad X_{Mg_2Al_4Si_5O_{18}} + X_{Fe_2Al_4Si_5O_{18}} = 1,$$

equations (33) and (34) define the composition of garnet and cordierite co-existing with Al_2SiO_5 and quartz at any particular pressure and temperature. The Gar-Als-Crd loop in Figure 16 is the graphical representation of the control on garnet and cordierite composition exercised by reactions among the four minerals (R13, R14). Furthermore, as Figure 16 illustrates, as temperature increases the compositions of garnet and cordierite coexisting with Al_2SiO_5 and quartz change only in a very specific manner controlled by the buffer reactions. The systematic change in composition of Fe-Mg solid solution minerals in progressively metamorphosed pelitic schists (A. B. Thompson, 1976) may be simply regarded as the result of the buffering of mineral composition by the prograde mineral reactions.

Compositions of Fe-Mg minerals participating in continous reactions change with increasing temperature in a manner prescribed by $T-X_{mineral}$ loops such as those shown in Figure 16. The $T-X_{mineral}$ loops are analogous to $T-X_{fluid}$ curves which represent the "continuous" devotalization reactions in Figures 1, 6, and 11. In both situations a mineral reaction buffers the composition of either a fluid solution (Figs. 1, 6, 11) or a solid solution (Figs. 15, 16). The isobaric $T-X_{mineral}$ loops in Figure 16 intersect at a particular temperature, and their intersection generates a "discontinuous" reaction. The isobaric "discontinuous" reaction not only occurs at a unique temperature, but also involves Fe-Mg solid solutions that have unique compositions. The intersection of the $T-X_{mineral}$ loops at "discontinuous" reactions are analogous to the intersection of $T-X_{fluid}$ curves in Figure 1 at isobaric invariant points. Reaction at isobaric invariant points on a $T-X_{CO2}$ diagram occurs at a unique temperature and involves a fluid solution of unique composition. In both situations mineral equilibria define isobaric invariant conditions at which mineral reactions control or buffer the compositions of either fluid or solid solutions.

Because of differences in bulk composition between different beds of an interlayered sequence of pelitic schists, different "continuous" buffer reactions may simultaneously occur during metamorphism. With reference to Figure 16, for example, some samples may contain garnet + cordierite + Al_2SiO_5 + quartz while other samples may contain biotite + cordierite + muscovite + garnet + quartz. At any temperature below that of the "discontinuous" reaction, "continuous" reactions buffer the compositions of garnet and cordierite to different values (Fig. 16). Differences in mineral compositions in samples of pelitic schist from the same outcrop but with different "continuous" reaction assemblages therefore can be interpreted in terms of the buffering of the composition of mineral solutions by mineral reactions. This situation is directly analogous to observed differences in the chemical potentials of volatile species between different samples from the same outcrop (*cf*. study by Rumble, 1978) which are interpreted in terms of the buffering of the composition of fluid solutions by mineral reactions.

When mineral-fluid equilibria involve both solid and fluid solutions, the added variance often complicates the simple relationships shown on $T-X_{CO_2}$ phase diagrams constructed for systems without solid solution. In theoretical analysis of such systems, it is common practice to separate the compositional variables. This may be done, for example, by constructing isobaric $T-X_{mineral}$ sections at constant mole fraction of species in the fluid (*cf*. A. B. Thompson, 1976; Rice, 1980) or by constructing isobaric $T-X_{fluid}$ sections valid for particular mineral compositions (*cf*. Rice, 1980). Although this separation of variables is desirable for graphical representation, it allows one to lose sight of the interrelationship between temperature and the compositional variables.

In some systems, the solid solutions exhibit exchange of anions which are also part of the mixed volatile phase (e.g., OH-F exchange in hydrous silicates). As an example, consider the following equilibrium involving humite minerals:

$$2 \text{ chondrodite} = \text{clinohumite} + \text{periclase} + H_2O$$

$$2(Mg_4Si_2O_8Mg(OH)_2) = Mg_8Si_4O_{16}Mg(OH)_2 + MgO + H_2O \ . \qquad (R15)$$

In the fluorine-free system, this equilibrium corresponds to a simple dehydration reaction. When fluorine is present, however, the compositions of coexisting chondrodite and clinohumite are governed by the following OH-F exchange equilibrium:

$$OH\text{-chondrodite} + F\text{-clinohumite} = F\text{-chondrodite} + OH\text{-clinohumite}. \qquad (R16)$$

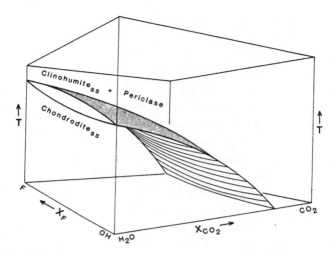

Figure 17. Schematic isobaric $T - X_{CO_2} - X_F$ diagram showing the shape of a reaction surface for an equilibrium such as (R12). The $T - X_F$ plane corresponds to a continuous loop such as illustrated on Figure 16. The $T - X_{CO_2}$ plane shows the dehydration boundary for the fluorine-free reaction.

By analogy with Fe-Mg reactions, (R15) coupled with (R16) defines an (OH-F) "continuous" reaction. Figure 17 illustrates the shape of the isobaric divariant reaction surface in $T-X_{CO_2}-X_F$ space for an equilibrium like (R15). Because X_F in chondrodite is greater than X_F in coexisting clinohumite, the temperature of equilibrium increases with increasing fluorine content. The upper surface of the loop corresponds to clinohumite composition whereas the lower surface represents the composition of chondrodite.

Because fluorine is not only partitioned between (OH-F) solid solution phases but also between minerals and the mixed-volatile phase, the fluid in equilibrium with a fluorine-bearing assemblage of minerals must be characterized by a finite fugacity of HF. The fugacity of HF in equilibrium with an individual humite-group mineral is governed by an exchange equilibrium of the following type:

$$OH\text{-humite} + 2HF = F\text{-humite} + 2H_2O \ . \tag{R17}$$

Equation (1) for an exchange equilibrium such as (R17) can be written in the form:

$$ln\left[\frac{f_{H_2O}}{f_{HF}}\right] = \frac{-\Delta\bar{H}_r}{RT} + \frac{\Delta\bar{S}_r}{R} - \frac{\Delta\bar{V}_s(P-1)}{RT} - ln\left[\frac{a_{F\text{-Hu}}}{a_{OH\text{-Hu}}}\right] \ . \tag{35}$$

From this expression, it is apparent that at constant P and T the fugacity and activity variables are not independent but may vary only within the constraint of equation (35).

An equilibrium assemblage consisting of chondrodite solid solution, clinohumite solid solution, periclase and fluid is associated with a number of intensive variables. These variables, however, are not all independent, but are simultaneously constrained by the energy-balance equations corresponding to reactions (R15), (R16) and (R17). In fact, at fixed P and T, coexistence of the three minerals results in one degree of freedom; if the composition of one of the humite minerals is fixed, then the activities of the solid components and the fugacities of the volatile components are uniquely determined. Alternatively, for a particular value of OH/F for the bulk mineral-fluid system, the compositions of the three mineral phases and the fluid are uniquely determined at a particular pressure and temperature by equilibrium relations based on reactions (R15)-(R17). Coexisting chondrodite, clinohumite, periclase, and H_2O-CO_2-HF fluid is an example therefore of a system in which mineral reactions simultaneously buffer the composition of mineral solutions as well as the fluid solution.

We can now examine the isobaric "progressive metamorphism" of a chondrodite rock starting at temperatures below the equilibrium surface in Figure 17. As heat is added to this rock, its temperature will increase until the equilibrium surface is reached. The point of intersection on the equilibrium surface will be determined by the bulk composition of the reactant assemblage (i.e., F/F + OH) ratio in chondrodite and $H_2O/(H_2O + CO_2)$ ratio in the pore fluid). Under conditions of internal buffering (i.e., mineral reaction without infiltration), further addition of heat will produce clinohumite and periclase; the composition of the clinohumite will be governed by (R16) while composition of fluid is governed by (R15) and (R17). With increase in temperature phase compositions will follow the equilibrium surface, with both chondrodite and clinohumite becoming progressively more fluorine-rich as the fluid becomes more H_2O-rich. At any point along the reaction surface, the activities of components in the humite minerals, as well as the fugacities of H_2O, CO_2 and HF will be buffered. Reaction will continue, and all intensive variables will be controlled, until chondrodite is consumed. At this point, clinohumite will have the same F/(F + OH) ratio as the initial chondrodite.

In systems with additional, linearly independent reactions, continuous reaction surfaces such as the one shown in Figure 17 intersect to form (OH-F) discontinuous reactions. The isobarically univariant discontinuous reactions correspond to isobaric invariant points in the $T-X_{CO_2}$ plane for the pure OH (fluorine-free) equilibria (Rice, 1980). Although the discontinuous reactions are polythermal, they should form distinct isograds in the field, since rocks of different bulk composition will contain the discontinuous assemblage with identical or very similar mineral compositions. Progressive metamorphism accompanied by internal buffering (and little or no infiltration) should therefore be characterized by zones of isobarically divariant, continuous reaction assemblages separated by isograds corresponding to the discontinuous reaction assemblages. One should observe a systematic change in mineral proportions and mineral compositions across the divariant zones. Mineral compositions in the discontinuous (isograd) assemblages should be identical, or nearly so, in a variety of different bulk compositions. Observation of such relationships may be taken as diagnostic of the internal control of mineral and fluid composition by buffering during metamorphism. Although data are sparse, petrologic studies involving (OH-F) minerals in mixed volatile systems summarized by Rice (1980) indicate that internal control of mineral and fluid compositions plays an important role in the metamorphism of fluorine-bearing rocks.

Infiltration and the external control of mineral composition

Many studies of mineral equilibria in metamorphic rocks are consistent with the internal control of mineral composition by buffering. In some cases, however, the composition of minerals may be controlled by an agent or process external to the rock under consideration. The situation may occur, for example, when rocks are continuously infiltrated during metamorphism by externally-derived fluid. Certain compositional parameters of minerals then adjust themselves to be in equilibrium with the external reservoir of fluid.

The external control of mineral composition by infiltration, rather than internally by buffering, has been recognized in a number of isotopic studies of metamorphic and altered igneous rocks. The basic evidence is uniformity in isotopic composition for a particular mineral in samples of rock that initially contained the mineral with a different composition prior to metamorphism or alteration. If internal processes, such as devolitilization reactions, can be sufficiently characterized to demonstrate that they could not be responsible for the observed shift in isotopic composition, then conventionally the change in isotopic composition is attributed to infiltration. The uniformity in isotopic composition of the same mineral in different samples is interpreted as the adjustment of the mineral in all samples to equilibrium with the infiltrating fluid, i.e., external control of mineral composition by infiltration.

311

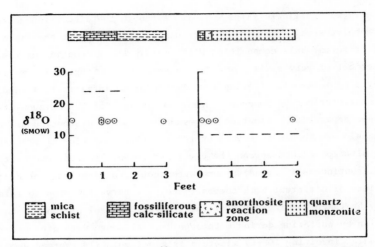

Figure 18. Oxygen isotopic composition (δ^{18}O, SMOW) of quartz separated from rock samples collected along two traverses through the outcrop illustrated in Figure 12 at Beaver Brook, New Hampshire. Quartz from each of the four lithologic types exposed has the same isotopic composition within error of measurement. From Rumble et al. (1982).

The external control of mineral isotopic composition is illustrated by the study of Rumble et al. (1982). They investigated an outcrop of inter-bedded silicified Devonian brachiopods and shale now metamorphosed to calc-silicate rock and mica schist (Fig. 12). The metasediments are cut by a syn-metamorphic quartz monzonite dike. Quartz in unmetamorphosed silicified Devonian brachiopods has isotopic composition δ^{18}O = +24$\pm$2 °/oo (Savin and Epstein (1970). Quartz in unaltered quartz monzonites has composition δ^{18}O = +10$\pm$3% (Taylor, 1967).

Quartz in each of the rock types has identical post-metamorphic composition, within error of measurement δ^{18}O = +14.5$\pm$0.2°/oo (Fig. 18). The uniformity in the isotopic composition of quartz is consistent with equilibration of all rocks during metamorphism with a pervasive infiltrating fluid. Rumble et al. quantitatively evaluated the effect of metamorphic devolitilization reactions on isotopic composition and concluded that internal reactions could account for, at most, 10-20% of the inferred changes in isotope composition that occurred during metamorphism. They attributed the change in δ^{18}O to infiltration of the outcrop by four to six rock volumes of H_2O-rich fluid whose isotopic composition must have been heavier than fluid initially in equilibrium with quartz monzonite, but lighter than fluid initially in equilibrium with the calc-silicate rock. The identical quartz compositions in each sample were interpreted as an isotopic composition imposed on quartz in all rocks by the external H_2O fluid.

Mineral composition controlled by combined infiltration and buffering

As in the discussion of the controls of metamorphic fluid composition, we now consider the control of mineral composition by a process that combines buffering with infiltration. Rocks are considered to be in communication with an external fluid reservoir of limited size. Mineral composition is controlled by an interplay between mineral reactions (which internally control mineral composition by buffering) and infiltration (which serves to further modify the composition of certain minerals through mineral-fluid reaction).

The control of mineral composition by a combination of buffering and infiltration is not often recognized because the combined processes produce mineral assemblages that are very similar to those produced by buffering phenomena alone. For example, either with buffering alone or with combined buffering and infiltration, mineral equilibria exist in rocks which buffer the concentration of certain components in mineral solid solution. As with the consideration of the control of fluid composition by combined buffering and infiltration, modal data turn out to be important in identifying the control of mineral composition by combined processes.

A recent example is an investigation of the controls on plagioclase composition during progressive regional metamorphism of impure carbonate rocks from south-central Maine by Ferry (1982b). As noted above (Ferry, 1980a), studies of reaction progress based on modal data indicate that this terrain was infiltrated by large quantities of H_2O-rich fluid, but that the buffering capacity of most rocks was maintained throughout the metamorphic event. Fluid composition was controlled by an interplay between the infiltration and the buffer reactions. At the lowest grades studied, the carbonate rocks contain albite (An_{0-2}) while at the highest grades many rocks contain anorthite (An_{90-98}). Part of the observed change in plagioclase composition may readily be attributed to several prograde mineral reactions that involve the components $NaAlSi_3O_8$ and $CaAl_2Si_2O_8$ dissolved in plagioclase solid solution. The principal reactions (with increasing metamorphic grade) include:

$$\text{muscovite + ankerite + quartz + ilmenite + HCl + } H_2O$$
$$= \text{biotite + calcite + } CaAl_2Si_2O_8 \text{ + NaCl + } CO_2 \qquad \text{(R18)}$$

$$\text{biotite + calcite + quartz + } NaAlSi_3O_8 \text{ + HCl}$$
$$= \text{calcic amphibole + sphene + } CaAl_2Si_2O_8 \text{ + KCl + } CO_2 \text{ + } H_2O \qquad \text{(R19)}$$

313

$$\text{amphibole + calcite} + CaAl_2Si_2O_8 + HCl + H_2O$$
$$= \text{zoisite + sphene + quartz} + NaCl + CO_2 + H_2 \qquad (R20)$$

and
$$\text{amphibole + calcite + quartz} + HCl$$
$$= \text{diopside + sphene} + CaAl_2Si_2O_8 + NaCl + CO_2 + H_2O \ . \qquad (R21)$$

A fifth reaction, which proceeded throughout the progressive sequence, is:

$$2NaAlSi_3O_8 + CaCO_3 + 2HCl$$
$$= CaAl_2Si_2O_8 + 4SiO_2 + CO_2 + H_2O + 2NaCl \ . \qquad (R22)$$

At grades higher than those at which reaction (R18) occurred, these mineral equilibria (or combinations of them) buffered the chemical potentials of $CaAl_2Si_2O_8$ and/or $NaAlSi_3O_8$ in plagioclase solid solution according to equilibrium relations like equations (1) and (25).

If plagioclase composition were exclusively controlled by internal buffering reactions (i.e., no infiltration), then the composition and modal amount of plagioclase in a rock would have been determined by the bulk rock chemistry and the progress of mineral reactions involving components of plagioclase solid solution. Modal analysis indicates that the rocks, however, contain far more $CaAl_2Si_2O_8$ (in plagioclase solid solution) than can be accounted for by reactions (R18)-(R22) operating in an isochemical system. Examination of reactions (R18) to (R22) reveals that all equilibria evolve a mixed volatile fluid containing not only CO_2 and H_2O, but also NaCl or KCl. The control of plagioclase composition during metamorphism was closely analogous to the control of metamorphic fluid composition by combined buffering and infiltration. During progressive metamorphism, mineral reactions [e.g., (R18)-(R22)] internally buffered the chemical potentials of the components $NaAlSi_3O_8$ and $CaAl_2Si_2O_8$ in plagioclase. As these reactions proceeded, however, Na was incorporated into the infiltrating fluid and removed from the local system. This process resulted in an effective subtraction of albite component from the bulk-rock composition. Whole-rock chemical analyses of the samples studied by Ferry (1982b), in fact, reveal that the high-grade specimens are depleted in Na relative to their low-grade equivalents. Removal of an element comprising the albite component allowed mineral reactions to buffer the composition of plagioclase to much higher mole fractions of anorthite than would have been possible if the bulk composition remained constant (*cf.* Fig. 15). As with the case of fluid solutions controlled by a combination of buffering and infiltration (Table 2), the composition of plagioclase in the rocks studied by Ferry (1982b) cannot exclusively be explained by mineral reactions in a system closed to mass transfer to and from external sources. Phase composition and modal abundance can only be explained by both mineral reactions and addition or subtraction of mass (Na) from the rocks through infiltration.

Summary

Many petrologic studies of metamorphic rocks have found systematic changes in mineral composition with increasing grade. The systematic changes can often be explained by the control on mineral composition exerted by prograde mineral equilibria. We stress that this control of mineral composition is simply an example of the buffering of phase composition which is rigorously analogous to the manner in which mineral reactions commonly buffer fluid composition during metamorphism. The same process that often internally controls metamorphic fluid composition also internally controls mineral composition. The similarity of terms involving the f_i and the a_j in equation (1) may be considered as the formal mathematical statement that mineral reactions exert a constraint on the composition of mineral solid solutions that is exactly similar to the constraint they exert on fluid solutions.

We also note, however, that mineral compositions are not always internally controlled during metamorphism exclusively by buffer reactions. In particular, the isotopic composition of minerals in some metamorphic rocks appears to have been controlled by infiltration independent of the local mineral reactions. In other cases, mineral composition is the result of an interplay between the phenomena of internal buffering and infiltration. While it seems that buffering is the most important control on the prograde evolution of metamorphic mineral compositions, infiltration with or without buffering is a factor that should be given consideration as well.

BUFFERING, HEAT FLOW, AND THE CONTROL OF TEMPERATURE

General statement

In the previous two sections we have explored the capacity of mineral equilibria to buffer both mineral and fluid composition. Compositional variables that refer to a fluid phase (the f_i) and compositional variables that refer to mineral phases (the a_j) enter into equation (1). Mathematically, the process of buffering may be considered as the constraint on these compositional variables exercised by equation (1). Temperature, however, is also a variable in equation (1). Thus, by analogy, mineral equilibria that define relations like (1) are capable of buffering not only fluid and mineral

composition, but also temperature. The buffering of temperature can be
readily expressed by equation (1) through consideration of an equilibrium
relation among minerals in which $RT(\Sigma_i \nu_i \ln f_i + \Sigma_j \nu_j \ln a_j) = 0$ (e.g., consideration of an equilibrium among pure volatile-free substances). In this
case, equation (1) reduces to:

$$\Delta \bar{H}_r - T\Delta \bar{S}_r + \Delta \bar{V}_r (P-1) = 0 \ . \tag{36}$$

At constant pressure, the mineral equilibrium defines a unique temperature,

$$T_{eq} = \frac{\Delta \bar{H}_r + \Delta \bar{V}_r (P-1)}{\Delta \bar{S}_r} \ . \tag{37}$$

The minerals in equilibrium will buffer temperature at T_{eq} through their reaction relationship. For example, if heat is added to the assemblage, the
reaction will proceed in the direction of positive $\Delta \bar{H}_r$: the heat will be
consumed as chemical work and temperature will remain fixed at T_{eq} until
one or more of the reactants is consumed. If heat is extracted from the
assemblage, the opposite will occur: the reaction will buffer temperature
at T_{eq} by progressing in the direction of negative $\Delta \bar{H}_r$.

As a specific example, consider coexisting kyanite and andalusite which
define the equilibrium relation:

$$\Delta \bar{H}_r - T\Delta \bar{S}_r + P\Delta \bar{V}_r = 1052 - 2.224T + 0.17P = 0 \tag{38}$$

(Holdaway, 1971). The two minerals may be in equilibrium at P = 3000 bars,
T = 440°C. The coexisting kyanite and andalusite at 3000 bars buffer temperature at 440°C. If heat is added, kyanite will react to form andalusite
and temperature will remain at 440°C until all kyanite is consumed. If heat
is extracted, andalusite will react to form kyanite and temperature will
remain at 440°C until all andalusite is consumed.

Simple calculations can be used to quantitatively evaluate the capacity
of mineral reactions to buffer temperature. Consider again the equilibrium
between kyanite and andalusite at 3000 bars, 440°C. At these pressure-
temperature conditions, 1 cm^3 kyanite (= 0.0227 moles kyanite) can exist
in equilibrium with andalusite. If heat is added to the kyanite, andalusite
forms and the mineral reaction buffers temperature at 440°C. The amount of
heat required to convert one mole kyanite to one mole andalusite is 1586 cal.
The 1 cm^3 kyanite, as it reacts to form andalusite, will buffer temperature
until all kyanite is consumed, i.e., until 0.0227 x 1586 = 36 cal are added
to the original 1 cm^3 mineral. The buffer capacity of the kyanite-andalusite
reaction may therefore be represented as ~36 cal/cm^3 of reactants (the exact

value depends on pressure of reaction). The heat capacity of kyanite at 440°C is 43.86 cal/degree-mole = 0.995 cal/degree-cm^3 (Robie et al., 1978). Thus, the kyanite-andalusite reaction can buffer temperature against an addition of heat which, in the absence of reaction, would raise the temperature of kyanite by 36°C.

The buffer capacity of common metamorphic mineral reactions may be semi-quantitatively compared by dividing $\Delta\bar{H}_r$ for the reaction by the molar volume of the reactants. For typical dehydration reactions, the buffer capacity is in the range 50-150 cal/cm^3; for decarbonation reactions, approximately 200-400 cal/cm^3; and for mixed-volatile (CO_2-H_2O) reactions, approximately 10-100 cal/cm^3. The volumetric heat capacity of minerals (and hence rocks) does not vary much about an average value of $\sim$0.9 cal/degree-cm^3. Typical devolatilization reactions can be expected to consume an amount of heat as chemical work that would otherwise (in an unreactive rock) raise the temperature by up to several hundred degrees.

An example of the buffering of temperature

The buffering of temperature by mineral reactions has not been extensively explored by metamorphic petrologists. The phenomenon may be demonstrated, however, by a reexamination of mineral equilibria among zoisite, anorthite, calcite, quartz, tremolite, diopside, and CO_2-H_2O fluid, which was discussed earlier. The six minerals coexist at an isobaric invariant point in the presence of CO_2-H_2O fluid. At 3500 bars pressure, the equilibrium conditions of the invariant point are T = 481°C, X_{CO_2} = 0.094. If heat is added to or extracted from a rock containing these minerals in equilibrium with fluid, reaction will proceed in such a way that temperature is held constant. The quantitative relation between reaction progress and the amount of heat added to or extracted from the assemblage may be easily evaluated. In quadrant I of Figure 13, net reaction produces anorthite, calcite, and diopside. Reaction may be described in terms of the progress variables ξ_6 and ξ_9. The relation between the amount of heat, Q, added to rocks which plot in quadrant I and the progress variables is:

$$Q(cal) = 16,804\ \xi_9 + 67,765\ \xi_6 . \tag{39}$$

Numerical coefficients are $\Delta\bar{H}$ for reactions (R9) and (R6), respectively, at 3500 bars, 481°C. Equation (39) defines straight lines in quadrant I of the ξ_6-ξ_9 diagram, which relate values of ξ_6 and ξ_9 for any particular rock to the amount of heat needed to produce the observed extent of reaction. The straight lines calculated from equation (39) are the dashed contours.

317

Numerical values on the contours refer to kcal heat/(1000 cm^3 metamorphosed rock). A positive sign for calculated Q indicates that heat was added to rocks as reactions (R6) and (R9) progressed at the invariant point. Because all dashed contours in quadrant I are labelled with positive values of Q, all rocks with measured values of ξ_6 and ξ_9 which plot in quadrant I must have had heat added to them while anorthite, calcite, and diopside crystallized. Rocks will not plot in quadrant I if heat was extracted from them as reaction occurred at the isobaric invariant point.

The remaining quadrants II, III and IV were contoured with lines of constant heat in a manner analogous to the exercise for quadrant I. In quadrant II, the dashed contours are a solution to:

$$Q(cal) = 67,665 \; \xi_6 - 16,804 \; \xi_7 \; . \tag{40}$$

In quadrant III, the dashed contours are a solution to:

$$Q(cal) = -67,665 \; \xi_{10} - 16,804 \; \xi_7 \; . \tag{41}$$

In quadrant IV, the dashed contours are a solution to:

$$Q(cal) = 16,804 \; \xi_9 - 67,665 \; \xi_{10} \; . \tag{42}$$

In equations (39)-(42) the numerical coefficients are $\Delta \bar{H}$ for the reaction indicated by the subscript of the associated progress variable at the pressure and temperature of the invariant point.

The dashed "0" contour divides Figure 13 into two parts. Along the "0" contour, which runs through quadrants II and IV, mineral reaction can take place without either addition of heat to rock or extraction of heat from it. Infiltration alone drives mineral reaction along the dashed "0" contour. Northeast of the dashed "0" contour, reaction occurs only with the addition of heat to rock; southwest of the "0" dashed contour, reaction occurs only with extraction of heat from rock.

Figure 13 demonstrates the large capacity of the assemblage zoisite-calcite-anorthite-quartz-tremolite-diopside-fluid to buffer temperature. For certain combinations of reaction progress, up to 200 cal/cm^3 rock can be absorbed or liberated as chemical work. This amount of heat, if used to change the temperature of the rocks in the absence of reaction, would result in temperature changes of over 200°C.

Returning to the study of Ferry (1982a), 14 rock samples containing the assemblage zoisite-calcic plagioclase-calcite-quartz-calcic amphibole-diopside have been plotted in Figure 13 on the basis of their reaction history and modes. The samples were all collected from the same outcrop, and it is highly unlikely that, at any point in time, the samples were ever more than a fraction of a degree in temperature from each other. Figure 13 indicates that most samples consumed 2-15 cal heat/cm^3 during the time that zoisite and diopside crystallized. Curiously, four samples suggest that small amounts of heat were extracted from them while zoisite and diopside crystallized. For all samples, a measurable amount of heat was added to or extracted from rocks without change in temperature. The heat was consumed or liberated not by a change in temperature of the rock, but through mineral reaction. The mineral reactions thus resisted changes in temperature that otherwise would have occurred due to heat-rock interaction. In this manner, the reactions may be considered to have buffered temperature. The maximum amounts of heat involved in reaction are equivalent to the amount of heat that, without reaction, would have raised the temperature of the rocks by 30°-40°C.

Reactions involving isobaric univariant and isobaric invariant mineral assemblages may be analyzed in a fashion similar to that in Figure 13 for reactions among zoisite, calcite, anorthite, quartz, tremolite, diopside, and fluid. Values of $\Delta \bar{H}_r$, however, must be known or be estimated at the pressure-temperature conditions of reaction. Because $\Delta \bar{H}_r \neq 0$ for almost all reactions, most reactions must have some capacity to buffer temperature. The buffering of temperature by mineral reactions during metamorphism therefore should be a common phenomenon.

Discussion

Buffering of temperature and fluid composition during metamorphism. Figure 13 emphasizes that the buffering of temperature and fluid composition are cooperative processes: in response to heat-rock and/or fluid-rock interactions, mineral-fluid reactions in isobaric invariant assemblages prohibit changes in both temperature and fluid composition until at least one of the mineral reactants is exhausted. Furthermore, the buffering of temperature and the buffering of fluid composition are processes that apparently reinforce each other. For example, with reference to Figure 13, the larger the amount of heat added to a rock containing zoisite + calcite + anorthite + quartz + tremolite + diopside, the larger is the volume of fluid that the rock can buffer to the composition of fluid in equilibrium with the six minerals. Conversely, rocks have a greater capacity to buffer temperature if fluid-rock interactions accompany heat-rock interactions.

The simultaneous buffering of temperature and fluid composition by mineral reactions can be viewed as a consequence of equation (1). Both temperature and fluid composition (as the f_i) are variables constrained by the equation. As emphasized earlier, the process of buffering can simply be regarded as the constraint that equation (1) exerts upon intensive variables. Because the equation represents a constraint which acts both on temperature and on the f_i, mineral equilibria simultaneously buffer temperature and fluid composition. Equation (1) is a mathematical statement that temperature and fluid composition act as analogous variables during metamorphism. This has long been recognized and is an unwritten foundation of the use of T-X_{CO_2} diagrams in metamorphic petrology. Although it is commonly accepted that mineral equilibria may buffer fluid composition, usually not much attention is given to the possibility that mineral equilibria may buffer temperature as well. The analogy between temperature and fluid composition emphasizes that these same mineral equilibria must be considered, in addition, as temperature buffers. Whenever mineral reactions involving volatile components buffer fluid composition, they also buffer temperature and vice versa. The discussion here is intended to unify the concepts of the buffering of temperature and fluid composition and to highlight the observation that they are closely integrated phenomena during metamorphic events.

Controls of temperature during metamorphism. Earlier we discussed two extreme mechanisms for the control of fluid composition during metamorphism. The control of temperature during metamorphism may similarly be viewed from two extremes. At one extreme, temperature is controlled by regional heat flow exclusive of any modification by mineral reactions. At the other extreme, temperature might be considered as controlled by the internal liberation or consumption of heat by mineral reactions exclusive of any interaction between the rock and external heat sources or heat sinks.

The following algebraic representation of the heat budget of a metamorphic rock illustrates the quantitative relationship between these two extreme controls:

$$Q_{ext} = c\Delta T + \Sigma_i \Delta \bar{H}_i^{P,T} \xi_i \tag{43}$$

where c is the volumetric heat capacity of the rock, ΔT is the change in temperature that occurs during metamorphism, $\Delta \bar{H}_i^{P,T}$ is the enthalpy of reaction for reaction i at the pressure-temperature conditions of metamorphism, and ξ_i is the progress of reaction i. Q_{ext} is the amount of heat added to the rock from an external heat reservoir or extracted from the rock and added to the external reservoir. At one extreme, metamorphic temperature is controlled exclusive of mineral reactions when $\Sigma_i \Delta \bar{H}_i^{P,T} \xi_i = 0$ and $Q_{ext} = c\Delta T$ (i.e., metamorphism without mineral reactions). At the other extreme, one can consider metamorphic temperature to be controlled exclusively by mineral reactions when $\Sigma_i \Delta \bar{H}_i^{P,T} \xi_i = -c\Delta T$ (i.e., $Q_{ext} = 0$; no exchange of heat between the rock and an external heat reservoir). In general, for metamorphic events, both $c\Delta T > 0$ and $\Sigma_i \Delta \bar{H}_i \xi_i > 0$. Thus $Q_{ext} > 0$. Consequently, neither extreme mechanism controls metamoprhic temperature. The temperature of a rock during metamorphism is a result of the interplay between (a) heat added to or extracted from the rock by interaction with an external heat reservoir and (b) heat internally consumed or liberated by mineral reactions. We note the parallel between this sort of control on temperature and the manner in which the molecular constituency of metamorphic fluid is controlled by an interplay between molecules introduced into the rock by infiltration and molecules liberated or consumed internally by mineral-fluid reactions.

BUFFERING AND THE CONTROL OF PRESSURE

We have explored the capacity of mineral equilibria to buffer mineral composition, fluid composition, and temperature. Pressure, however, is also an intensive variable in equation (1). Because equation (1) identifies a constraint exercised on pressure by a mineral equilibrium, the mineral assemblage can therefore be considered to act as a pressure buffer. The buffering of pressure can be readily expressed by equation (1) through the consideration again of an equilibrium among pure volatile-free substances. In this case, equation (1) reduces to equation (36) above. At constant temperature, the mineral equilibrium defines a unique pressure, P_{eq}:

$$P_{eq} \simeq \frac{T \Delta \bar{S}_r - \Delta \bar{H}_r}{\Delta \bar{V}_r} \; . \tag{44}$$

The minerals in equilibrium will buffer pressure at P_{eq} through their reaction relationship. If, for example, compressive work is done on the assemblage, the reaction will progress in the direction of negative $\Delta \bar{V}_r$. The compressive work will be converted into chemical work, and pressure will remain fixed at P_{eq} until one or more of the reactants is consumed.

It is questionable whether compressive work is ever a significant driving force behind mineral reactions during metamorphism, and consequently questionable whether the buffering of pressure by mineral reactions ever occurs in nature. We have introduced the concept of pressure buffering principally to emphasize the equivalence of the variables f_i, a_j, T, and P in equation (1), and to emphasize that the same mineral reaction may be capable of buffering several of these intensive variables simultaneously during metamorphism.

CONCLUDING REMARKS

Chemical thermodynamics identifies those intensive variables that characterize the state of a system at equilibrium. In the consideration of metamorphism, the set of intensive variables of principal interest include pressure, temperature, mineral composition, and fluid composition. When minerals in a rock are in a reaction relationship during metamorphism, they define relationships like equation (1), which serve as constraints that act upon two or more of the intensive variables. Mineral reactions therefore may be considered as processes that control the values of these variables during a metamorphic event. The control of intensive variables by chemical reactions is directly analogous to the phenomena of buffering described in standard chemistry texts. We have devoted this chapter to demonstrating, through numerous case studies, that mineral reactions do indeed buffer intensive variables during metamorphism. The buffering of fluid composition has become well established since the pioneering studies of H. J. Greenwood. The control of mineral composition by mineral reactions has long been recognized, but not often described by the concept of buffering. More recent studies provide evidence that mineral reactions have a large capacity to buffer temperature, and indeed have done so during metamorphic events. While in principle, mineral reactions may buffer pressure, no case study has yet demonstrated that this process is important during metamorphism.

Hopefully, this chapter has convinced the reader that there is a common theme that must run through any consideration of what controls intensive variables during metamorphism. Buffering reactions serve as internal controls within rocks which, wholly or in part, determine the values of fluid composition, mineral composition, temperature, and possibly even pressure during metamorphism. Furthermore, mineral reactions also serve as buffers in the sense that mineral parageneses resist attempts to change those intensive variables that may be influenced by agents or processes external to the particular rock under consideration.

Buffering is not the only control on intensive variables governing metamorphism, and we have identified infiltration as a second process of general interest. Numerous petrologic and geochemical studies indicate that infiltration of metamorphic rock by aqueous fluid may control or influence the concentration of components in mineral solid solutions and coexisting fluid solutions. If heat flow can be considered as an honorary form of infiltration, then a form of infiltration may also be considered to partially control temperature during metamorphic events. Thus there is a second common theme which runs through consideration of what controls intensive variables during metamorphism: namely, infiltration. The flow of fluid and heat through rocks during metamorphism serves as an external control which, wholly or in part, determines fluid and mineral composition and temperature.

Buffering and infiltration do not normally act alone; in fact, they more often function in concert. Metamorphic fluid composition is, in general, a result of volatiles released by local mineral reactions and molecules introduced into, or removed from, a rock by infiltration. Mineral compositions, in general, are a result of mineral components produced or consumed by heterogeneous mineral reactions and components (or species) introduced into, or removed from, rocks by infiltration. Similarly, temperature is a result of heat produced or consumed by mineral reactions, and heat introduced into, or removed from, rocks by heat flow. Thus, in general, intensive variables can be considered as controlled during metamorphism by an interplay between local buffering phenomena and infiltration. This is not to say, of course, that in any given terrain or at any point in time, one or the other process might not dominate. Buffering and infiltration cannot be regarded as the only controls of intensive variables during metamorphism. We nevertheless believe that a unified model of metamorphism conveniently emerges from consideration of the similarities by which buffer reactions and infiltration act as controls on fluid composition, mineral composition, and temperature.

ACKNOWLEDGMENTS

This chapter has profited greatly from the constructive comments of D. Rumble, F. S. Spear, J. Selverstone, and S. A. Rawson who reviewed the manuscript at various stages. Support from the National Science Foundation in the form of Grants EAR-7922877 (JMR) and EAR-8020567 (JMF) is most gratefully acknowledged. JMF further acknowledges support from Research Corporation in the form of a Cottrell Grant.

Albee, A.L. (1965) Phase equilibria in three assemblages of kyanite-zone pelitic schists, Lincoln Mountain quadrangle, Vermont. J. Petrol. 6, 246-301.

Allen, J.M. (1978) Calc-silicate equilibria in a remetamorphosed aureole. Trans. Am. Geophys. Union 59, 407.

_____ and J.J. Fawcett (1982) Zoisite-anorthite-calcite stability relations in H_2O-CO_2 fluids at 5000 bars: an experimental and SEM study. J. Petrol. 23, 215-239.

Bowman, J.R. and E.J. Essene (1982) P-T-X(CO_2) conditions of contact metamorphism in the Black Butte aureole, Elkhorn, Montana. Am. J. Sci. 282, 311-341.

_____, _____, and J.R. O'Neil (1980) Origins and evolution of metamorphic fluids in dolomitic marbles, Elkhorn, Montana. Trans. Am. Geophys. Union 61, 391.

Bottinga, Y. and P. Richet (1981) High pressure and temperature equation of state and calculation of the thermodynamic properties of gaseous CO_2. Am. J. Sci. 281, 615-661.

Bucher-Nurminen, K. (1981) The formation of metasomatic reaction veins in dolomitic marble roof pendants in the Bergell intrusion (Province Sondrio, Northern Italy). Am. J. Sci. 281, 1197-1223.

Burnham, C.W., J.R. Holloway, and N.F. Davis (1969) *Thermodynamic Properties of Water to 1000°C and 10,000 bars.* Geol. Soc. Am. Spec. Paper 132, 96 pp.

Butler, P., Jr. (1969) Mineral compositions and equilibria in the metamorphosed iron formation of the Gagnon Region, Quebec, Canada. J. Petrol. 10, 5-101.

Carmichael, D.M. (1970) Intersecting isograds in the Whetstone Lake area, Ontario. J. Petrol. 11, 147-181.

Cermignani, C. and G.M. Anderson (1973) Origin of a diopside-tremolite assemblage near Tweed, Ontario. Canadian J. Earth Sci. 10, 84-90.

Chatterjee, N.D. (1971) Phase equilibria in the Alpine metamorphic rocks of the environs of the Dora-Maira-Massif, Western Italian Alps. N. Jahrb. Mineral. Monat. 114, 181-245.

Chinner, G.A. (1960) Pelitic gneisses with varying ferrous/ferric ratios from Glen Clova, Angus, Scotland. J. Petrol. 1, 178=217.

Crawford, M.L., D.W. Kraus, and L.S. Hollister, (1979) Petrologic and fluid inclusion study of calc-silicate rocks, Prince Rupert, British Columbia. Am. J. Sci. 279, 1135-1159.

Erdmer, P. (1981) Metamorphism at the northwest contact of the Stanhope pluton, Quebec Appalachians: mineral equilibria in interbedded pelite and calc-schist. Contrib. Mineral. Petrol. 76, 109-115.

Evans, B.W. and V. Trommsdorff (1970) Regional metamorphism of ultramafic rocks in the central Alps: paragenesis in the system CaO-MgO-SiO_2-H_2O. Schweiz. Mineral. Petrogr. Mitt. 50, 481-507.

_____ and _____ (1974) Stability of enstatite + talc, and CO_2 metasomatism of metaperidotite, Val d'Efra, Lepontine Alps. Am. J. Sci. 274, 274-296.

Ferry, J.M. (1976a) Metamorphism of calcareous sediments in the Waterville-Vassalboro area, south-central Maine: mineral reactions and graphical analysis. Am. J. Sci. 276, 841-882.

_____ (1976b) P, T, f_{CO_2}, and f_{H_2O} during metamorphism of calcareous sediments in the Waterville-Vassalboro area, south-central Maine. Contrib. Mineral. Petrol. 57, 119-143.

_____ (1979) A map of chemical potential differences within an outcrop. Am. Mineral. 64, 966-985.

_____ (1980a) A case study of the amount and distribution of heat and fluid during metamorphism. Contrib. Mineral. Petrol. 71, 373-385.

_____ (1930b) A comparative study of geothermometers and geobarometers in pelitic schists from south-central Maine. Am. Mineral. 65, 720-732.

_____ (1981) Petrology of graphitic sulfide-rich schists from south-central Maine: an example of desulfidation during prograde regional metamorphism. Am. Mineral. 66, 908-930.

_____ (1982a) An example of the buffering of temperature by mineral reactions during prograde regional metamorphism. Am. J. Sci. (submitted).

_____ (1982b) Mineral reactions and element migration during metamorphism of calcareous sediments from the Vassalboro Formation, south-central Maine. Am. Mineral. (submitted).

Frost, B.R. (1982) Contact metamorphic effects of the Stillwater Complex: the concordant iron formation: a discussion of the role of buffering in metamorphism of iron formation. Am. Mineral. 67, 142-149.

Fyfe, W.S., F.J. Turner, and J. Verhoogen (1958) *Metamorphic Reactions and Metamorphic Facies.* Geol. Soc. Am. Memoir 73, 259 pp.

324

Ghent, E.D., D.B. Robbins, and M.Z. Stout (1979) Geothermometry, geobarometry, and fluid compositions of metamorphosed calc-silicates and pelites, Mica Creek, British Columbia. Am. Mineral. 64, 874-885.

Glassley, W.E. (1975) High grade regional metamorphism of some carbonate bodies: significance for the orthopyroxene isograd. Am. J. Sci. 275, 1133-1163.

Gordon, T.M. and H.J. Greenwood (1971) The stability of grossularite in H_2O-CO_2 mixtures. Am. Mineral. 56, 1674-1688.

Grambling, J.A. (1981) Kyanite, andalusite, sillimanite, and related mineral assemblages in the Truchas Peak region, New Mexico. Am. Mineral. 66, 702-722.

Greenwood, H.J. (1962) Metamorphic reactions involving two volatile components. Carnegie Inst. Wash. Year Book 61, 82-85.

_____ (1967) Wollastonite: stability in H_2O-CO_2 mixtures and occurrence in a contact metamorphic aureole near Salmo, British Columbia, Canada. Am. Mineral. 52, 1669-1680.

_____ (1967) Mineral equilibria in the system MgO-SiO_2-H_2O-CO_2. In P.H. Abelson, Ed., Researches in Geochemistry, II. John Wiley and Sons, New York, pp. 542-549.

_____ (1975) Buffering of pore fluids by metamorphic reactions. Am. J. Sci. 275, 573-593.

Grew, E.S. (1978) Oxygen and water gradients in granulite-facies migmatitic pelitic gneisses from Molodezhaya Station, Antarctica (67°40'S, 40°50'E). Geol. Soc. Am. Abstr. Prog. 10, 412.

Guidotti, C.V. (1970) The mineralogy and petrology of the transition from the lower to upper sillimanite zone in the Oquossoc area, Maine. J. Petrol. 11, 277-336.

Hewitt, D.A. (1973) The metamorphism of micaceous limestones from south-central Connecticut. Am. J. Sci. 273A, 444-469.

Holdaway, M.J. (1971) Stability of andalusite and sillimanite and the aluminum silicate phase diagram. Am. J. Sci. 271, 97-131.

Hoy, T. (1976) Calc-silicate isograds in the Riondel area, southeastern British Columbia. Canadian J. Earth Science 13, 1093-1104.

James, H.L. and A.L. Howland (1955) Mineral facies in iron-and-silica-rich rocks. Bull. Geol. Soc. Am. 66, 1580-1581.

Jansen, J.H.B., A.H. Kraats, H. Rijst, and R.D. Schuiling (1978) Metamorphism of siliceous dolomites at Naxos, Greece. Contrib. Mineral. Petrol. 67, 279-288.

Kerrick, D.M. (1974) Review of metamorphic mixed-volatile (H_2O-CO_2) equilibria. Am. Mineral. 59, 729-762.

Kerrick, D.M., K.E. Crawford, and A.F. Randazzo (1973) Metamorphism of calcareous rocks in three roof pendants in the Sierra Nevada, California. J. Petrol 14, 303-325.

Korzhinskii, D.S. (1959) Physicochemical Basis of the Analysis of the Paragenesis of Minerals. Trans. Consultants Bureau, New York.

_____ (1967) On thermodynamics of open systems and the phase rule (a reply to the second critical paper of D.F. Weill and W.S. Fyfe). Geochim. Cosmochim. Acta 30, 829-835.

Kranck, S.H. (1961) A study of phase equilibria in a metamorphic iron formation. J. Petrol. 2, 137-184.

Labotka, T.C., D.T. Vaniman, and J.J. Papike (1982) Contact metamorphic effects of the Stillwater Complex, Montana: the concordant iron formation: a reply to the role of buffering in metamorphism of iron formation. Am. Mineral. 67, 149-153.

Melson, W.G. (1966) Phase equilibria in calc-silicate hornfels, Lewis and Clark County, Montana. Am. Mineral. 51, 402-421.

Misch, P. (1964) Stable association wollastonite-anorthite and other calc-silicate assemblages in amphibolite-facies crystalline schists of Nanga Parbat, Northwest Himalayas. Beitr. Mineral. Petrol. 10, 315-356.

Moore, J.N. and D.M. Kerrick (1976) Equilibria in siliceous dolomites of the Alta aureole, Utah. Am. J. Sci. 276, 502-524.

Morgan, B.A. (1979) Petrology and mineralogy of eclogite and garnet amphibolite from Puerto Calsello, Venezuela. J. Petrol. 11, 101-145.

Newton, R.C., J.V. Smith, and B.F. Windley (1980) Carbonic metamorphism, granulites and crustal growth. Nature 288, 45-50.

Osberg, P.H. (1974) Isochemical metamorphism, south-central Maine. Trans. Am. Geophys. Union 55, 450.

Rice, J.M. (1977a) Contact metamorphism of impure dolomitic limestone in the Boulder aureole, Montana. Contrib. Mineral. Petrol.59, 237-259.

_____ (1977b) Progressive metamorphism of impure dolomitic limestone in the Marysville aureole, Montana. Am. J. Sci. 277, 1-24.

_____ (1980) Phase equilibria involving humite minerals in impure dolomitic limestones: Part II. Calculated stability of chondrodite and norbergite. Contrib. Mineral. Petrol. 75, 205-223.

Rumble, D. (1974) The Gibbs phase rule and its application to geochemistry. J. Wash. Acad. Sci. 64, 199-208.

_____ (1978) Mineralogy, petrology, and oxygen isotopic geochemistry of the Clough Formation, Black Mountain, New Hampshire, U.S.A. J. Petrol. 19, 317-340.

_____, J.M. Ferry, T.C. Hoering, and A.J. Boucot (1982) Fluid flow during metamorphism at the Beaver Brook fossil locality, New Hampshire. Am. J. Sci. 282, 886-919.

Savin, S.M. and S. Epstein (1970) The oxygen isotopic composition of coarse-grained sedimentary rocks and minerals. Geochim. Cosmochim. Acta 34, 323-329.

Shaw, D.M. (1956) Geochemistry of pelitic rocks. Part III: major elements and general geochemistry. Bull. Geol. Soc. Am. 67, 919-934.

Shmonov, V.M. and K.I. Shmulovich (1974) Molar volumes and the equation of state for CO_2 at 100-1000°C and 2000-10000 bar. Akad. Nuak SSSR Dokl. 217, 205-209.

Skippen, G.B. and D.M. Carmichael (1977) Mixed-volatile equilibria. *In* H.J. Greenwood, Ed., *Short Course in Application of Thermodynamics to Petrology and Ore Deposits*. Mineral. Assoc. of Canada, Vancouver, pp. 109-125.

Spear, F.S. (1977) Phase equilibria of amphibolites from the Post Pond Volcanics, Vermont. Carnegie Inst. Wash. Year Book 76, 613-619.

_____ (1981) $\mu(H_2O)-\mu(CO_2)-X(Fe-Mg)$ relations in amphibolite assemblages. Geol. Soc. Am. Abstr. Prog. 13, 559.

Suzuki, K. (1977) Local equilibrium during the contact metamorphism of siliceous dolomites in Kasuga-mura, Japan. Contrib. Mineral. Petrol. 61, 79-89.

Taylor, H.P., Jr. (1967) Oxygen isotope studies of hydrothermal mineral deposits. *In* H.L. Barnes, Ed., *Geochemistry of Hydrothermal Ore Deposits*. Holt, Rinehart, and Winston, New York, pp. 109-142.

Thompson, A.B. (1976) Mineral reactions in pelitic rocks: I. Prediction of P-T-X(Fe-Mg) phase relations. Am. J. Sci. 276, 401-424.

Thompson, J.B., Jr. (1970) Geochemical reaction and open systems. Geochim. Cosmochim. Acta 34, 529-551.

Trommsdorff, V. (1972) Change in T-X during metamorphism of siliceous dolomite rocks of the central Alps. Schweiz. Mineral. Petrog. Mitt. 52, 567-571.

_____ and B.W. Evans (1969) The stable association enstatite-forsterite-chlorite in amphibolite facies ultramafics of the Lepontine Alps. Schweiz. Mineral. Petrog. Mitt. 49, 325-332.

_____ and _____ (1974) Alpine metamorphism of peridotitic rocks. Schweiz. Mineral. Petrog. Mitt. 14, 333-352.

Valley, J.W. and E.J. Essene (1980) Calc-silicate reactions in Adirondack marbles: the role of fluids and solid solution: Summary. Bull. Geol. Soc. Am. 91, 114-117.

Vidale, R.J. and D.A. Hewitt (1973) "Mobile" components in the formation of calc-silicate bands. Am. Mineral. 58, 991-997.

Walther, J.V. and P.M. Orville (1980) Rates of metamorphism and volatile production and transport in regional metamorphism. Geol. Soc. Am. Abstr. Prog. 12, 544.

Weill, D.F. and W.S. Fyfe (1967) On equilibrium thermodynamics of open systems and the phase rule (a reply to D.S. Korzhinskii). Geochim. Cosmochim. Acta 31, 1167-1176.

Williams-Jones, A.E. (1981) Thermal metamorphism of siliceous limestones in the aureole of Mount Royal, Quebec. Am. J. Sci. 281, 673-696.

Zen, E-an (1961) Mineralogy and petrology of the system $Al_2O_3-SiO_2-H_2O$ in some pyrophyllite deposits of North Carolina. Am. Mineral. 46, 52-66.

_____ (1963) Components, phases, and criteria of chemical equilibrium in rocks. Am. J. Sci. 261, 929-942.

Chapter 8
STABLE ISOTOPE FRACTIONATION during METAMORPHIC
DEVOLATILIZATION REACTIONS
D. Rumble III

INTRODUCTION

The methods of stable isotope geochemistry have been used in the study of contact and regionally metamorphosed rocks

(1) to estimate metamorphic temperatures;

(2) to determine the nature and extent of fluid-rock interaction during metamorphism and to deduce the source of fluids; and

(3) to infer the sedimentary or igneous precursors of metamorphosed rocks.

The emphasis in this volume is on methods of studying fluid-rock interactions in regionally metamorphosed terranes. A brief review is given of the methods of stable isotope geothermometry as well as the methods of assessing the degree of attainment of isotope equilibrium during metamorphism.

DEFINITIONS AND TECHNIQUES

Stable isotope analyses are reported in the "δ" notation where δ is defined by

$$\delta = 1000 \left[\frac{R_{sample}}{R_{standard}} - 1 \right]$$

and R is the isotope abundance ratio of the rarer to the more common isotope (i.e., D/H, $^{18}O/^{16}O$, $^{13}C/^{12}C$; the symbol "D" refers to the stable isotope of hydrogen, ^{2}H, called deuterium). It may be seen that δ values give deviations in the isotope ratios of samples from a given standard. Absolute abundance ratios are not reported because they cannot be measured as precisely as relative differences. A discussion of the various standards used in stable isotope analysis is given by Friedman and O'Neil (1977).

The partitioning of isotopes between pairs of fluid species or minerals is:

$$\alpha_{A-B} = R_A/R_B \qquad \text{or, in } \delta \text{ notation,} \qquad \alpha_{A-B} = \frac{1 + \delta_A/1000}{1 + \delta_B/1000} \, ,$$

where A and B refer to a given pair of minerals or fluid species. In order to obtain a useful approximation, take the natural logarithm of both sides of the preceding equation

$$\ln \alpha = \ln(1 + \delta_A/1000) - \ln(1 + \delta_B/1000).$$

Now $\delta/1000$ is always less than 1.0; therefore, the following relation holds, approximately,

$$\ln(1 + \delta_A/1000) \simeq \delta_A/1000 \ ,$$

and similarly for δ_B. By substitution one obtains

$$\Delta_{A-B} \equiv \delta_A - \delta_B \simeq 1000 \ln \alpha_{A-B} \ .$$

The accuracy of the approximation decreases as both the values of δ_A and δ_B and the difference between them increases above 10.0 (Friedman and O'Neil, 1977; Table 1, p. KK2).

The stable isotope composition of samples is routinely measured in a dual inlet, double- or triple-collector mass-spectrometer (Nier, 1947; McKinney et al., 1950). The various correction factors that must be taken into account to obtain accurate results have been described by Craig (1957), Deines (1970), and Mook and Grootes (1973). Gas samples are prepared from rocks and minerals for mass spectrometric analysis by a variety of techniques depending on the isotopes and minerals which are to be analyzed (Hoefs, 1980, p. 23-59).

ISOTOPE EXCHANGE REACTIONS

Isotope partitioning and geothermometry

The partitioning of light stable isotopes between pairs of minerals or fluid species is governed by exchange reactions (Urey, 1947) such as illustrated in the following examples:

$$\tfrac{1}{2}Si^{18}O_2 + \tfrac{1}{4}Fe_3{}^{16}O_4 = \tfrac{1}{2}Si^{16}O_2 + \tfrac{1}{4}Fe_3{}^{18}O_4$$

for a pair of minerals, or,

$$\tfrac{1}{2}Si^{18}O_2 + H_2{}^{16}O = \tfrac{1}{2}Si^{16}O_2 + H_2{}^{18}O$$

for a mineral and a fluid species, or,

$$\tfrac{1}{2}C^{18}O_2 + H_2{}^{16}O = \tfrac{1}{2}C^{16}O_2 + H_2{}^{18}O$$

for a pair of fluid species. Fractional stoichiometric coefficients are used because of the convention of writing the reactions in terms of one exchangeable atom.

The equilibrium constant, K, for an oxygen isotope exchange reaction between a pair of minerals such as quartz and magnetite that have only one crystallographically distinct site for oxygen is:

$$K = \frac{\left(x_{16,Q}^{2}\right)^{1/2}\left(x_{18,Mt}^{4}\right)^{1/4}}{\left(x_{18,Q}^{2}\right)^{1/2}\left(x_{16,Mt}^{4}\right)^{1/4}}$$

where the $X_{i,j}$ give the atomic fraction of ^{18}O or ^{16}O in the oxygen site of quartz (Q) or magnetite (Mt), the whole number exponents are the multiplicities of the oxygen sites in the mineral formulas, and the fractional exponents are the stoichiometric coefficients of the isotope exchange reaction (cf. Thompson, 1969; Kerrick and Darken, 1975). By substituting the atomic fraction of ^{16}O and ^{18}O

$$X_{16,Q} = \left(\frac{^{16}O}{^{18}O + {}^{16}O}\right)_{Q}$$

and similarly for the other $X_{i,j}$ one obtains:

$$K = \alpha = (^{18}O/^{16}O)_{Mt}/(^{18}O/^{16}O)_{Q} \ .$$

Urey (1947) and Bottinga (1969) give derivations of the isotopic equilibrium constant for isotope exchange between molecular substances such as the fluid species CO_2 and H_2O. It is unlikely that α_{A-B} is a function of δ_A or δ_B at natural abundance levels because mutually substituting isotopes are similar in their thermodynamic properties.

Theoretical studies show that the equilibrium constant, $K = \alpha$, for oxygen isotope exchange between pairs of silicate or oxide minerals is a linear function of $1/T^2$, where T is in degrees Kelvin, at crustal temperatures (Bottinga and Javoy, 1977; Javoy, 1977; O'Neil, 1977). Many of the experimental studies of α compiled by Friedman and O'Neil (1977) reflect the predicted $1/T^2$ dependence. Experimental investigation of D/H partitioning between micas and amphiboles gives a $1/T^2$ dependence for α (Suzuoki and Epstein, 1976). Because isotopic substitution has such a small effect on the molar volume of a mineral or fluid species, α is probably independent of P. Experimental studies over a range of 1 to 20 kbar confirm that there is no measurable dependence of α on P (Hoering, 1961; Clayton et al., 1975). In some cases α is a function of the chemical composition of minerals and fluids, such as $^{18}O/^{16}O$ in albite-anorthite solid solutions (O'Neil and Taylor, 1967) and in aqueous salt solutions (Truesdell, 1974),

329

and D/H in phlogopite-annite solid solutions (Suzuoki and Epstein, 1976).

The fractionation of oxygen isotopes is usually chosen for geothermometric studies of metamorphic rocks for the following reasons (for a discussion of carbon isotope geothermometry see Valley and O'Neil, 1981):

(1) Oxygen-bearing minerals are ubiquitous;

(2) It has been recognized that metamorphic minerals may be ranked according to a systematic order of enrichment in ^{18}O, and furthermore, that partitioning between minerals decreases in magnitude more or less regularly with increasing metamorphic grade (Garlick and Epstein, 1967; Taylor and Coleman, 1968; Shieh and Taylor, 1969a,b; Rye et al., 1976).

(3) Experimental studies have shown that α is strongly affected by changing T; the effects of changing chemical composition can be calibrated experimentally (see references in Friedman and O'Neil, 1977).

Recent reviews of oxygen isotope geothermometry have shown that there are problems in the areas of (1) accuracy of experimental calibrations and (2) retrograde isotope exchange between adjacent minerals (Bottinga and Javoy, 1975, 1977; Deines, 1977). The significance of oxygen isotope temperature estimates in a specific case must be evaluated not only by analyzing more than one pair of minerals but also by taking into account recrystallization history and results from cation-mineral geothermometers.

Attainment of isotope equilibrium during metamorphism

Conclusions concerning the nature and extent of stable isotope equilibration during metamorphism are influenced by (1) the criteria used to test for attainment of equilibrium, (2) the specific isotope in question, and (3) the spatial scale over which measurements have been made. These factors are discussed in the framework of two contrasting types of isotope equilibrium. The two concepts are presented below in stark contrast in order to vividly highlight their differences. In nature, however, both types of equilibration probably take place simultaneously, although one or the other may predominate as determined by specific circumstances.

The first concept of isotopic equilibrium considers equilibration of minerals by isotope exchange with an all-pervading, infiltrating fluid of uniform isotopic and chemical composition (Taylor et al., 1963). If the infiltration of such fluid continued in sufficient quantity over a long enough period of time, then each mineral species would become isotopically

homogeneous despite whatever differences in isotopic composition may have
existed prior to metamorphism. Such a process conceivably could lead to
isotope equilibration over distances of hundreds of kilometers during metamorphism (Shieh and Schwarcz, 1974).

The concept of local isotopic equilibration between adjacent mineral
grains stands in contrast to the model which involves pervasive, infiltrating
fluid. Equilibration would be brought about either by isotope exchange between minerals via a small amount of static pore fluid or directly between
minerals in mutual contact. The minerals of a mineralogically and isotopically homogeneous rock layer could achieve exchange equilibrium by this
mechanism but would not necessarily reach equilibrium with the minerals of
intercalated rocks of different mineralogic and isotopic composition. The
process of grain-to-grain exchange reactions could lead to equilibration on
the scale of individual sedimentary beds or on the scale of a single intru-
sive or volcanic unit but probably would not result in isotopic homogenization of different rock types even within a single outcrop (Anderson, 1967).

A criterion for isotopic equilibration with infiltrating fluid (the
first model discussed above) is to test by analysis whether or not a particular mineral is homogeneous throughout an outcrop. It is particularly important in the application of this criterion to have some assurance that the
mineral in question was isotopically heterogeneous prior to metamorphism.
With this criteria, Taylor et al. (1963) showed that quartz from kyanite-
grade mica schist was uniform in composition to within $\pm$ 0.1^{o}/oo over an
area measuring several hundred meters across, despite observed variations
in the chemical and modal composition of the rocks studied. Infiltration
of a fluid of homogeneous isotopic composition is indicated by the results
of Taylor et al. (1963). In a study of biotite-zone marbles, however, it
was found that both δ^{18}O and δ^{13}C values of calcite were uniform for a distance of 30 m along a single bedding unit, but, upon crossing a bedding
plane, the values changed by 0.5 and 0.3^{o}/oo ($\pm$ 0.1^{o}/oo), respectively, over
a distance of 2 cm (Sheppard and Schwarcz, 1970). In a 10-meter outcrop of
granulite facies quartzite, marble, amphibolite, and granitic gneiss, quartz
varied from 10.7^{o}/oo in quartzite to 17.7^{o}/oo in marble (Anderson, 1967).
The data of Sheppard and Schwarcz (1970) and Anderson (1967) show that the
rocks did not equilibrate with a pervasive infiltrating fluid of uniform
isotopic composition. A comparative investigation of two areas, one of
interbedded quartzite and quartz-mica schist and another of intercalated
calc-silicate and mica-schist showed that oxygen isotope homogenization

331

between contiguous rock layers was positively correlated with the amount of devolatilization which occurred during metamorphism. Thorough isotopic homogenization occurred over distances of meters in rocks in which devolatilization reactions released 20%, by weight, of CO_2. In the absence of such devolatilization, homogenization did not occur between adjacent rocks of different mineralogic and isotopic composition (Rumble et al., 1982). Rye et al. (1976) similarly found that ^{18}O depletion of impure marbles was greatest in beds where decarbonation reactions had produced calc-silicate minerals. The results of Rumble et al. (1982) and Rye et al. (1976) show that infiltration is facilitated by a mechanism for enhancing permeability such as devolatilization reactions. The efficacy of the criterion of mineral homogeneity in detecting isotopic homogenization during metamorphism is strongly dependent on the quality of the evidence of premetamorphic heterogeneity. If the mineral under investigation was already isotopically homogeneous prior to metamorphism then its measured postmetamorphic homogeneity has little bearing on the question of isotopic communication during metamorphism. Failure to adequately assure premetamorphic heterogeneity could lead to an overestimation of the size of the volume over which isotopic equilibration took place during metamorphism.

Tests of grain-to-grain isotope exchange equilibrium (the second model discussed above) may be made by analyzing two or more pairs of minerals from the same rock in order to determine whether isotope partitioning gives concordant temperature estimates (Clayton and Epstein, 1961; Deines, 1977). Application of this criterion has shown that many metamorphic rocks attained isotope equilibrium during metamorphism (Hoernes and Friedrichsen, 1978, Fig. 4, p. 308; Hoernes and Hoffer, 1979, Fig. 2, p. 379). The same test reveals, however, a lack of concordancy and, therefore, failure to attain equilibrium fully in rocks subjected to more than one episode of prograde metamorphism or to retrograde metamorphism (Hoernes and Friedrichsen, 1978, Fig. 5, p. 312; Hoernes and Hoffer, 1979, Fig. 3, p. 384, Hoernes and Friedrichsen, 1980, Figs. 9 and 10, p. 30). A disadvantage of the concordant temperature method is that it implicitly assumes experimental calibrations of isotope partitioning are accurate. One could be led to an erroneous conclusion of isotope disequilibrium in a rock when, in fact, an imperfect experimental study was at fault.

In an alternative method of testing for the attainment of grain-to-grain isotopic equilibrium, analyses of pairs of minerals from an outcrop are plotted on an exchange equilibrium diagram (Fig. 1). The use of such diagrams

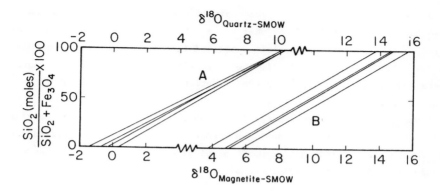

Figure 1. Oxygen isotope exchange diagram for coexisting quartz and magnetite. The group of tie-lines labeled A, from Mt. Finish, New Hampshire, shows poor evidence of equilibration because tie-lines are not mutually parallel. The array of tie-lines labeled B, from Wilmot Mtn., Vermont, give evidence of close approach to isotope exchange equilibrium because tie-lines are mutually parallel.

is common in phase equilibrium studies of ion exchange (Orville, 1963); their use for isotopic exchange is justified because of the close similarity between the thermodynamic treatment of isotope exchange and ion exchange equilibrium (Rumble, 1978). For outcrops where isotope exchange equilibrium has been attained, tie-lines connecting coexisting minerals plot as parallel lines. Failure to achieve equilibrium results in cross-cutting tie-lines (Fig. 1). A disadvantage of the exchange equilibrium diagram is that one could be led to ascribe an array of tie-lines to peak metamorphic conditions when what was really observed was retrograde equilibrium.

If all of the methods of testing for isotope equilibration are imperfect then what is one to do? An answer to the problem is to apply all of the methods to the same suite of samples. In this way the advantages of one method may cancel out the disadvantages of another. It is also strongly recommended that isotopic analyses be evaluated in relation to geologic and recrystallization history. Comparison of data on isotopic equilibration with that on chemical equilibrium can lead to mutually reinforcing conclusions concerning the degree and scale of equilibrium (Rumble, 1978).

STABLE ISOTOPE FRACTIONATION BETWEEN FLUIDS AND MINERAL ASSEMBLAGES

The preceding discussion concerned isotope exchange between single pairs of minerals or fluid species. It is obvious, however, that metamorphic fluids contain several species and that rocks usually consist of more

than one kind of mineral. It is necessary, therefore, to focus attention
on isotope fractionation effects that are a consequence of the multi-mineral,
multi-species nature of metamorphic rocks and fluids. The topics to be discussed in this section include (1) the role of "lever-rule" or mass balance
effects and (2) the control of isotope fractionation by the species composition of fluids.

Lever-rule effects are isotope fractionation effects caused by changing
modal proportions of minerals during a chemical reaction. These effects can
cause changes in the isotopic composition of minerals regardless of changes
in temperature or other factors. Consider, for example, the reaction

$$Fe_2SiO_4 \quad + \quad SiO_2 \quad = \quad 2FeSiO_3$$
$$fayalite \quad + \quad quartz \quad = \quad ferrosilite$$

taking place well within the stability field of ferrosilite in a closed
system so that the bulk $^{16}O-^{18}O$ content of the system cannot change. The
reactants are present in the correct stoichiometric ratio so that both
quartz and fayalite are exhausted and the rock consists entirely of ferrosilite upon completion of reaction. The relative order of enrichment in
^{18}O of these minerals, given in the accompanying bar graph, is quartz (Q,
highest), ferrosilite (Fs, intermediate), and fayalite (Fa, lowest). It is
assumed that isotopic equilibrium is maintained throughout the course of reaction and that the reaction proceeds isothermally so that Δ_{Fa-Q} and Δ_{Fs-Q}
are constant.

```
                        Fs
        o-----------o-----------o
        Fa   (2)         (1)    Q
```

$$\delta^{18}O \text{ increasing} \longrightarrow$$

Suppose that $\delta^{18}O$ of the bulk rock plots at (1), between ferrosilite and
quartz. As reaction proceeds and the amount of ferrosilite present increases while the amounts of quartz and fayalite decrease, the $\delta^{18}O$ values
of all three minerals will increase. If the reaction proceeds to completion,
the $\delta^{18}O$ value of ferrosilite will attain the bulk rock value at (1). As an
alternate example consider the consequences if the bulk rock $\delta^{18}O$ value were
at (2), between fayalite and ferrosilite. In this case, the $\delta^{18}O$ values of
all three minerals would decrease until reactants were exhausted. The $\delta^{18}O$
of ferrosilite would grow closer and closer to that of the bulk rock at (2).

Under equilibrium conditions, isotopes are partitioned among various
fluid species. The proportions of species change according to the identity

and composition of the minerals with which fluids are in equilibrium and as
a function of the P,T conditions to which rocks and fluids are subjected.
These two facts have important consequences for the fractionation of isotopes
that accompanies devolatilization reactions. The dehydration of muscovite
gives an example of how the fractionation of deuterium and hydrogen could be
controlled by the species composition of fluids. The relative partitioning
of D-H at isotopic equilibrium between H_2O (most enriched in D), muscovite
(M, intermediate), and CH_4 (least enriched) is illustrated in the accompanying bar graph.

$$CH_4 o-----------o-----------oH_2O$$
$$M$$

$$\delta D \text{ increasing} \longrightarrow$$

Suppose that muscovite breaks down by the reaction

$$KAl_3Si_3O_{10}(O(H,D))_2 + SiO_2 = KAlSi_3O_8 + Al_2SiO_5 + (H,D)_2O$$
$$\text{muscovite} \quad + \text{quartz} = \text{K-feldspar} + \text{sillimanite} + \text{fluid}$$

If product $(H,D)_2O$ escapes from the reaction site without re-equilibrating
with reactant or product minerals, the remaining muscovite will be progressively depleted in D. This process corresponds to Rayleigh distillation.
In contrast, consider the combined redox-dehydration reaction

$$2KAl_3Si_3O_{10}(O(H,D))_2 + 2SiO_2 + 2C = 2KAlSi_3O_8 + 2Al_2SiO_5 + C(H,D)_4 + CO_2$$
$$\text{muscovite} + \text{quartz} + \text{graphite} = \text{K-feldspar} + \text{sillimanite} + \text{fluid}$$

If $C(H,D)_4$ escapes from the rock without exchanging isotopes with the
minerals, the remaining muscovite will be *enriched* in D.

The isotope fractionation caused by mass balance effects can be expressed mathematically for the purpose of computation. The bulk or mean
isotopic composition of an assemblage of minerals is given, to a close approximation, by the equation (Craig, 1953, footnote, p. 87).

$$\delta^{18}O_{WR} = \sum_i X_i \, \delta^{18}O_i \tag{1}$$

where $\delta^{18}O_{WR}$ is the whole rock $\delta^{18}O$ value, X_i is the atomic fraction of
oxygen in mineral i relative to the total amount of oxygen is all the minerals of the assemblage, and $\delta^{18}O_i$ is the $\delta^{18}O$ value of mineral i. For
convenience it will be assumed that the mineral assemblages of interest are
quartz-bearing and are in isotope exchange equilibrium. The expression

$$\delta^{18}O_i = \delta^{18}O_Q - \Delta_{Q-i}$$

can be substituted in equation (1).

$$\delta^{18}O_{WR} = \delta^{18}O_Q - \sum_{i \neq Q} X_i \Delta_{Q-i} \qquad (2)$$

where $\delta^{18}O_Q$ is the $\delta^{18}O$ value of quartz, X_i is defined as for equation (1), and $\Delta_{Q-i} = \delta^{18}O_Q - \delta^{18}O_i$ ($\simeq 1000 ln\ \alpha_{Q-i}$). Note that the summation does *not* include quartz. By a similar procedure, the equation for bulk fluid composition is

$$\delta^{18}O_{FL} = \delta^{18}O_Q - \sum_f X_f \Delta_{Q-F} \qquad (3)$$

where $\delta^{18}O_{FL}$ is the $\delta^{18}O$ value of bulk fluid, X_f is the atomic fraction of oxygen in species f relative to the total amount of oxygen contained in all the species of the fluid, and Δ_{Q-f} is defined as for equation (2). If equation (3) is subtracted from equation (2), an expression for isotope fractionation between bulk fluid and whole rock may be obtained

$$\Delta_{WR-FL} = \sum_f X_f \Delta_{Q-f} - \sum_{i \neq Q} X_i \Delta_{Q-i}\ , \qquad (4)$$

where

$$\Delta_{WR-FL} = \delta^{18}O_{WR} - \delta^{18}O_{FL}\ .$$

In the application of equation (4) to rocks, the terms X_i are computed from modal and chemical analyses of minerals. The quantities X_f are determined by a two-step procedure: (1) the species composition of metamorphic fluid is calculated thermodynamically from mineral equilibria (Ferry & Burt Ch. 6, this volume), or fluid composition is determined by measurements of fluid inclusions; then (2) values of X_f are computed. The Δ_{Q-f}, and Δ_{Q-i} terms are functions of temperature and are independent of pressure, fluid species composition, and proportions of minerals. The values of isotope partitioning between many pairs of minerals and fluid species over a wide range of T have been compiled by Friedman and O'Neil (1977).

An example of the dependence of isotope fractionation on fluid composition is shown in Figure 2. It may be seen that Δ_{Q-FL} for $^{18}O-^{16}O$ fractionation varies from +1.8 for pure H_2O, to −0.5 at $X_{CO_2} = 0.18$, to −5.6 for pure CO_2 (off enlarged diagram). Important studies of the influence of fluid composition on fractionation of C and S isotopes have been presented by Ohmoto (1972), Rye and Ohmoto (1974), Ohmoto and Rye (1979), and Deines (1980). The most important conclusion to be drawn from the preceding discussion is that unique interpretations of isotopic analyses in terms of

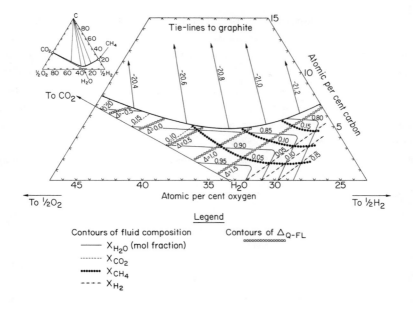

Figure 2. Phase equilibrium diagram of the system C-O-H at 600°C and 3.5 kbar. Contours of Δ_{Q-FL} show fractionation of $^{18}O-^{16}O$ between quartz and fluid. Tie-lines from graphite to fluid are labeled with appropriate values of $\log_{10} f_{O_2}$.

competing hypotheses of isotope fractionation cannot be made without data on fluid composition.

MODELS OF FLUID-ROCK INTERACTION AND THEIR ISOTOPIC EFFECTS

The purpose of this final section is to synthesize concepts previously discussed and to demonstrate how the combined methods of petrology and stable isotope geochemistry may be used to characterize fluid-rock interactions. Mathematical models of C-H-O isotope fractionation during metamorphic reactions and during infiltration of fluids through rocks have been developed by the writer. A unique feature of the models is that different hypotheses of reaction mechanism such as that of equilibrium fluid composition (Ferry, 1980) or constant porosity have been incorporated. With hypotheses of reaction mechanism it is possible to compute the isotopic evolution of model fluid-rock systems as a function of reaction progress. The model calculations may be compared to isotopic and petrologic data from rocks to evaluate which of several competing hypotheses of fluid-rock interaction is most nearly correct.

337

The computation scheme uses the principle of conservation of mass to calculate isotope effects for small increments of reaction progress. Temperature, pressure, and porosity may be varied in any specified way throughout the course of reaction. It is assumed that during each increment of reaction, chemical and isotopic equilibrium is maintained between minerals and fluid. Equilibrium does not exist, however, between minerals participating in successive increments of reaction. By this means, irreversible chemical and isotopic reactions may be approximated by a series of "infinitesimal" increments of reversible reactions, a concept pioneered in geochemistry by Helgeson (1968). The phrases "reaction progress," "extent of reaction," and "amount of reaction" are used interchangeably in this paper. "Extent of reaction" is defined, following Prigogine and Defay (1954, p. 10), as equal to the increase in gram formula weight units of a phase produced during a given reaction divided by its stoichiometric coefficient in the reaction. Note that in Figures 3 and 4 values of the abscissa, GFW garnet (gram formula weight garnet), should be divided by 3.0, the stoichiometric coefficient of garnet in the reaction, to obtain the normalized extent of reaction.

Three models of fluid-rock interactions have been investigated: (1) Rayleigh distillation with chemical reaction between minerals; (2) infiltration alone, without chemical reaction between minerals; and (3) combined infiltration and Rayleigh distillation with chemical reaction between minerals. All three types of fluid-rock interaction are likely to occur to some extent during metamorphism. The results of the computations described below show that it may be possible to determine which process was dominant in the evolution of individual rock samples by measuring petrologic and isotopic properties of the rocks.

Methods of computation

Rayleigh distillation. During Rayleigh distillation rocks interact only with fluids generated internally by devolatilization reactions between the rocks' minerals. The conditions of Rayleigh distillation require that once fluid is generated it is expelled immediately from the rock and not allowed to interact with the rock again. These conditions are likely to be met under some metamorphic conditions for two reasons: (1) The low porosity of metamorphic rocks, as observed at the surface, suggests that volatile products are expelled as fast as they are formed; and (2) once expelled, buoyancy forces would drive fluids through fissures towards the Earth's surface, preventing reequilibration. Rayleigh distillation for isotopic fractionation in a two-phase system is described by the following equation (Broecker and Oversby, 1971, p. 166):

$$\delta_{initial} - \delta_{final} = 1000(F^{\alpha-1} - 1) \tag{5}$$

where $(\delta_{initial} - \delta_{final})$ gives the change in reactant composition, F is the fraction of unconverted reactant, and α refers to the partitioning of isotopes between product and reactant. Equation (5) is inadequate to deal with devolatilization reactions, however, because it is valid only for constant α. The value of α changes continuously throughout the course of an isothermal, multi-phase, multi-component devolatilization reaction in response to changing proportions of minerals as may be seen by inspecting equation (4), above. Fractionation, α, would also change during a polythermal, polybaric reaction because of changing proportions of both fluid and mineral species. Accordingly, a numerical scheme was devised to calculate the effects of Rayleigh distillation taking into account all of the factors influencing α.

Computation of Rayleigh distillation begins with a rock consisting of specified amounts of reactant minerals and a specified whole rock isotopic composition. Temperature, pressure, and porosity are also given and may be constant or change in a predetermined manner throughout the course of reaction. A small increment of a given devolatilization reaction takes place with mineral and fluid species at chemical and isotopic equilibrium. The magnitude of the reaction increment is controlled by the constraint that the volume of volatile products does not exceed the specified porosity. The rock now consists of product and reactant minerals together with a small amount of fluid phase with which it is in equilibrium. Assuming the reaction involves quartz, the value of $\delta^{18}O$ (quartz) is calculated with the equation:

$$\delta^{18}O_Q = \delta^{18}O_{WR} + \sum_i \Delta_{Q-i} X_i' + \sum_f \Delta_{Q-f} X_f' \tag{6}$$

[for derivation, see equations (2) and (3), above] where the subscripts Q and WR refer to quartz and whole rock, respectively; X_i' is the atomic fraction of oxygen in mineral "i" relative to the total oxygen in minerals *plus* fluid; X_f' is the atomic fraction of oxygen in fluid species "f" relative to the total oxygen in fluid *plus* minerals; and Δ_{Q-i} and Δ_{Q-f} are defined above. The next step is to remove the fluid products of reaction and compute a new value of $\delta^{18}O_{WR}$ with the equation:

$$\delta^{18}O_{WR} = \sum_i X_i \, \delta^{18}O_i$$

where X_i is the atomic fraction of oxygen in mineral "i" relative to the total oxygen in minerals *alone* and $\delta^{18}O_i = \Delta_{i-Q} + \delta^{18}O_Q$. At this point, another increment of devolatilization takes place and computation returns to equation (6). Reiteration continues until the desired extent of reaction progress is reached. Test calculations of isothermal Rayleigh distillation in an isothermal two-phase system (water-steam) in which α is constant show agreement of $\pm 0.1^o/oo$ for up to 70% completed reaction and $\pm 0.5^o/oo$ for 90% completed reaction between calculations performed using methods described in this study and the analytic form of Rayleigh's equation with constant α (see equation 5).

Infiltration. Isotopic infiltration is a process whereby the minerals in rock exchange isotopes with fluid flowing through the rock. By this means a rock may come to be in isotopic equilibrium with an external reservoir of fluid. Such a process is likely to occur to at least some extent during metamorphism because the volatile species released by metamorphic reactions must pass through other rocks in order to reach the Earth's surface. The effects of infiltration on the isotopic composition of rocks which do not experience devolatilization reactions is computed as follows: Consider a rock with specified modal mineralogical composition, a given whole rock isotopic composition and empty pores. Pressure, temperature, and porosity are specified and may be constant or change in a predetermined way during infiltration. To these pores add a small, specified amount of H_2O and CO_2 with specified isotopic composition not initially in equilibrium with the rock. The fluid is allowed to come to isotopic equilibrium with the minerals of the rock. In the case in which infiltration occurs without devolatilization reactions, the H_2O and CO_2 are added in the correct ratio to be in equilibrium with the mineral assemblage in the rock. Now compute the mean isotopic composition of the system consisting of whole rock plus fluid using the equation:

$$\delta^{18}O_{(system)} = \sum_f X'_f \, \delta^{18}O_f + X'_{WR} \, \delta^{18}O_{WR} \tag{7}$$

where X'_f is the atomic fraction of oxygen in fluid species "f" relative to the total oxygen in fluid *plus* minerals, and X'_{WR} is the atomic fraction of oxygen in all the minerals in the rock relative to total oxygen in rock *plus* fluid. Then $\delta^{18}O_Q$ is calculated with the equation:

$$\delta^{18}O_Q = \delta^{18}O_{(system)} + \sum_f \Delta_{Q-f} X'_f + \sum_i \Delta_{Q-i} X'_i \qquad (8)$$

where X'_f is defined as in equation (7) and X'_i is the atomic fraction of oxygen in mineral "i" relative to the total oxygen in minerals *plus* fluid. At this point, the previously infiltrated, small increment of fluid is removed from the rock. A new value of $\delta^{18}O_{WR}$ is computed from the equation:

$$\delta^{18}O_{WR} = \sum_i X_i \Delta^{18}O_i \qquad (9)$$

where X_i is the atomic fraction of oxygen in mineral "i" relative to the total amount of oxygen in minerals, alone. A new increment of fluid is added to the rock, computation returns to equation (7), and reiteration continues until the desired degree of infiltration and consequent change in $\delta^{18}O_{WR}$ has been achieved. Test calculations of infiltration into a rock *not* undergoing devolatilization using the computer model of this study give results in agreement with fluid-rock ratios computed using Taylor's (1977, p. 524) "open system" or "single-pass" model.

Combined Rayleigh distillation and infiltration. A rock undergoing a devolatilization reaction may also often experience infiltration because such reactions should enhance permeability (Fyfe, Price, and Thompson, 1978, p. 309). Experiments have shown that permeability is increased by increasing pore pressure and increasing porosity (Brace, 1972). Devolatilization reactions would enhance permeability in two ways: (1) Such reactions would maintain or increase fluid pressure to values equal to lithostatic pressure; and (2) reactions would tend to increase porosity because the volume occupied by solid mineral products is usually less than that of reactants.

Computation of combined distillation and infiltration is made with equations (7)-(9) and an additional equation derived from Ferry's (1980) equation relating reaction progress and water-rock ratio. The calculations consider the specific situation in which rock is infiltrated by pure H_2O fluid while minerals undergo dehydration-decarbonation reactions. Equilibrium fluid composition is related to the number of moles of H_2O introduced into the rock by infiltration and the number of moles H_2O and CO_2 produced by the reaction according to:

$$n^{rxn}_{CO_2} = \frac{X^{eq}_{CO_2}(n^{rxn}_{H_2O} + n^{ext}_{H_2O})}{1 - X^{eq}_{CO_2}} \qquad (10)$$

where $n^{rxn}_{CO_2}$ and $n^{rxn}_{H_2O}$ are the number of moles of CO_2 and H_2O,

341

respectively, released by devolatilization reaction, $n^{ext}_{H_2O}$ is the number of moles of H_2O added from an external reservoir, and $X^{eq}_{CO_2}$ is the mole fraction of CO_2 in fluid in equilibrium with the mineral assemblage. Equation (10) serves the purpose of relating the amount of H_2O introduced by infiltration ($n^{ext}_{H_2O}$) to the amount of reaction progress (as measured by $n^{rxn}_{CO_2}$). The changes in amounts of reactant and product minerals are calculated from $n^{rxn}_{CO_2}$ with the stoichiometric coefficients of the reaction. The amount of infiltrated H_2O, $n^{ext}_{H_2O}$, is introduced in sufficiently small amounts that the volume of H_2O added to the rock does not exceed its specified porosity. Pressure and temperature can be changed in any specified manner during reaction.

Computation begins with a rock of given modal mineralogical composition and given isotopic composition. A small amount of pure H_2O with specified $\delta^{18}O$ is infiltrated into the rock from an external reservoir of fluid. This fresh fluid sweeps away any old, volatile reaction products. The new fluid makes it possible for reaction to proceed because the H_2O is initially not in either chemical or isotopic equilibrium with the mineral assemblage. Decarbonation reaction begins and continues until an equilibrium fluid composition has been attained (equation 10). Isotopic equilibration takes place at the same time. The pore fluid, which now consists of molecules introduced from an external reservoir and molecules generated internally by a decarbonation reaction, is flushed out of the rock by the next increment of infiltrating fluid. The computation is repeated until the desired extent of reaction is achieved.

Results of computations

Model calculations have been carried out for (1) Rayleigh distillation, (2) infiltration, and (3) combined infiltration and Rayleigh distillation. The dehydration-decarbonation reaction

$$2Ca_2Al_3Si_3O_{12}(OH) + 5CaCO_3 + 3SiO_2 = 3Ca_3Al_2Si_3O_{12} + 5CO_2 + H_2O \qquad (R1)$$
$$\text{clinozoisite} \quad + \text{calcite} + \text{quartz} = \quad \text{grossular} \quad + \text{fluid}$$

has been chosen for illustrative purposes in all ensuing calculations. The reaction has a steep positive slope where it is stable on the H_2O-rich side of a T-X_{CO_2} diagram. The position of the equilibrium curve shifts to CO_2-richer compositions with increasing temperature and shifts to H_2O-richer compositions with increasing pressure. At higher temperature reaction (R1) becomes metastable with respect to the reaction clinozoisite + CO_2 =

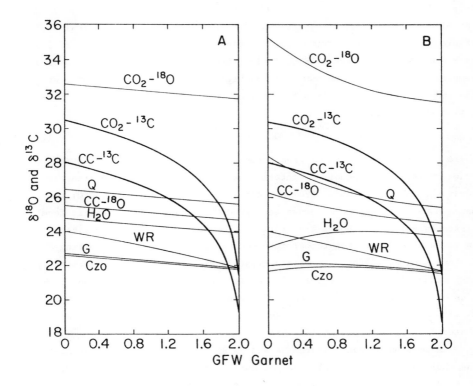

Figure 3. Graph of $\delta^{18}O$ and $\delta^{13}C$ versus reaction progress (gram formula weight units of garnet) for the reaction clinozoisite (Czo) + calcite (Cc) + quartz (Q) = garnet (G) + H_2O + CO_2. Heavy lines show $\delta^{13}C$ values. Both 3A and 3B show isotope fractionation caused by Rayleigh distillation of H_2O and CO_2. Initial values of $\delta^{18}O$ and $\delta^{13}O$ for whole rock (WR) were 24.0 (SMOW) and 28.0°/oo (arbitrary value), respectively. Initial amounts of quartz, calcite, and clinozoisite were 2.1, 3.4, and 1.4, respectively, gram formula weight units. Reactants are nearly exhausted and reaction completed at GFW garnet = 2.0. Figure 3A is for isothermal, isobaric conditions, 600°C and 3.5 kbar. Figure 3B is for isobaric, polythermal conditions, 3.5 kbar and 400°-600°C such that T(°C) = 400 + 198.97 G - 16.98 G^2 - 32.49 G^3 + 8.12 G^4 where G is gram formula weight units of garnet (equation 11).

anorthite + calcite + H_2O (Kerrick and Ghent, 1979). In all the calculations illustrated in Figures 3-5, the initial amounts of clinozoisite, calcite, and quartz were 1.4, 3.4, and 2.1 gram formula weight units, respectively. These amounts were chosen so that reactants would be exhausted and the rock would consist of 2.0 gram formula weight units of garnet upon completion of reaction.

Rayleigh distillation. The effects of Rayleigh distillation on ^{18}O and ^{13}C values of minerals and whole rock participating in reaction (R1) are shown in Figure 3. The horizontal axis gives the number of gram formula weight units of grossular formed during reaction. Values of $\delta^{18}O$ and $\delta^{13}C$

for minerals, fluid species, and whole rock are shown on the vertical scale. Figure 3A presents an isotope fractionation path for the conditions of 3.5 kbar and 600°C. In Figure 3B, reaction takes place at 3.5 kbar and over a temperature interval from 400°C at the beginning of reaction to 600°C at the end of reaction. The equation giving temperature as a function of reaction progress is

$$T(°C) = 400 + 198.97G - 16.98G^2 - 32.49G^3 + 8.12G^4 \qquad (11)$$

where G equals the number of gram formula weight units of garnet formed by reaction (R1). The equation dictates that the rate of reaction with temperature, dG/dT, increases from a minimum at low temperature (400°C) to a maximum value at the peak temperature (600°C); however, the specific form of the equation was chosen arbitrarily. The rate of reaction is likely to increase with increasing temperature because potentially rate-limiting factors such as diffusivity within crystals are known to be accelerated by temperature increases. The initial whole rock $\delta^{18}O$ value of 24°/oo (SMOW) lies within the range of ancient calcareous rocks. The initial value of δ^{13}(28°/oo), however, was chosen so that $\delta^{18}O$ and $\delta^{13}C$ values could be presented conveniently on the same graph.

Inspection of both Figures 3A and 3B shows that $\delta^{18}O$ values of minerals and whole rock decrease only 1-2°/oo during reaction. Values of $\delta^{13}C$, however, decrease by a large amount. Note that in Figure 3B the vertical spacing between the $\delta^{18}O$ curves of minerals or fluid species changes (H_2O versus calcite is a clear example), whereas in Figure 3A the curves are mutually parallel. The changes in spacing reflect changing partitioning of ^{18}O and ^{16}O between minerals and fluid species, a consequence of the polythermal conditions of Figure 3B.

The disparity in behavior between $\delta^{18}O$ and $\delta^{13}C$ during Rayleigh distillation may be explained by considering how many oxygen atoms versus how many carbon atoms are converted by reaction from constituents of a mineral to fluid species. For every 11 atoms of oxygen lost from the rock as H_2O and CO_2, 36 atoms remain behind in garnet (for three formula units of garnet, see reaction R1). In contrast, all of the carbon atoms in the rock are converted from mineral constituents to fluid species upon completion of reaction. At the portion of the plotted curve in Figure 3 where $\delta^{13}C$ decreases most rapidly, the amount of residual calcite is very small and is rapidly disappearing. A similar disparity between the behavior of $\delta^{18}O$ and δD is found in calculations of dehydration reactions.

Much larger depletions in ^{18}O than those shown in Figure 3 are predicted with an alternative model of Rayleigh distillation presented by Lattanzi et al. (1980). In their model no oxygen isotope exchange occurs between calcite and other minerals in the rock during devolatilization (ibid., Fig. 3 caption, p. 896). Under these conditions, for every two atoms of oxygen expelled from the rock as CO_2 only one atom remains behind (for one formula unit of calcite). This model can be tested by analyzing calcite as well as all other minerals present in order to determine whether or not calcite is in oxygen isotope exchange equilibrium with the other minerals.

Infiltration. Results of calculations show that there are no petrographic or isotopic criteria for easily distinguishing cases of either infiltration, alone, or combined infiltration and distillation in natural samples. It has been found that infiltration of a rock which has already experienced a devolatilization reaction during an earlier metamorphic event produces a rock whose petrographic and isotopic properties are similar to those of a rock that has undergone simultaneous infiltration and devolatilization reaction, provided that all other conditions are the same. For this reason, the discussion of the computed effects of infiltration will be presented in the next section on combined infiltration and distillation.

Combined infiltration and Rayleigh distillation. Figures 4A and 4B show the results of computing combined infiltration by pure H_2O and distillation for reaction (R1). The conditions for Figure 4A are 3.5 kbar and 600°C; for Figure 4B they are 3.5 kbar and 400°-600°C. The equation relating temperature to reaction progress (equation 11) is the same for Figure 4B as it is for Figure 3B. A value of $12.5^{o}/oo$ $\delta^{18}O$ (SMOW) was chosen for infiltrating H_2O because it is sufficiently out of oxygen isotopic equilibrium with the rock that effects of infiltration may be seen clearly. This value lies within the field of metamorphic waters over the temperature range of interest (Taylor, 1979). The volume of infiltrating water added at each increment of reaction was specified as 1% of the volume of solid minerals present during the increment, consistent with the low porosity of metamorphic rocks. From a comparison of Figures 3 and 4 it may be seen that infiltration can have a far greater effect on $\delta^{18}O$ values than Rayleigh distillation, alone. The capacity of infiltration to change a rock's $\delta^{18}O$ is limited only by (1) the amount by which infiltrating fluid is *not* in isotopic equilibrium with the rock being infiltrated, and (2) the

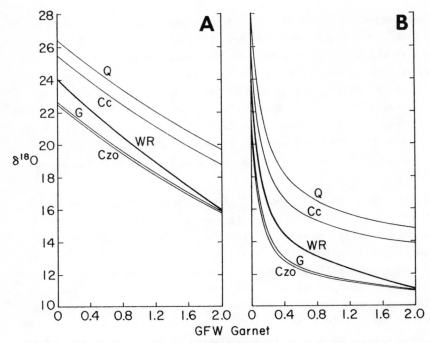

Figure 4. Plots of isotope fractionation caused by combined Rayleigh distillation and infiltration by pure H_2O for the reaction clinozoisite (Czo) + quartz (Q) + calcite (Cc) = garnet (G) + H_2O + CO_2. Heavy lines show whole rock values. Infiltrating H_2O equals 12.5°/oo, $\delta^{18}O$ SMOW. Amount of water added at each increment of reaction determined by (1) equilibrium X_{CO_2} and (2) constant porosity of 1%. Initial $\delta^{18}O_{WR}$ and $\delta^{13}C_{WR}$ and initial amounts of Cc, Q, and Czo as in Figure 3. Figure 4A shows isothermal, isobaric conditions of 600°C and 3.5 kbar. Figure 4B shows isobaric, polythermal conditions of 3.5 kbar and 400°-600°C as in Figure 3B (equation 11). Mineral thermodynamic data from Ferry (1976); fugacity coefficients of fluid species from Kerrick and Jacobs (1981).

amount of infiltrating fluid. The $\delta^{13}C$ versus reaction progress curves, not shown in Figures 4A and 4B, are identical, respectively, to those in Figures 3A and 3B because the infiltrating fluid was assumed to be pure H_2O.

A comparison of Figures 4A and 4B reveals that the $\delta^{18}O$ versus extent of reaction curves are affected greatly by thermal history. Figure 4A, depicting isothermal, isobaric infiltration and reaction, shows that the mineral assemblage did not attain oxygen isotopic equilibrium with infiltrating H_2O even upon completion of reaction. In Figure 4B, however, it may be seen that equilibrium with infiltrating fluid was approached more closely and earlier during the course of reaction. The attainment of isotopic equilibrium in Figure 4B may be seen where the mineral curves begin to flatten out. Complete equilibration with infiltrating fluid would be shown by mineral curves becoming horizontal so that values of $\delta^{18}O_i$ were constant with increasing reaction progress.

The controlling influence of thermal history on isotope fractionation
paths has three causes: (1) the temperature dependence of isotope partitioning, α; (2) the temperature dependence of isotope fractionation between
bulk fluid and whole rock, Δ_{WR-FL}, due to changing proportions of fluid
species (equation 4); and (3) the temperature dependence of water-rock
ratios computed from equation (10). The third process listed is dominant
under the conditions of Figure 4. Inspection of equation (10) shows that
for small values of $X^{eq}_{CO_2}$, larger values of $n^{ext}_{H_2O}$ are required in order
to achieve a given amount of reaction progress (i.e., to achieve a given
amount of $n^{rxn}_{CO_2}$). Under the isothermal, isobaric conditions of Figure 4A,
$X^{eq}_{CO_2}$ has a constant value of 0.14. In Figure 4B, however, $X^{eq}_{CO_2}$ changes
from 0.02 at 400°C to 0.14 at 600°C. At the smaller values of $X^{eq}_{CO_2}$, large
amounts of external H_2O result in a closer approach to oxygen isotopic
equilibrium. For larger values of $X^{eq}_{CO_2}$, smaller quantities of externally
derived H_2O are required for a given amount of reaction progress. Smaller
volumes of external H_2O lead to a lesser degree of isotopic equilibration
between infiltrating H_2O and mineral assemblage. The overall water-rock
ratio (by volume) for Figure 4A is 1.9; for Figure 4B it is 7.7. Water-rock
ratios were computed by summing the values of $n^{ext}_{H_2O}$ in equation (10).

Isotope fractionation paths are shown for two geothermal gradients in
Figures 5A and 5B. The geothermal gradient for Figure 5A is 100°C/km, a
value appropriate to near-surface contact metamorphism. A gradient of
25°/km, representative of deep-seated Barrovian regional metamorphism, was
used to compute Figure 5B. Temperature variation with extent of reaction was
given by equation (11) for both Figures 5A and 5B. Pressure was specified as
a linear function of temperature and ranged from 1.1 to 1.7 kbar in Figure 5A
and from 4.5 to 6.8 kbar in Figure 5B. The equilibrium value of X_{CO_2} follows
a curving path for the reaction (R1) through $P-T-X_{CO_2}$ space as determined by
the intersection of the chosen geothermal gradient and the divariant equilibrium surface. The other conditions of reaction were the same as in previous
models of combined infiltration and distillation.

It may be seen that selecting different geothermal gradients has a
noticeable effect on computed oxygen isotope fractionation paths. In Figure
5A, the rock did not attain equilibrium with infiltrating H_2O during reaction. The rock shown in Figure 5B, however, approached isotopic equilibrium
with infiltrating H_2O much more closely for a given extent of reaction.
Note that in both Figures 5A and 5B, the temperature interval over which
reaction occurred is the same but the pressure range is different. One is
led to deduce that pressure has a controlling effect on the path of stable

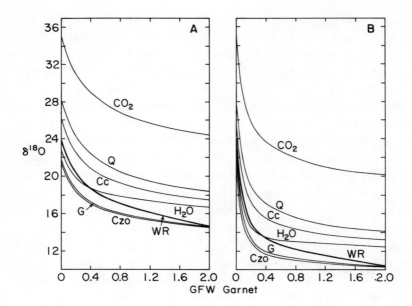

Figure 5. Plots of $\delta^{18}O$ versus gram formula weight (GFW) units of garnet formed by the reaction clinozoisite (Czo) + quartz (Q) + calcite (Cc) = garnet (G) + CO_2 + H_2O. Heavy curves show whole rock (WR) $\delta^{18}O$ values. In both figures T varied from 400°C at the start of reaction to 600°C at the end of reaction (equation 11). In Figure 5A, P varied from 1.1 to 1.7 kbar along a geothermal gradient of 100°/km (contact metamorphism). In Figure 5B, P varied from 4.5 to 6.8 kbar along a geothermal gradient of 25°C/km (Barrovian regional metamorphism). Initial value of $\delta^{18}O$ for whole rock was 24.0°/oo. Infiltrating H_2O equals 12.5°/oo $\delta^{18}O$ SMOW. Initial amounts of quartz, calcite, and clinozoisite were 2.1, 3.4, and 1.4 gram formula weight units, respectively. Reactants are nearly exhausted and reaction completed at GFW garnet = 2.0. Thermodynamic data on minerals from Ferry (1976); P-V-T properties of fluids from Kerrick and Jacobs (1981); and isotope partitioning from Friedman and O'Neil (1977).

isotope fractionation. Experimental studies have shown, however, that isotope partitioning between pairs of minerals is pressure independent (Hoering, 1961; Clayton et al., 1975). This apparent discrepancy requires explanation.

There are two related processes that give rise to the pressure dependence of isotope fractionation paths:

(1) One of these controlling factors is isotope fractionation between a multi-species fluid and a multi-mineral rock. During the computation of Figure 5, isotope partitioning varied with temperature but was not a function of pressure. The isotope fractionation between bulk fluid and whole rock, however, changed throughout the course of reaction (equation 4) as a function of pressure, temperature, and extent of reaction.

(2) The other controlling factor is the pressure dependence of water-rock ratios computed with equation (10). The dependence of the quantity of

externally derived H_2O, $n^{ext}_{H_2O}$, on $X^{eq}_{CO_2}$ has been discussed above. Equilibrium X_{CO_2} varies from 0.02 to 0.33 in Figure 5A and from 0.004 to 0.06 in Figure 5B. In order to achieve a given extent of reaction where $X^{eq}_{CO_2}$ is small, a large amount of external H_2O is required to dilute the CO_2 produced by mineral reaction. For larger values of $X^{eq}_{CO_2}$, smaller amounts of H_2O must be infiltrated. The rock subjected to a Barrovian gradient has approached isotopic equilibrium with infiltrating fluid more closely because it has experienced larger volumes of infiltrating H_2O (11.9 volumetric water-rock ratio at the end of reaction in Fig. 5B). For a contact metamorphic gradient, however, equilibrium was not achieved with infiltrating H_2O because the water-rock ratio at the end of reaction was much less, 3.3 by volume (Fig. 5A).

Considering the combined distillation-infiltration models (Figs. 4-5) more generally, it becomes apparent that there is a way of verifying the assumptions regarding initial conditions and reaction mechanisms. Given a rock for which both premetamorphic, mineralogical, chemical, and isotopic compositions and post-metamorphic compositions are known, it is possible to make two independent estimates of fluid-rock ratio during metamorphism. Ferry (1980) has shown that fluid-rock ratios can be estimated from petrographic modal analysis (see equation 10) of devolatilization reactions. Fluid-rock ratios can also be computed from changes in isotopic composition that have occurred as a consequence of metamorphism (Taylor, 1977). The models of combined infiltration and distillation presented above bring these two methods together because the models, in effect, give isotopic fluid-rock ratio (vertical axis, Figs. 4 and 5) as a function of petrographic fluid-rock ratio (horizontal axis). Taking the values of premetamorphic and post-metamorphic mineralogical and isotopic composition as fixed, it is possible to change the relationship between the two fluid-rock ratios by changing the equations relating reaction progress, temperature, and pressure (equation 11). Thus, it is possible to find a specific model of a specific rock for which the petrographic fluid-rock ratio equals the isotopic fluid-rock ratio. If no such agreement can be found no matter how many adjustments are made to the model, then one would be led to suspect that the rock did not experience combined distillation and infiltration during metamorphism.

CONCLUSIONS

The most important general conclusion to be drawn in this chapter concerns the *inter*dependence of petrology and stable isotope geochemistry in the study of metamorphic rocks. It is impossible to gain a unique interpretation of fluid-rock interaction from stable isotope analyses alone without supporting data on the species composition of metamorphic fluids. But this is a true interdependence, not a one-way dependence, for stable isotope analyses are required to identify the origin of externally derived fluids.

By way of recapitulation, it is informative to make a practical list of the data needed to understand fluid-rock interactions during metamorphism:

(1) Chemical compositions of coexisting minerals combined with mineralogical thermodynamic data and P-V-T properties of fluids are needed for geothermometry, geobarometry, and calculation of metamorphic fluid compositions.

(2) The chemical data as well as petrographic modal analyses and systematic study of textures and fabrics must be used to identify specific metamorphic reactions and to measure their extent.

(3) Stable isotopic analyses of whole rocks and coexisting minerals are required for geothermometry and to estimate the isotopic composition of fluid species.

(4) Data on the *pre*-metamorphic mineralogical, chemical, and isotopic composition of rocks must be deduced from sedimentological, stratigraphic, or other information on geologic history. Pre-metamorphic properties of metasedimentary rocks may be obtained by studying their unmetamorphosed stratigraphic and sedimentological equivalents. Pre-metamorphic properties of meta-igneous rocks may be determined by investigating unmetamorphosed igneous equivalents. The initial conditions of a model calculation are fully as important as the final conditions. In many cases, the application of model calculations may be strictly limited by the difficulty of making accurate estimates of pre-metamorphic properties.

Collection of such detailed information on metamorphic rocks requires a great deal of effort. The question arises as to whether or not the potential benefits to be realized are commensurate with the effort expended. This question can be answered by listing potential benefits (*cf.* Rice and Ferry, this volume):

(1)   A distinction can be made between rocks that have interacted predominantly with fluids generated internally by devolatilization reactions and rocks that have interacted predominantly with fluids derived from an external reservoir.

(2)   Quantitative estimates of fluid-rock ratios can be made from isotopic analyses and from petrographic modal analysis. The capability of making two independent estimates is quite important because it provides a means of verifying fluid-rock ratios.

(3)   Rocks can be ranked according to their relative permeabilities (Rumble et al., 1982).

(4)   The most exciting potential benefit of these methods is the mapping of fluid-rock ratios in relation to geologic structures such as bedding, unconformities, faults, pluton-wall rock contacts, isograds, and the like. Such maps of the fluid circulation systems of metamorphic belts will probably lead to a new, higher level of understanding of the dynamic process of metamorphism.

ACKNOWLEDGMENTS

M.D. Barton, M.T. Einaudi, J.M. Ferry, R.T. Gretogy, T.C. Hoering, and H.S. Yoder, Jr. contributed to this study through discussion and criticism. L.W. Finger provided invaluable advice on computing methods. Preparation of this report was supported in part by National Science Foundation, grant EAR 7919769.

NOTE ADDED IN PROOF

An excellent recent review of isotopic geothermometry is "Isotopic Thermometry," by R.N. Clayton (1981) *In* Newton, R.C., A. Navrotsky, and B.J. Wood, Eds., *Thermodynamics of Minerals and Melts*, Springer-Verlag, New York, pp. 85-109.

CHAPTER 8 REFERENCES

Anderson, A.T. (1967) The dimensions of oxygen isotopic equilibrium attainment during prograde metamorphism. J. Geol. 75, 323-332.

Bottinga, Y. (1969) Calculated fractionation factors for carbon and hydrogen isotope exchange in the system calcite-carbon dioxide-graphite-methane-hydrogen-water vapor. Geochim. Cosmochim. Acta 33, 49-64.

_____ and M. Javoy (1973) Comments on oxygen isotope geothermometry. Earth Planet. Sci. Lett. 20, 250-265.

_____ and _____ (1975) Oxygen isotope partitioning among the minerals in igneous and metamorphic rocks. Revs. Geophys. Space Phys. 13, 401-418.

Brace, W.F. (1972) Pore pressure in geophysics, in H.C. Heart et al., eds., *Flow and Fracture in Rocks*. Geophys. Monogr. Am. Geophys. Union 16, 265-273.

Broecker, W.S. and V.M. Oversby (1971) *Chemical Equilibrium in the Earth*, McGraw-Hill, New York.

Clayton, R.N. and S. Epstein (1961) The use of oxygen isotopes in high-temperature geological thermometry. J. Geol. 69, 447-452.

_____, J.R. Goldsmith, K.J. Kavel, T.K. Mayeda and R.C. Newton (1975) Limits on the effect of pressure on isotopic fractionation. Geochim. Cosmochim. Acta 39, 1197-1201.

Craig, H. (1953) The geochemistry of the stable carbon isotopes. Geochim. Cosmochim. Acta 3, 53-92.

_____ (1957) Isotopic standards for carbon and oxygen and correction factors for mass-spectrometric analysis of carbon dioxide. Geochim. Cosmochim. Acta 12, 133-149.

Deines, P. (1970) Mass spectrometer correction factors for the determination of small isotopic composition variations of carbon and oxygen. Internat. J. Mass Spec. Ion Phys. 4, 283-295.

_____ (1977) On the oxygen isotope distribution among mineral triplets in igneous and metamorphic rocks. Geochim. Cosmochim. Acta 41, 1709-1730.

_____ (1980) The carbon isotopic composition of diamonds: relationship to diamond shape, color, occurrence, and vapor composition. Geochim. Cosmochim. Acta 44, 943-961.

Ferry, J.M. (1976) P, T, f_{CO_2} and f_{H_2O} during metamorphism of calcareous sediments in the Waterville-Vassalboro area, south-central Maine. Contrib. Mineral. Petrol. 57, 119-143.

_____ (1980) A case study of the amount and distribution of heat and fluid during metamorphism. Contrib. Mineral. Petrol. 71, 373-385.

Friedman, I. and J.R. O'Neil (1977) Compilation of stable isotope fractionation factors of geochemical interest. U. S. Geol. Survey, Prof. Paper 440-KK.

Fyfe, W.W., N.J. Price and A.B. Thompson (1978) *Fluids in the Earth's Crust*. Elsevier, New York, 383 p.

Garlick, G.D. and S. Epstein (1967) Oxygen isotope ratios in coexisting minerals of regionally metamorphosed rocks. Geochim. Cosmochim. Acta 31, 181-214.

Helgeson, H.C. (1968) Evaluation of irreversible reactions in geochemical processes involving minerals and aqueous solutions -- I. Thermodynamic relations. Geochim. Cosmochim. Acta 32, 853-877.

Hoefs, J. (1980) *Stable Isotope Geochemistry*. Springer-Verlag, Berlin.

Hoering, T.C. (1961) The physical chemistry of isotopic substances. Carnegie Inst. Wash. Year Book 60, 201-207.

Hoernes, S. and H. Friedrichsen (1978) Oxygen and hydrogen isotope study of the polymetamorphic area of the northern Otztal-Stubai Alps (Tyrol). Contrib. Mineral. Petrol. 67, 305-315.

_____ and _____ (1980) Oxygen and hydrogen isotopic composition of alpine and pre-alpine minerals of the Swiss Central Alps. Contrib. Mineral. Petrol. 72, 19-32.

_____ and E. Hoffer (1979) Equilibrium relations of prograde metamorphic mineral assemblages. Contrib. Mineral. Petrol. 68, 377-389.

Javoy, M. (1977) Stable isotopes and geothermometry. J. Geological Soc. 133, 609-636.

Kerrick, D.M. and E.D. Ghent (1979) $P-T-X_{CO_2}$ relations of equilibria in the system CaO-Al$_2$O$_3$-SiO$_2$-CO$_2$-H$_2$O, in V.A. Zharikov, V.I. Fonarev and V.A. Kori Kovskii, eds., *Problems of Physicochemical Petrology*, Akademii Nauk, SSSR, 32-52.

_____ and G.K. Jacobs (1981) A modified Redlich-Kwong equation for H$_2$O, CO$_2$ and H$_2$O-CO$_2$ mixtures. Am. J. Sci. 281, 735-767.

_____ and L.S. Darken (1975) Statistical thermodynamic models for ideal oxide and silicate solid solutions with application to plagioclase. Geochim. Cosmochim. Acta 39, 1431-1442.

Lattanzi, P., D.M. Rye and J.M. Rice (1980) Behavior of ^{13}C and ^{18}O in carbonates during contact metamorphism at Marysville, Montana. Am. J. Sci. 280, 890-906.

McKinney, C.R., J.M. McCrea, S. Epstein, H.A. Allen and H.C. Urey (1950) Improvements in mass spectrometers for the measurement of small differences in isotope abundance ratios. Revs. Scientific Instruments 21, 724-730.

Mook, W.G. and P.M. Grootes (1973) The measuring procedure and corrections for the high-precision mass-spectrometric analysis of isotopic abundance ratios. Inter. J. Mass Spectrometry Ion Phys. 12, 273-298.

Nier, A.O. (1947) A mass spectrometer for isotope and gas analysis. Revs. Scientific Instruments 18, 398-411.

Ohmoto, H. (1972) Systematics of sulfur and carbon isotopes in hydrothermal ore deposits. Econ. Geol. 67, 551-579.

_____ and R.O. Rye (1979) Isotopes of sulfur and carbon, *in* H.L. Barnes, ed., *Geochemistry of Hydrothermal Ore Deposits*, 2nd ed. John Wiley, New York, 509-567.

O'Neil, J.R. (1977) Stable isotopes in mineralogy. Phys. Chem. Minerals 2, 105-123.

_____ and H.P. Taylor (1967) The oxygen isotope and cation exchange chemistry of feldspars. Am. Mineral. 52, 1414-1437.

Orville, P.M. (1963) Alkali ion exchange between vapor and feldspar phases. Am. J. Sci. 261, 201-237.

Prigogine, I. and R. Defay (1954) *Chemical Thermodynamics*. Longmans, London, 543 p.

Rumble, D. (1978) Mineralogy, petrology, and oxygen isotopic geochemistry of the Clough Formation, Black Mountain, Western New Hampshire, USA. J. Petrol. 19, 317-340.

_____, J.M. Ferry, T.C. Hoering and A.J. Boucot (1982) Fluid flow during metamorphism at the Beaver Brook fossil locality. Am. J. Sci. 282, 886-919.

Rye, R.O. and H. Ohmoto (1974) Sulfur and carbon isotopes and ore genesis: a review. Econ. Geol. 69, 826-842.

_____, R.D. Schuiling, D.M. Rye and J.B.H. Jansen (1976) Carbon, hydrogen, and oxygen isotope studies of the regional metamorphic complex at Naxos, Greece. Geochim. Cosmochim. Acta 40, 1031-1049.

Sheppard, S.M.F. and H.P. Schwarcz (1970) Fractionation of carbon and oxygen isotopes and magnesium between coexisting metamorphic calcite and dolomite. Contrib. Mineral. Petrol. 26, 161-198.

Shieh, Y.N. and H.P. Schwarcz (1974) Oxygen isotope studies of granite and migmatite, Grenville province of Ontario, Canada. Geochim. Cosmochim. Acta 38, 21-45.

_____ and H.P. Taylor (1969) Oxygen and carbon isotope studies of contact metamorphism of carbonate rocks. J. Petrol. 10, 307-331.

_____ and _____ (1969) Oxygen and hydrogen isotope studies of contact metamorphism in the Santa Rosa range, Nevada. Contrib. Mineral. Petrol. 20, 306-356.

Suzuoki, T. and S. Epstein (1976) Hydrogen isotope fractionation between OH-bearing minerals and water. Geochim. Cosmochim. Acta 40, 1229-1240.

Taylor, H.P., Jr. (1977) Water/rock interactions and the origin of H_2O in granitic batholiths. J. Geol. Soc. 133, 509-558.

_____ (1979) Oxygen and hydrogen isotope relationships in hydrothermal mineral deposits, *in* H.L. Barnes, ed., *Geochemistry of Hydrothermal Ore Deposits*, 2nd ed. John Wiley, New York, 236-277.

_____ and R.G. Coleman (1968) $^{18}O/^{16}O$ ratios of coexisting minerals in glaucophane-bearing metamorphic rocks. Geol. Soc. Am. Bull. 79, 1727-1756.

_____, A.L. Albee and S. Epstein (1963) $^{18}O/^{16}O$ ratios of coexisting minerals in three assemblages of kyanite-zone pelitic schist. J. Geol. 71, 513-522.

Thompson, J.B. (1969) Chemical reactions in crystals. Am. Mineral. 54, 341-375.

Truesdell, A.H. (1974) Oxygen isotope activities and concentrations in aqueous salt solutions at elevated temperatures. Earth Planet. Sci. Lett. 23, 386-396.

Urey, H.C. (1947) The thermodynamic properties of isotopic substances. J. Chem. Soc. 562-581.

Valley, J.W. and J.R. O'Neil (1981) $^{13}C/^{12}C$ exchange between calcite and graphite: a possible thermometer in Grenville marbles. Geochim. Cosmochim. Acta 45, 411-419.

Chapter 9
COMPOSITIONAL ZONING and INCLUSIONS in METAMORPHIC MINERALS
R.J. Tracy

INTRODUCTION

The quantitative study of chemical zoning in igneous and metamorphic minerals is a relatively new field which originated in the early 1960's because of
the introduction of the electron probe microanalyzer. Before this time, it had
been recognized that chemical zoning probably existed in minerals as the under-
lying cause of such well-known phenomena as color zoning (e.g., pyroxenes,
amphiboles, tourmalines), intracrystalline refractive index variation (e.g.,
garnets) and variation in extinction angle (e.g., plagioclase). However, there
was no readily available method of performing microchemical analyses to determine the spatial variation of composition within mineral grains until the advent
of the microprobe. The earliest published microprobe studies of zoning in metamorphic minerals deal almost exclusively with zoned garnets (Atherton and Edmunds,
1966; Banno, 1965; Chinner, 1962; DeBethune et al., 1965; Evans and Guidotti,
1966; Harte and Henley, 1966; Hollister, 1966). Most of these early papers are
largely descriptive except for Harte and Henley (1966) which proposed several
qualitative models for garnet zoning, and by Hollister (1966) which presented
an important quantitative model that will be discussed in some detail below.
Many studies in the years since 1966 have extended the measurement of chemical
zoning profiles to a much greater variety of metamorphic minerals (see Table 1),
and have presented a considerable number of theoretical analyses dealing with
the development and preservation of zoning.

The goal of this review is to summarize the research of the last twenty
years on all aspects of zoning in metamorphic minerals and, more importantly
in light of the purpose of this volume, to develop a coherent picture of zoned
minerals as critical indicators of the processes of crystal growth and mass
transfer in metamorphic rocks. Mineral zoning is a clear indicator of disequilibrium and provides one important opportunity for application of increasingly sophisticated models of the kinetics of metamorphic processes.

Two types of zoning will be discussed. The first is *growth zoning*, zoning
developed because of a continuous or discontinuous change in composition of
material supplied to the growing surface of the crystal. An obvious requirement
of this type of zoning is slow volume diffusion so that the interior of the

Table 1. Examples of metamorphic minerals that show zoning.

Mineral	Elements	Reference
Garnet	Fe, Mg, Mn, Ca	Numerous
Staurolite	Fe, Mg, Ti, Al, Si	Hollister (1970)
Chloritoid	(Fe, Mg)	Ribbe (1980)
Al-silicates	Fe^{3+}, Ti	Chinner et al. (1969)
Zircon	(color)	Malcuit and Heimlich (1972)
Tourmaline	Fe, Mg	Chinner (1965)
Phengite	Fe, Mg, Al, Si	Boulter and Raheim (1974)
Phlogopite	Fe, Mg	Rimsaite (1970)
Annite	Fe, Mg, Ti	Kwak (1981)
Hypersthene	Mg, Al, Si	Obata (1980)
Diopside	Mg, Al, Si	Tracy et al. (1978)
Omphacite	Mg, Na, Ca, Al	Carpenter (1980)
Gedrite	Mg, Na, Al, Si	Robinson et al. (1982)
Sodic Amphibole	Ca, Na, Fe, Si	Holland and Richardson (1979)
Plagioclase	Ca, Na	Nord et al. (1978)
Cordierite	Fe, Mg	Tracy and Dietsch (1982)
Spinel	Fe, Mg, Cr, Al	Frost (1976)
Siderite	Fe, Mn, Mg	Jones and Ghent (1971)

crystal is effectively isolated from the rest of the rock. It is typically inferred that bell-shaped profiles for Mn and/or Ca in garnets which display euhedral crystal faces are indicative of growth zoning. The possibility of modification or obliteration of growth zoning through intracrystalline diffusion will be discussed.

A second type of metamorphic zoning may be called *diffusion zoning*, and it differs from growth zoning in that diffusion zoning is imposed upon a pre-existing crystal that may or may not have been homogeneous. This type of zoning is developed through intracrystalline diffusion that is driven by reaction of the crystal's surface either with an adjacent mineral grain or with the matrix in general through the medium of an intergranular fluid or film. Metamorphic diffusion zoning may develop during either heating or cooling of the rock, and may be the source of importance evidence on reaction rates and rates of retrograde processes.

This review cannot go into kinetics in any detail, but there will be extensive reference to the literature on the kinetics of metamorphic processes. It is hoped that the reader will be left with an appreciation of the importance of kinetics to the development and preservation of zoning, and with some understanding of how zoning can be used to produce quantitative kinetic models. Relatively little of the terminology of diffusion will appear in this review, but several definitions are in order. *Diffusion* is the transport of matter driven by a chemical potential gradient (which may reflect either a compositional gradient or a temperature gradient). *Volume diffusion* or *intracrystalline diffusion* is transport through the crystal lattice which requires lattice defects (typically point defects) in order to operate. *Intergranular diffusion*, on the other hand, is transport along grain boundaries and free surfaces, or along surface defects or dislocations. This type of diffusion may include transport of species through an interstitial fluid. *Interdiffusion* is the specific counter-diffusion of two or more species, that is, balanced transport in opposite directions. A typical cause of interdiffusion would be the mass balance constraint imposed by stoichiometry in the case of volume diffusion in a binary compound. The reader is referred to Manning (1974) and Freer (1981) for more detailed discussions of diffusion mechanisms, and to Brady (1975a,b) for a discussion of the theory of diffusion applied to metamorphic rocks.

GROWTH ZONING

Continuous growth zoning of garnet

Models of continuous growth zoning of garnet. The first studies of metamorphic mineral zoning, and the majority of studies since, have concentrated on garnet, undoubtedly because it is the most commonly occurring zoned metamorphic mineral. Many of the early garnet zoning studies involved garnets from pelitic rocks at grades lower than sillimanite grade and concluded that the zoning originated through fractionation processes operating during garnet growth. This conclusion was largely based upon the higher concentrations of manganese found in the cores of garnet porphyroblasts whose outer margins were depleted in Mn and appeared to consist of well-formed crystal faces. It was suggested (Harte and Henley, 1966; Chinner, 1965) that since garnet was known to fractionate Mn relative to other minerals, the bell-shaped Mn profiles in garnets were due to depletion of Mn in the rock through concentration in the first garnet nuclei to form. Successive layers of garnet would then have a depleted reservoir of manganese to draw upon, and would therefore become progressively less manganiferous. Hollister (1966) quantified the above suggestion by proposing that the depletion of manganese in the rock could be modelled as a Rayleigh fractionation or distillation process, similar to the application of Rayleigh fractionation to the distribution of trace elements between crystals and liquid. The basic assumption that Hollister made in order to apply the Rayleigh model was that the fractionation factor was a constant during the garnet growth process. As Hollister notes, this assumption requires that the fractionated element (Mn in this case) be so low in concentration both in the

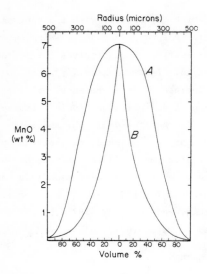

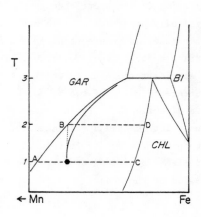

Figure 1 (to the left). Profiles of Mn concentration in garnet from Kwoiek, British Columbia (data of Hollister, 1966). Profile A is for Mn concentration versus radius (top scale) while profile B is for Mn concentration versus volume (bottom scale). Note that the radius profile seems to suggest high concentration of Mn in the entire garnet, but the volume profile indicates that more than half of the volume of garnet is characterized by less than 1 wt % MnO.

Figure 2 (to the right). Schematic Fe-Mn pseudobinary T-X diagram showing reaction loops for chlorite (+ quartz) → garnet, chlorite (+ muscovite) → biotite, and biotite → garnet (+ musco- vite) (these are only hypothetical reactions, and not necessarily complete). Hexagon indi- cates initial proportions of garnet and chlorite. For explanation of other symbols, see text.

reservoir and in the crystal that Mn/(Mn+Mg+Fe) is approximately equal to Mn/(Mg+Fe). The garnet to which Hollister applied the model was one from Kwoiek, British Columbia; the manganese concentration profile of garnet is illustrated in Figure 1. Hollister calculated a composition profile using the Rayleigh model that agreed closely with the measured profile for the Kwoiek garnet. He ascribed the minor disagreement between calculated and measured profiles to errors in the assumption that the fractionation factor was constant. Specifi- cally, potential difficulties with Hollister's analysis are: (a) the fractiona- tion factor would be expected to change with changing temperature, and (b) the fractionation factor depends upon the exact mineralogic composition of the reser- voir or matrix, especially on the relative proportions of matrix minerals that are in direct reaction relation with the garnet (chlorite and biotite in the case of the Kwoiek garnet).

It is perhaps unfortunate that Hollister chose to describe his model in terms of the Rayleigh distillation process, because the model has thereby pro- voked criticism by petrologists who prefer a more specific mechanistic alternative that may be called the reaction partitioning model. Kretz (1973) and Trczienski (1977) have described this model, which is inherent in the treatment of a number of other authors. The reaction partitioning model proposes that the zoned mineral grows while its surface composition is controlled by some multivarient equilibrium with one or more reactant minerals. Figure 2 illustrates a sample reaction in which chlorite is the principal Fe-Mn-Mg reactant mineral and garnet is the product mineral:

$$(Fe,Mg,Mn)_{4.5}Al_3Si_{2.5}O_{10}(OH)_8 + 2SiO_2 = 1.5 \ (Fe,Mg,Mn)_3Al_2Si_3O_{12} + 4 \ H_2O \qquad (1)$$

Chlorite   Quartz   Garnet

The Fe-Mn join is of course only one pseudobinary join in a Fe-Mg-Mn pseudoternary system. For the bulk composition shown by the hexagon, the compositions of coexisting garnet and chlorite at temperature T_1 are given by points A and C. Given *homogeneous and heterogeneous* equilibrium, we would expect that a temperature increase from T_1 to T_2 would cause the garnet composition to change from A to B and the chlorite composition to change from C to D. Therefore, at T_2 the chlorite would be totally consumed and garnet would have a homogeneous composition of B.

If, however, we specify that the material crystallized as garnet loses communication with the matrix (i.e., there is no longer homogeneous equilibrium), then the "effective bulk composition" (EBC) is depleted in the material now sequestered in the garnet interiors. In our example, this will drive the EBC toward higher Fe/Mn since the material removed as garnet interiors has a higher Mn/Fe than the initial bulk composition (IBC). The path for the EBC that is shown in Figure 2 must have this shape, but the exact path is determined by the efficiency of the fractionation of Mn by the garnet. The graphical similarity between this behavior and the case of igneous fractional crystallization is rather striking; the only difference is that metamorphic fractionation takes place with rising temperature while igneous fractionation occurs upon cooling. There are a couple of interesting implications to the metamorphic fractionation process. The first is that garnet zoning may encompass a much more extensive compositional path than in an equilibrium case; garnet composition may in fact go all the way past B to the termination of the T-X loop at T_3 in Figure 2. The second is that chlorite may never be consumed completely in the reaction as long as the chlorite coexisting with garnet has Mn/Fe>0, as in Figure 2.

Kretz (1973) has pointed out a possible additional complexity in interpreting a diagram such as Figure 2. We have so far assumed that heterogeneous equilibrium is maintained (i.e., *reactive* mineral compositions lie on an equilibrium T-X loop) but not homogeneous equilibrium in garnet, and the EBC therefore changes. Kretz suggested that slow volume diffusion in a reactive phase, chlorite for example, could allow only the outer edges of chlorite grains to react and equilibrate with the garnet edges. This would result in both garnet and chlorite grains with Mn-richer cores. Kretz argues that continued chlorite consumption would eventually lead to breakdown of the Mn-rich cores, and this would be reflected by an increase in Mn at the very edge of the garnet. Such a reversal in Mn zoning, in apparently growth-zoned and unresorbed garnets, has been reported by a number of authors (e.g., Hollister, 1966; DeBethune et al., 1968; Drake, 1968; Mueller and Schneider, 1971).

Atherton (1968) presented a model similar to Hollister's, but based on a treatment of zone refining developed by Pfann. The equation that Atherton gives is

$$C = kC_o(1-g)^{k-1} \qquad (2)$$

where C is concentration of an element in crystallizing solid, C_o is concentration of that element in the whole system before crystallization begins, k is the distribution coefficient and g is the modal amount of the crystallizing solid. (The original Pfann application of this model to zone refining involved two phases, solid and melt, in a binary alloy.) For comparison, the Rayleigh fractionation equation given by Hollister (1966) is

$$M_G = \lambda M_o(1-W^G/W^o)^{\lambda-1} \qquad (3)$$

where M_G is the weight fraction of an element in the garnet edge, M_o is the weight fraction of that element in the initial rock, λ is the fractionation factor, and W^G and W^o are the weights of crystallized garnet and the initial system, respectively. (2) and (3) are apparently the same equation, which is not surprising since zone refining is a Rayleigh fractionation-type process.

It should be apparent that the difference between the Hollister and Atherton models and the "reaction partitioning" model is really only a semantic one. The fractionation factors, λ or k, of Hollister and Atherton are simply empirical factors that express the bulk fractionation of an element between garnet and all the reacting phases in the rock. The Hollister and Atherton models may not be as physically correct as the "reaction partitioning" model, but they have been shown by their adherents to be capable of accurate predictions of garnet zoning. One potential pitfall of these two models, however, is that they are nominally isothermal if a constant λ or k is used, while growth of a large garnet porphyroblast may take place over a considerable temperature

interval. The temperature dependence of k has been taken into account by
Atherton for one zoned garnet from Scotland by calculating the profile using
more than one value for k (Atherton, 1968, Fig. 6). On the other hand, Hollister
(1966, 1969) argues that there may not *be* much of a temperature interval in the
growth of garnets in British Columbia. The garnets may have grown rapidly and
nearly isothermally in the rapidly heated rocks of a thermal aureole; if the
rocks were heated more rapidly than reaction kinetics could keep up with, then
reactions would have been overstepped and ultimately would have proceeded with
greatly enhanced rates. Loomis (1972, 1975) has proposed a similar mechanism
for rocks in the aureole of the Ronda peridotite.

If one accepts that the "reaction partitioning" model correctly describes
garnet zoning (or that of any other zoned mineral), then it follows that a change
in the assemblage of reactant minerals may cause an abrupt change in the trend of
zoning in the garnet. For example, suppose that the zoned garnet produced in
the reaction already illustrated in Figure 2 was heated to and above T_3. Above
T_3 biotite replaces chlorite as the principal Fe-Mg-Mn reactant, and the two-
phase loop above T_3 reflects a different Fe-Mn partitioning between garnet and
the new reactant mineral. A zoned garnet that has grown during the progress of
both reactions will show a decided kink in the zoning profile of Mn. It is even
possible that the higher-temperature loop could allow virtually no change in the
Fe/Mn of garnet. Thompson (1976) reviewed the topological possibilities of T-X
loops and gives examples of a number of garnet-producing reactions which may be
useful in understanding zoning in garnets produced by this mechanism.

Compositional zoning in garnets from greenschist grade to kyanite grade.
Compositional mapping of garnets is an alternative to profiles for displaying
the zoning pattern. If it is assumed that garnet is zoned in a radially symmetrical pattern, then a single radial profile should be sufficient. Zoning
in many garnets, however, is not radially symmetrical and a single profile in
one of these garnets might give an erroneous impression of the overall zoning
pattern. A good example is the zoning displayed by a large (about 8 mm) garnet
in pelitic schist from Gassetts, Vermont (Thompson et al., 1977). Many garnets
from this terrain show rotation during growth (Rosenfeld, 1970) and this rotation produces an asymmetry in the compositional distribution. Figure 3 shows
contour maps for mole percent of garnet components in the Gassetts garnet.
Zoning irregularities may be due either to rotation about a sub-horizontal axis
during growth, or to coalescence of a number of small garnet nuclei (coinciding
with the highs in Mn and Ca) followed by relatively smooth growth around the
coalesced center. A random profile through a garnet as complexly zoned as the

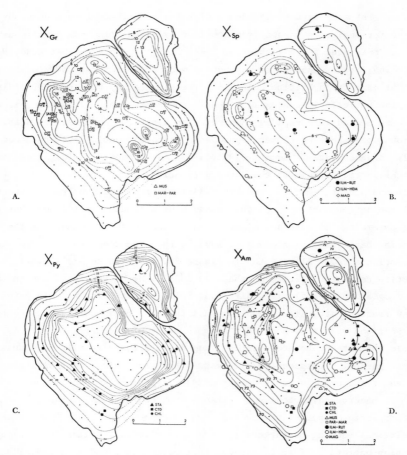

Figure 3. Composition "maps" of garnet from Gassetts Schist (Thompson et al., 1977) showing composition contours and locations of inclusions. Scale bar is two mm long. $X_{Gr} = 100$ Ca/(Ca +Mn+Mg+Fe); $X_{Sp} = 100$ Mn/(Ca+Mn+Mg+Fe); $X_{Py} = 100$ Mg/(Ca+Mn+Mg+Fe); $X_{Am} = 100$ Fe/(Ca+Mn+Mg+Fe). Symbol identities are shown in the figure legends. Hachures on contour lines point toward lower values of X_{Am} in (D). Dots indicate probe analysis points for garnet.

Gassetts garnet would likely be very difficult to interpret. In fact, it is possible that the several cases of oscillatory zoning in garnet that have been reported are actually cases of rotation and/or coalescence.

Another example of an irregularly zoned garnet is shown in Figure 4, based upon unpublished data of the author on a garnet-kyanite-staurolite schist from central Massachusetts. This cross-section shows zoning suggestive of rotation about an approximately east-west axis, although the outermost part of the garnet appears to be concentrically zoned. A very interesting feature of this garnet, however, is the reversal in Mg zoning, a feature which occurs in a number of other garnets in rocks of this grade. Most kyanite-staurolite-grade garnets

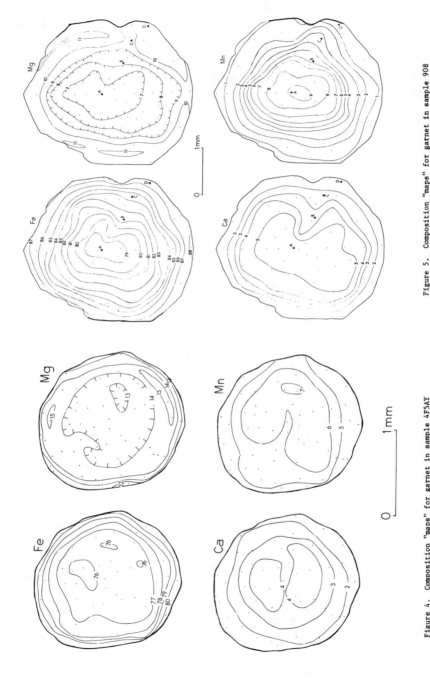

Figure 4. Composition "maps" for garnet in sample 4F5AY (Tracy et al., 1976), kyanite-staurolite grade schist from Massachusetts. Note the non-concentric zoning of Ca and Mn; for explanation see text. Dots indicate probe analysis points.

Figure 5. Composition "maps" for garnet in sample 908 (Tracy et al., 1976), kyanite-staurolite grade schist from Massachusetts. Dots indicate probe analysis points. Triangles are analyzed inclusions of ilmenite (see Fig. 13). Note the reversal in Mg zoning near the garnet edge.

show an increase outward in Mg and in Mg/Fe as the Mn and Ca decrease; in some garnets this trend continues right to the margin and in others the Mg reversal occurs. It is unclear whether this abrupt reversal is a prograde or retrograde effect, and further study is clearly needed. Another variation in kyanite-staurolite-grade garnet zoning is shown in Figure 5, a composition map of garnet in sample 908 (Tracy et al., 1976). In contrast to the previous two garnets, zoning of all elements in nicely concentric, but 908 also shows the edge reversal in Mg.

A third method of displaying the zoning trends of garnets is on ternary diagrams whose apices are the zoned elements; the two most useful of these diagrams appear to be Fe-Mg-Mn and Fe-Mg-Ca. Diagrams of this sort have been used by Brown (1969), Tracy et al. (1976) and others. Figure 6 consists of Fe-Mg-Mn and Fe-Mg-Ca diagrams showing trends in a number of growth-zoned garnets from rocks whose maximum grade was kyanite-staurolite or lower. The data are from a wide variety of geological locales, but it is striking how similar most of the trends are. The core-to-rim decrease in Mn and Ca is apparent in all the trends, as is the general tendency to constant or slightly increasing Mg/Fe from core to rim. These diagrams nicely display the reversal in Mg zoning that

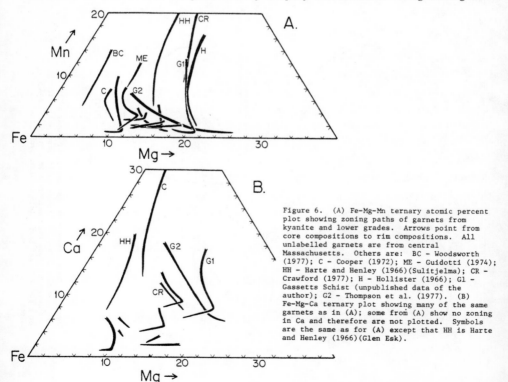

Figure 6. (A) Fe-Mg-Mn ternary atomic percent plot showing zoning paths of garnets from kyanite and lower grades. Arrows point from core compositions to rim compositions. All unlabelled garnets are from central Massachusetts. Others are: BC – Woodsworth (1977); C – Cooper (1972); ME – Guidotti (1974); HH – Harte and Henley (1966)(Sulitjelma); CR – Crawford (1977); H – Hollister (1966); G1 – Gassetts Schist (unpublished data of the author); G2 – Thompson et al. (1977). (B) Fe-Mg-Ca ternary plot showing many of the same garnets as in (A); some from (A) show no zoning in Ca and therefore are not plotted. Symbols are the same as for (A) except that HH is Harte and Henley (1966)(Glen Esk).

364

was noted in Figures 4 and 5.

An important advantage of diagrams such as those in Figure 6 is that they may be used to characterize the reactions by which garnet was being produced and to deduce changes in the reactant assemblage. For example, an interpretation of trend H in Figure 6A would be as follows: at the beginning of its growth, the garnet strongly fractionates Mn (as noted above). If it is assumed that the reactants are low in Mn, then the Fe/Mg ratio of the material being supplied to the growing garnet may be predicted at any point in its growth by taking a tangent to the garnet zoning trend at that point and projecting to the Fe-Mg edge of the diagram. (In this sense, the Fe-Mg-Mn diagram should obey the same sorts of mass-balance rules that apply to fractional crystallization in ternary liquidus diagrams.) The curved path for garnet H in Figure 6A therefore records Mg enrichment in the material supplied to the garnet's growth surface. This trend, in turn, is consistent with a garnet-producing reaction represented by a T-X(Fe-Mg) loop which moves to higher X_{Mg} with increasing T. Presumably, a smooth and continuous path such as that shown by garnet H represents garnet growth during a single reaction. Assemblage data given by Hollister (1966) suggest that this reaction is

$$\text{Chlorite + Staurolite + Quartz = Garnet.} \qquad (4)$$

On the other hand, garnets with kinked paths such as G1 (from the Gassetts schist, unpublished data of the author) appear to have grown during a more complex reaction history. Much of the zoning path of garnet G1 is similar to that of garnet H, suggesting that it might represent growth during a reaction similar to (4), which is consistent with the mineral assemblage in the Gassetts schist. If the "tail" toward Fe-enrichment in G1 is a prograde growth feature, it represents garnet formation in a new reaction, in which the material supplied to the garnet is rapidly depleted in Mg; it is very difficult to formulate such a reaction, however. (The alternative is that the "tail" resulted from a minor amount of retrograde requilibration. A minor resorption of the garnet edge accompanied by production of a more magnesian phase such as chlorite or biotite would drive the garnet compositions in the directions shown by the "tails" in G1 and several other garnets in Figure 6A.)

The Fe-Mg-Ca diagram in Figure 6B can be interpreted in much the same way as the Mn diagram. Crawford (1977) and Banno and Kurata (1972) have discussed in some detail the potential complexities of Ca zoning in garnets, some of which arise because of the possibility that there may be phases in a rock that are more calcic than garnet, e.g., epidote, plagioclase and apatite. But as Figure

6B shows, the zoning of Ca in lower-grade garnets is relatively systematic, in contrast to the data we will see later for higher-grade garnets.

Compositional zoning in garnets from the sillimanite + muscovite zone. Garnets in rocks whose metamorphic grade is slightly higher than the ones previously described show zoning patterns that are quite different (e.g., Fig. 7).

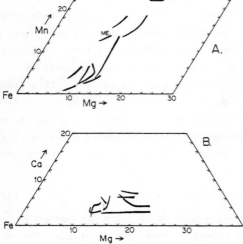

Figure 7. (A) Fe-Mg-Mn ternary atomic percent plot of sillimanite+muscovite-grade garnets. Arrows point from core to rim. Unlabelled garnets are from central Massachusetts; ME – Guidotti (1974). (B) Fe-Mg-Ca ternary plot of the same garnets as in (A), excluding those that show no Ca zoning. All are from central Massachusetts.

All of these garnets are from sillimanite or sillimanite-staurolite-bearing assemblages and all coexist with muscovite rather than K-feldspar. While these garnets typically show a core-to-rim decrease in Mn similar to the previous group of garnets, the trend toward lower Mn is accompanied by an increase in the ratio Fe/Mg. In most of the garnets of this grade that have been examined, no evidence of zoning has been found in the interiors which is similar to that in the typical lower-grade garnets. There are several possible explanations for this behavior: (a) the sillimanite-grade garnets have entirely grown under conditions near peak metamorphic conditions and therefore never had zoning typical of lower grade; (b) as a corollary of the preceding, complete recrystallization of lower-grade garnets under sillimanite-grade conditions obliterated any evidence of lower-grade zoning; and/or (c) the higher temperatures in the sillimanite zone allowed volume diffusion to operate at efficient rates which modified original growth zoning. The likelihood of the first two processes is extremely difficult to assess; if they occur at all they are probably dependent upon exact P-T-time metamorphic path as well as upon strain rate and deformational history. The third process has been proposed by Anderson and Buckley (1973), Anderson and Olimpio (1977) and Woodsworth (1977) and certainly appears valid for garnets metamorphosed to sillimanite + K-feldspar grade conditions as will be discussed below. Anderson and Olimpio (1977) argue that volume diffusion may be important even at grades as low as staurolite or staurolite-kyanite, and suggest that the classic bell-shaped Mn profiles (as in Fig. 1) are in fact modified after garnet growth.

Ca zoning in sillimanite-grade garnets, as shown in Figure 7B, follows no consistent pattern and in fact suggests that several different processes may have been at work. The increase in Ca near the rims of several garnets likely reflects the involvement of some calcic phase such as plagioclase or epidote in the garnet-producing reaction.

It has already been pointed out that there are advantages to viewing garnet zoning trends on diagrams such as Figures 6 and 7, but it should also be noted that there is a disadvantage. The trends portrayed on Figures 6 and 7 give no information about the relative volumes of garnet of the different compositions. For example, it is possible that a large garnet could have most of its zoning concentrated at the rim, or, alternatively, a strongly zoned core could be surrounded by a wide homogeneous rim. For this reason, it is advisable to use either a profile or a map together with a ternary diagram in order to get the fullest appreciation of a garnet zoning pattern.

Many garnets from higher grade rocks have been studied, but their discussion will be deferred to a later section on diffusion zoning. It is now generally accepted that intracrystalline diffusion in garnet becomes important at the temperatures reached above those characteristic of the sillimanite zone, and it is therefore appropriate to discuss zoning in these higher grade garnets as originating through diffusion.

Discontinuous zoning in garnet

All of the garnets discussed in the previous section are continuously zoned, that is, there are no breaks or gaps in the zoning trends (although reversals may occur). Examples of garnets that do show breaks are those reported by Rosenfeld (1970) and Rumble and Finnerty (1974) in rocks from eastern Vermont. Figure 8 shows one such garnet and its measured zoning profile, which Rumble and Finnerty ascribe to polymetamorphism. Similar, apparently anomalous zoning has been reported by Albee (1968) and Brown (1969). Rumble and Finnerty (1974) suggest that the almadine-rich cores grew during Ordovician contact metamorphism and that the grossular and spessartine-rich

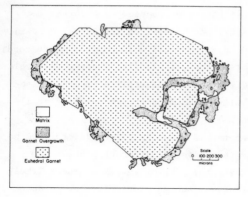

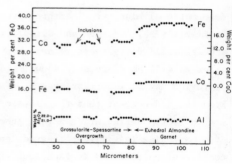

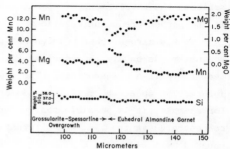

Figure 8. Sketch and composition profiles of
discontinuously zoned garnet from central
Vermont. From Rumble and Finnerty (1974).

overgrowths grew during a late Devonian lower-grade regional metamorphism which
pervasively retrograded the matrix of the rock. The break in zoning between
core and overgrowth is therefore an "unconformity" analagous to a stratigraphic
one.

It is worthwhile to note the pattern of zoning that occurs in the outer
part of the almandine-rich core shown in Figure 8. The significant increase in
Fe/Mg toward the edge of the core zone, coupled with the minor increase in Mn,
is consistent with some retrograde reequilibration of the outer part of the pre-
existing garnet, as noted in the previous section. This presumably predated the
overgrowth, but may have occurred at an early stage in the second metamorphism.
The very different composition of the overgrowth, especially its enrichment in
Ca and Mn, indicates that the later garnet-producing reaction was completely
different from the earlier one, in part because of the difference in grade and
in part due to the change in "effective bulk composition" caused by the isola-
tion of so much Fe in the refractory early garnets.

It appears likely from the extreme difference in core and overgrowth chemistry that the garnet in Figure 8 indeed shows a polymetamorphic "unconformity".

However, Thompson et al. (1977) suggested that a break in garnet zoning could be produced within a single metamorphic episode. Their mechanism involves garnet production in a continuous (multivariant) reaction in which a reactant is consumed entirely. At this point the rock leaves the T-X reaction loop (cf. Fig. 2) and garnet is no longer produced. At a later stage in the heating of the rock, a different garnet-producing reaction may be intersected and a different garnet composition overgrown on the earlier garnet. During the hiatus in garnet growth, the earlier garnet may either be resorbed slightly or remain inert.

Although no apparent examples are available, it is a further possibility that a compositional break such as that in the Vermont garnets could be smoothed out if the second metamorphism were at high enough grade that intracrystalline diffusion could operate. It is likely, however, that if the second metamorphism was lower-grade than the first, but still high enough for significant mass transfer by diffusion, garnet would actually be resorbed and diffusion zoning would operate at the margin, as described in detail in a later section.

Growth zoning in other minerals

Table 1 shows the variety of metamorphic minerals in which some kind of zoning has been observed. Zoning in minerals other than garnet and perhaps plagioclase appears to be rather unusual, however. Staurolite is probably the most familiar to petrologists, but zoned staurolites seem to be much more common than zoned garnets. Two papers on zoned staurolite include studies of staurolite from British Columbia (Hollister, 1970) and from Ireland (Smellie, 1974). Both staurolites are sector-zoned in Ti, Mg, Fe, Si and Al. Interestingly, both occurrences are from contact metamorphosed rocks in the aureoles of plutons, while most reports of staurolite in regional metamorphism do not indicate any zoning. This suggests that there is something special about the conditions of contact metamorphism which is conducive to the development of zoning; this point will be discussed in detail in a later section. Hollister (1970) does, however, report analyses of staurolites from a number of different locales which show some zoning. In general, the regional metamorphic staurolites show significant zoning only in Ti.

Figure 9A shows the zoning in staurolite from Kwoiek, British Columbia and Figure 9B portrays the different crystallographic sectors in staurolite (Hollister, 1970). Hollister presented several models for the development of sector zoning and concluded that one is more likely than the others. The preferred model assumes disequilibrium between the bulk of each sector and the matrix, but

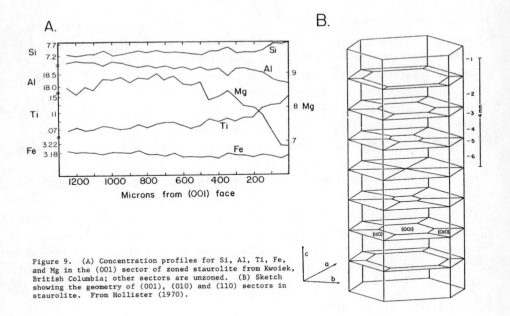

A.

B.

Figure 9. (A) Concentration profiles for Si, Al, Ti, Fe, and Mg in the (001) sector of zoned staurolite from Kwoiek, British Columbia; other sectors are unzoned. (B) Sketch showing the geometry of (001), (010) and (110) sectors in staurolite. From Hollister (1970).

chemical equilibrium between the surface layers of the sectors and the matrix. Homogenization of each sector depends upon the diffusion of cations from inner layers to outer layers, eliminating compositional gradients. Hollister suggested that rapid heating and rapid crystal growth isolated the inner growth layers faster than diffusion would equilibrate them, and therefore tended to preserve sector zoning. He further suggested that this is consistent with his observation that sector zoning in all elements is preserved in contact metamorphic staurolite, that only Ti is sector zoned in staurolite from regional metamorphic terrains with closely spaced isograds, and that no sector zoning is observed in regional metamorphic terrains with widely spaced isograds.

Tourmaline is another mineral of pelitic schists in which zoning has been observed. Many reports on the petrography of tourmaline-bearing rocks have noted that this mineral is color-zoned, but very little data are available on chemical zoning within tournaline crystals. Chinner (1962) and O'Connor (1973) have reported microprobe data which indicate that the rims of zoned tourmaline crystals have a lower value of Fe/Mg than the cores, but neither gives a zoning profile. The author has analyzed a zoned tourmaline from near Jamaica, in eastern Vermont, and these unpublished data are given in Figure 10. Large (up to 1 mm diameter) tourmalines commonly have cores that are blue-green in color and rims that are greenish-brown, with a rather abrupt boundary between the two

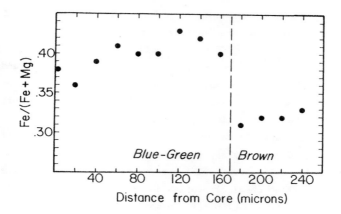

Figure 10.   Concentration profile in zoned tourmaline from Cambro-Ordovician
pelitic schist at kyanite-staurolite grade from near Jamaica, Vermont (unpublished probe data of
the author). This tourmaline has a blue-green core and a brown rim; the color change corresponds to the change in chemistry. No variation was found in any elements other than Fe and Mg.

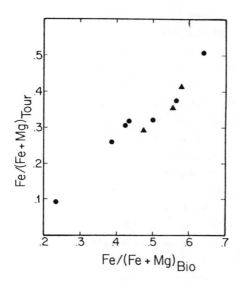

Figure 11.   Correlation of compositional variation of tourmaline and biotite in
schists ranging from staurolite to sillimanite grade. Filled circles - unpublished data of the
author from central Massachusetts; triangles - data from Waterbury, Connecticut and Philadelphia
(O'Connor, 1973).

371

zones. The only chemical variation is in Fe/(Fe+Mg) and the sense of this variation is consistent with the observations of Chinner and O'Connor. As Figure 10
shows, an abrupt change in Fe/(Fe+Mg) coincides with the color zoning. In addition, there appears to be a systematic zoning toward higher Fe/(Fe+Mg) within
each of the color zones. This zoning presumably reflects a systematic shift in
the Fe/(Fe+Mg) of the other minerals with which each successive surface layer
of the tourmaline was in equilibrium during growth. It has been observed that
tourmaline shows sympathetic compositional variation with other ferromagnesian
phases in rocks (O'Connor, 1973); the data for pelitic schists from southern
New England in Figure 11 serve to illustrate this relationship. The abrupt
break in the tourmaline chemical profile in Figure 10 may very well reflect
the growth (or recrystallization) of tourmaline in two different continuous
equilibria, in much the same manner that discontinuous zoning may develop in
garnet (see previous section).

Compositional zoning in minerals in mafic schists appear to be particularly
useful as chronicles of changing P-T conditions in rocks with histories of polymetamorphism. As an example, Laird and Albee (1981a) reported that epidote,
garnet, calcite, omphacite, and amphibole are zoned in mafic schists collected
from Tillotson Peak, northern Vermont. Zoning in pyroxene and amphibole were
characterized in some detail. Omphacites had $NaAlSi_2O_6$-, $NaFeSi_2O_6$-rich cores
and augite-rich rims. Some pyroxenes were set in a matrix of cryptocrystalline
material with composition $NaAlSi_3O_8$. Pyroxene zoning was interpreted in terms
of a metamorphic history in which the rims of an initial acmite + jadeite-rich
pyroxene later decomposed according to the reaction jadeite + quartz = albite.
Amphiboles had glaucophane-rich cores and actinolite-rich rims. Amphibole rims
systematically differed in composition from the cores by several amphibole substitution schemes; rims were lower in glaucophane ($NaAlCa_{-1}Mg_{-1}$) and Al-tschermak
($Al_2Mg_{-1}Si_{-1}$) contents but higher in Ti-tschermak ($TiAl_2Mg_{-1}Si_{-2}$), edenite
($NaAlSi_{-1}$), and ferric iron ($FeAl_{-1}$) contents (substitution schemes written
according to those developed by Thompson in Chapter 1 of this volume). Amphibole
zoning was interpreted in terms of a history in which the initial glaucophane-
rich amphibole later developed actinolite-rich rims through a continuous reaction involving combinations of actinolite, glaucophane, chlorite, omphacite,
plagioclase, garnet, epidote, quartz, sphene, magnetite, muscovite, carbonate,
and sulfide. Zoning in both pyroxene and amphibole were believed to record
changes in T and P during mineral growth. Pyroxene and amphibole cores grew
under high-pressure, low-temperature conditions (P = 9 $\pm$ 2kbar; T = 450°C).
Pyroxene and amphibole rims record a later retrograde event which was either

(a) at lower pressure and the same temperature, (b) at higher temperature and the same pressure, or (c) at lower pressure and temperature.

In a landmark paper Laird and Albee (1981b) extended consideration of chemical zoning in amphiboles in mafic schists to occurrences of the rock type world wide. They examined the composition of amphibole in the "common assemblage" amphibole + chlorite + epidote + plagioclase + quartz + Ti-phase $\pm$ carbonate $\pm$ K-mica $\pm$ ferric oxide, from numerous metamorphic terrains. Amphibole compositions were found to systematically differ in the assemblage in terrains metamorphosed under conditions of different facies series. Differences in amphibole composition were displayed on four different diagrams: (1) Na(M4) versus Na(A) + K; (2) [Al(VI) + Fe^{3+} + Ti + Cr] versus Al(IV); (3) Na(M4) versus [Al(VI) + Fe^{3+} + Ti + Cr]; and (4) Na/(Ca + Na) versus Al/(Al + Si). On each diagram amphiboles from different facies series plotted in different but slightly overlapping areas. The Na/(Ca + Na) versus Al/(Al + Si) diagram was preferred because the positions of amphibole compositions on it were not influenced by any of the assumptions used to calculate cation distributions of amphiboles from electron microprobe analyses. On the Na/(Ca + Na) versus Al/(Al + Si) diagram amphiboles from different facies series plot in different elongated areas which radiate outward from the origin. Amphiboles from high-pressure facies series terrains plot at the highest values of Na/(Na + Ca) for a reference value of Al/(Al + Si); amphiboles from the low-pressure facies series plot at the lowest values of Na/(Na + Ca); amphiboles from the intermediate pressure facies series plot at intermediate values of Na/(Na + Ca).

Laird and Albee (1981b) examined chemically zoned amphiboles in polymetamorphic mafic schists from numerous localities in Vermont. Amphibole compositions were used to infer P-T conditions during crystallization using the Na/(Na + Ca) versus Al/(Al + Si) diagram, and relative timing of the P-T conditions was inferred from the geometry of the zoning and overgrowths (core records oldest conditions; rims record the youngest). They inferred up to four metamorphic episodes for some samples. For example, in one sample, zoning and overgrowth patterns apparently record two early metamorphic episodes (one high-pressure and the other medium-pressure), possibly of Ordovician age, followed by two later low-pressure episodes of possible Devonian age. The use of zoned amphiboles as chronicles of metamorphic history in polymetamorphic terrains awaits further applications elsewhere.

There are, of course, a number of additional metamorphic minerals that display growth zoning, and their zoning generally follows the principles already discussed.

Inclusions in growth-zoned minerals

The presence of one or several kinds of matrix minerals as small inclusions in large metamorphic porphyroblasts is a commonly observed phenomenon. Quartz typically occurs as swarms of small inclusions in garnet, staurolite, chloritoid and cordierite porphyroblasts, giving rise to classic "sieve" texture (Spry, 1969). Oriented inclusions of quartz and other minerals have been used to deduce strain directions and magnitudes through analysis of rotations of garnets (Rosenfeld, 1970). Inclusions may also be used for geothermometry/geobarometry (Adams et al., 1975a,b). Besides quartz, minerals which commonly occur as inclusions include biotite, graphite, Fe-Ti oxides and zircon. The importance of inclusions in metamorphic minerals is that their identity, their distribution, and their relationship to chemical zoning in the host mineral may tell us much more about the processes of crystal growth in metamorphic rocks.

Spry (1969, pp. 169-180) gave a detailed discussion of the physical processes that may lead to the formation of inclusions. He noted that poikiloblastic crystals are usually suggestive of growth in a higher-energy environment than homogeneous crystals because of the considerably greater surface free energy in a poikiloblast (perhaps as much as an order of magnitude greater than an inclusion-free crystal of similar size). This energy consideration generally suggests faster growth of a poikiloblast relative to an inclusion-free crystal. There is, however, the possibility of significant variation in the energy of the host-inclusion interface due to different degrees of similarity between the structures of the two phases. If the structures are very dissimilar, the interfacial energy will be high and the host crystal will tend to reject the impurity unless the growth rate is very high. On the other hand, if the impurity has a structure which may loosely or tightly bond with the surface of the growing crystal, it may become adsorbed and is therefore more likely to become overgrown as an inclusion. Whether an impurity at the growth surface of a metamorphic mineral becomes an inclusion or not is therefore a rather delicate balance of kinetic factors. To summarize, Spry gave several common reasons for poikiloblastic texture to develop: (1) rapid growth of the host; (2) ability of the host lattice to absorb the strain of accommodating the impurity; (3) abundance of the impurity throughout the rock; (4) surface affinity (i.e., low surfacial energy) between host and inclusion; (5) difficulty of nucleation of the host.

Some qualitative observations about inclusions and hosts can be made, as Spry (1969) notes. The very common occurrence of abundant quartz inclusions in cordierite may be due to the fact that cordierite-quartz interfaces are of suitably low energy under most growth-rate conditions. A similar observation may be made about the abundance of quartz inclusions in lower-grade garnets, while higher grade garnets (above sillimanite grade) are notably poorer in inclusions. In this regard, it is interesting to note also that garnet-producing reactions at lower grade are typically quartz-producing, while at sillimanite grade and above, garnet-producing reactions are quartz-consuming (Thompson, 1976). Finally, as noted by Ramberg (1952), mica forms relatively high energy surfaces and is typically rejected by such phases as garnet, cordierite, and staurolite. Mica rinds around prophyroblasts of these minerals may represent accumulations of impurities rejected during porphyroblast growth.

It should be apparent that there is great potential for using observations on metamorphic poikiloblasts to make estimates of pressure-temperature-time relationships in different metamorphic terrains. Unfortunately, knowledge of important kinetic parameters such as crystal growth rates and mechanisms, intergranular diffusivities, and surficial energies between different phases are not well known enough to enable quantitative estimates to be made. In some cases, inclusions may be useful as indicators of the reaction history of a rock when used in conjunction with growth zoning in their host. Thompson et al. (1977) report a study of a large growth-zoned garnet which contains numerous inclusions of several different phases. Contour maps of the distribution of Fe, Mg, Mn and Ca in this garnet have already been given in Figure 3, which also shows the distribution of inclusions. The compositions of such included ferromagnesian phases as staurolite and chloritoid vary systematically from core to rim of the garnet and have been used to construct "paleo-AFM" diagrams for several lower-grade stages in the metamorphism of the rock (Fig. 12). It is

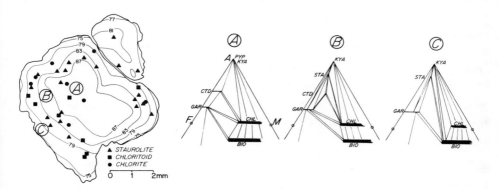

Figure 12. Composition "map" of Gassetts garnet (Thompson et al., 1977) showing contours of 100 Fe/(Fe+Mg) and locations of ferromagnesian inclusions. A, B, and C are zones of the garnet which represent the "paleo-AFM" topologies indicated. Direction of movement of three-phase triangles may also be deduced (see Thompson et al. for discussion).

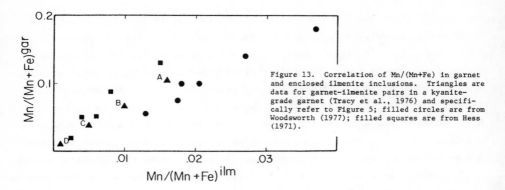

Figure 13. Correlation of Mn/(Mn+Fe) in garnet and enclosed ilmenite inclusions. Triangles are data for garnet-ilmenite pairs in a kyanite-grade garnet (Tracy et al., 1976) and specifically refer to Figure 5; filled circles are from Woodsworth (1977); filled squares are from Hess (1971).

remarkable that a number of phases which occur as inclusions in the garnet -- magnetite, rutile, chloritoid, margarite and epidote -- do not occur in the matrix outside the garnet (they completely reacted away). This garnet, in its preservation of growth zoning and unreacted included phases, serves as a "tape recording" of metamorphic reaction history.

A second example of the systematic behavior of growth-zoned garnet and its inclusions involves the partitioning of Fe and Mn between garnet and ilmenite inclusions. Tracy et al. (1976), Woodsworth (1977), and Thompson et al. (1977) have all noted the tendency of ilmenite in the interiors of growth-zoned garnets to have higher Mn/Fe. This systematic relationship is shown in Figure 13, which incorporates data on rocks from British Columbia, Massachusetts, and Vermont. The relationship has been interpreted as reflecting the depletion of Mn in ilmenite as garnet progressively sequestered Mn in its interior -- in other words, at any point the growth surface of the garnet and the matrix ilmenite were in equilibrium along a T-X loop such as that in Figure 2. The relationship should hold as long as the compositions of neither ilmenite nor adjacent garnet were altered by exchange processes or by intra-crystalline diffusion in the zoned garnet. Woodsworth (1977) points out con-vincingly that garnet homogenization through intracrystalline diffusion has operated extensively in the higher grade rocks (cordierite grade) which he studied, without any significant reequilibration between host garnet and ilmenite inclusions.

Summary: kinetic controls of growth zoning and inclusion formation

If homogeneous and heterogeneous equilibrium in rocks and minerals were always maintained, we should not even expect to see high-grade metamorphic rocks at the Earth's surface, much less find compositional zoning or inclu-sions in metamorphic minerals. The abundant evidence of disequilibrium around us, in the form of unstable or metastable mineral compositions, mineral assemblages, and textures, makes it imperative for us to try and understand the kinetics of the prograde and retrograde metamorphic processes that led to the formation and/or preservation of such disequilibrium features.

In order to understand the kinetics of growth zoning, we need to know three things: (a) nucleation rates, (b) crystal growth rates, and (c) the rate of volume diffusion within the zoned mineral. Nucleation rate can provide an important, though indirect, control on growth rate: the lower the nucleation rate, the farther material must be transported through the rock matrix to reach a crystal growth surface, and hence the slower the growth rate will be. It is possible to evaluate nucleation rate (compared to growth rate) as a rate-controlling step by examination of mineral size populations and chemistry, as discussed in detail by Kretz (1973). He points out that the expressions for nucleation rate and growth rate may be combined into a single expression for reaction rate. Kretz deduced the following equation for nucleation rate of garnet in gneiss from Yellowknife, Canada:

$$dn/dt = 2.8 \ k_1 t; \tag{5}$$

n is the number of nuclei, t is time, and k_1 is a nucleation rate constant. In order to make calculations using this equation he selected an arbitrary value for the rate constant, k_1. This rate constant is probably dependent upon many variable parameters including bulk composition, grain size, exact mineralogy, pressure, and temperature, among others.

Growth rate itself is rather complex and is dependent upon a number of factors. For one particular rock, Kretz (1973) has also deduced an expression for garnet growth rate which is based upon the estimation that garnet surface area increases linearly with time:

$$da/dt = k_2; \tag{6}$$

a is the surface area of all garnets in the rock, and k_2 is a growth rate constant. The rate constant in equation (6) is probably dependent upon as many parameters as the rate constant for nucleation, but perhaps the most important is the rate of supply of material to the garnet growth surface. To evaluate this, we need to know the diffusivities of the constituent components of the growing crystal, and especially the diffusivity of the most slowly diffusing component since this will be the rate-limiting factor. (It should be noted that the diffusing components are likely to be more complex than the simplest oxide components of the growing crystal.) Diffusion of components to the growth surface may be either by volume (lattice) diffusion

377

or by intergranular diffusion (Brady, 1975a,b). The relative importance of each will depend upon the temperature, since volume diffusion seems to be dominant at higher temperatures and intergranular diffusion dominant at lower temperatures (Putnis and McConnell, 1980; Freer, 1981). Other important factors entering into estimation of a growth rate constant are the kinetics and mechanism of the actual absorption process at the growth surface; Kirkpatrick (1981) discusses this process in the case of igneous crystallization.

It should be stressed that the rate laws of equations (5) and (6) were deduced by Kretz (1973) for one specific rock, and as he notes, do not necessarily apply in other cases. They may be more general, but that is a premise that needs testing in other similar rocks from other terrains. There is clearly much further work to be done in this area.

A final kinetic consideration is the process by which a compositional gradient can initially be produced in a growing crystal, and how that gradient can be preserved. Hollister (1970) addressed this point in his discussion of zoning in staurolite. He noted that the most pronounced zoning occurred in the (001) sector (Fig. 9B), the sector which must have grown more rapidly than the others in order to maintain an idioblastic (euhedral) crystal shape. This suggested to Hollister that the greater rate at which the (001) growth face moved away from already-absorbed interior material rendered volume diffusion less efficient in eliminating the compositional gradient between newly absorbed material and material in the interior of this sector as compared to the other sectors. Of course, the rate of volume diffusion should be approximately the same in all sectors.

The key, therefore, to the development of growth zoning is the kinetic balance between the rate at which new material (of slightly different composition) is added to the growing face of the crystal and the rate at which this new material is exchanged with previously added material through volume diffusion. It is presumed that the composition of material supplied to the crystal face is constantly changing because of some continuous reaction process. In cases where compositional zoning is expressed by only two cations, for example the common pair of Fe and Mg, the controlling kinetic parameter in volume diffusion is the interdiffusion coefficient. When more than two species diffuse, as can be the case in garnet, a more complicated set of diffusion coefficients which includes cross-coefficients is required (Loomis, 1978a,b). There can be a wide variation in the magnitudes of diffusion coefficients for different species or combinations of species (Freer, 1981).

The absence of growth zoning in most metamorphic minerals is almost certainly an indication that the diffusivities of their solid solution components are high relative to growth rates at the temperatures of growth. For most minerals, only conditions of considerable overheating and overstepping of reaction (analogous to undercooling in igneous systems) can produce growth rates great enough to allow growth zoning to form. On the other hand, the very common occurrence of growth zoning in garnet is a necessary consequence of the very *low* diffusivities in this mineral relative to almost any reasonable growth rate. It is probably fair to say that any unzoned garnet that grew at temperatures below about 550°C did so in the absence of any change in the composition of the material supplied to it during growth.

The relationship between growth zoning and inclusions in metamorphic minerals is an interesting one. It has been observed that zoned garnets (many references) and zoned staurolites (Hollister, 1970; Smellie, 1974) typically are full of inclusions, principally quartz. But in the author's experience, inclusions are also common in unzoned garnet and staurolite, leading to the conclusion that the kinetics of growth zoning and inclusion formation are somewhat different. A growth rate rapid enough to cause entrapment and inclusion of adsorbed grains of foreign material may not generally be rapid enough to counterbalance the rate of homogenization through volume diffusion, at least in staurolite. Thus the distinction between the presence of inclusions and the presence of inclusions *plus* zoning has the potential of being an important kinetic marker for certain metamorphic minerals.

Finally, the preservation of growth zoning after its initial formation may be very difficult to assess. Anderson and Olimpio (1977) suggest that some modification of growth zoning may occur through volume diffusion even in relatively low-grade garnets. Although diffusion through the garnet lattice appears to be a very slow process at temperatures below about 600°C (Anderson and Buckley, 1973), very slow cooling of a rock, after crystallization of the final mineral assemblage, may allow significant diffusion to occur, even in garnet. And as noted in an earlier section, interpretation of sillimanite-grade garnet zoning is complicated by the fact that the typical temperatures of sillimanite-grade metamorphism -- 575° to 650°C -- are at the point where volume diffusion over significant distances in garnet may take place in geologically reasonable times (Anderson and Buckley, 1973). It is certainly possible that many staurolites in regional terrains were originally zoned, but that the slow cooling after the peak of metamorphism allowed volume diffusion to operate and homogenize the crystals. There seems to be very little doubt that the zoning observed in garnets from metamorphic

379

grade higher than sillimanite (i.e., above the sillimanite + K-feldspar isograd) is largely the result of volume diffusion; this forms the basis of the
next section.

DIFFUSION ZONING

Diffusion zoning in garnet

The reader will have noticed that the highest grade garnets discussed
in the previous sections were of sillimanite grade, while many studies of
garnet zoning from higher grades have been reported. The reason for this is
that most, if not all, of the zoning seen in these high-grade garnets is the
result of post-growth volume diffusion, commonly during the cooling which
immediately follows the peak of metamorphism, or during a later, lower-grade
remetamorphism. Most authors seem to accept the possibility that some garnet
resorption occurs during a retrograde event, but in one terrain in central
New England it has been proposed that garnet *growth* during cooling after a
near-granulite grade metamorphism can lead to what may be described as "retrograde growth zoning" (Cyan and Lasaga, 1982). The following discussion will
outline the characteristics of diffusion zoning observed in a number of cases,
and will review the implications of zoning for the kinetics of high temperature metamorphic processes.

It has been proposed by a number of authors that garnets which are heated
to temperatures near or above the sillimanite + K-feldspar isograd (approximately 650°C) are homogenized when volume diffusion obliterates any pre-existing compositional zoning (Woodsworth, 1977; Blackburn, 1969; DeBethune et al.,
1975; Grant and Weiblen, 1971; Tracy et al., 1976; Anderson and Buckley,
1973). A transition over a narrow temperature interval, perhaps 75°C, from a
regime where growth zoning is preserved to one where it is erased by diffusion
may be understood in terms of the temperature dependence of the diffusion
coefficient D:

$$D_T = D_o \exp(-Q/RT) \tag{7}$$

In this equation, D_o is essentially D at infinite temperature, D_T is the
coefficient of diffusion at the T of interest, and Q is the activation energy
of diffusion. The exponential dependence of D on T implies that we might
expect a relatively abrupt transition in diffusion with increasing T, other
factors such as pressure and heating and cooling rates being roughly equal.
In any case, we will assume for purposes of this discussion that any metamorphic garnet which has been heated to 650°C or above will be roughly

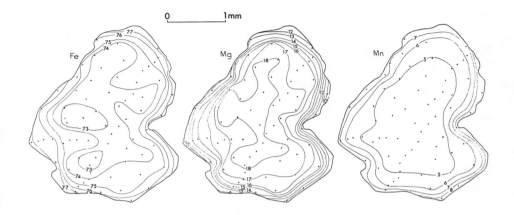

Figure 14. Composition "maps" of garnet in sample 933B (Tracy et al., 1976); the sample is from the lower part of the sillimanite + K-feldspar zone. Dots indicate analysis points. Ca zoning is not shown since it is less than one mole percent. The contours for each element are: 100Fe/(Fe+Mg+Mn+Ca), 100Mg/(Fe+Mg+Mn+Ca) and 100Mn/(Fe+Mg+Mn+Ca).

homogeneous in composition at these high temperatures.

We can use composition maps to display the distribution of cations in high grade garnets in the same way that we already have for lower grade garnets. Figure 14 shows maps of Fe-, Mg-, and Mn-distribution in a garnet from central Massachusetts that was described by Tracy et al. (1976); it comes from the lowest part of the sillimanite + K-feldspar zone. Calcium zoning is not shown because variation in Ca is so minor that contouring was not possible; this is typical of most garnets from high grades. The relatively homogeneous or weakly zoned core of the garnet gives way to pronounced zoning in all three cations near the rim. The sense of zoning of Mg and Mn is the reverse of that in the lower-grade, growth-zoned garnets, but zoning is generally concentric as in the case of growth zoning. Zoning of the sort shown in Figure 14 seems to be virtually restricted to garnets of the uppermost amphibolite facies; granulite facies garnets (to be discussed below) show zoning with similar trends but much different distribution. Garnets with zoning similar to that in Figure 14 have been described by a number of authors studying upper amphibolite facies rocks from geographically widespread terrains (e.g., Evans and Guidotti, 1966; Grant and Weiblen, 1971; Hollister, 1977; Loomis, 1976; Woodsworth, 1977; Tracy et al., 1976; O'Connor, 1973; Blackburn, 1969).

One puzzling feature of garnet zoning in upper amphibolite grade rocks is the increase of Mn approaching the garnet rim. A similar Mn increase has been observed in a number of lower-grade garnets and has been explained variously as the result of diffusion (Anderson and Olimpio, 1977), minor retrogression (Linthout and Westra, 1968), changing f_{O_2} (Tewhey and Hess, 1976) and Mn-enrichment in the last chlorite which breaks down to form garnet (Kretz, 1973). None of these explanations appears to have much relevance for the higher-grade garnets, and the preferred hypothesis involves the selective resorption of garnet edge in a retrograde process that consumed garnet and produced either biotite or cordierite (Grant and Weiblen, 1971; Tracy et al., 1976; Hollister, 1977). These two product minerals are both very poor in Mn and lower than garnet in Fe/(Fe + Mg); direct breakdown of garnet rim to produce biotite or cordierite should result in depletion of Mg and enrichment of

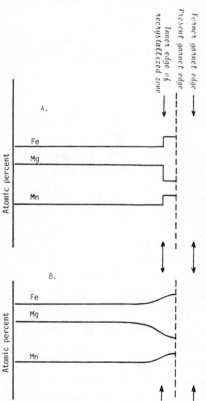

Fe and Mn in the residual garnet edge. The exact reaction mechanism is poorly understood at best, and is in some dispute, but one possible model would be as follows. The reaction would involve some degree of hydration, except in the unlikely case of production of anhydrous cordierite, and water therefore must be present or be introduced to the immediate region of the garnet rim. In addition, potassium would have to be added for biotite to be produced; potassium could be derived from K-feldspar in the matrix. Relatively rapid selective resorption of the garnet rim could involve removal only of some Fe and Mg (in a ratio different from that in which they are present in garnet) and the concomitant recrystallization of an appropriate amount of aluminosilicate matrix to form the product phase(s). This process could be accompanied by epitaxial recrystallization of the unresorbed garnet rim to produce the composition-distance relations shown in Figure 15, assuming that reaction and recrystallization rates were rapid relative to volume diffusion.

Figure 15. Schematic diagram of the edge of a high-grade garnet to illustrate one possible mechanism for development of rim zoning through operation of reaction (8). (A) shows the initial compositional discontinuity which develops immediately after resorption of the garnet edge. (B) shows the smoothed concentration profiles after volume diffusion has progressed for some time.

Figure 15 schematically illustrates the compositional discontinuity
that could be produced at the garnet rim as a result of selective resorption.
This discontinuity would be an adequate driving force for volume diffusion,
in which Fe and Mn would diffuse inward and Mg outward in an attempt to smooth
the discontinuity. Predictions of diffusion rates from published data (e.g.,
Freer, 1981) suggest that Fe and Mg should diffuse faster than Mn, and that
the profiles for Fe and Mg after some given time should be broader than the
profile for Mn. Reference to Figure 14 indicates roughly the following diffusivities: Mg > Fe > Mn, generally in accord with the prediction. It would
be desirable to try to quantify the above model for data such as that in
Figure 14, but the three-component diffusion problem is much more complicated
mathematically than a two-component treatment of interdiffusion (Loomis, 1978).

It is certainly qualitatively true that the extent of diffusion in the
case of high temperature garnet resorption/zoning must be controlled by the
sample's P-T-time path, and especially by the rate of cooling. Since the
distance of diffusion is dependent upon time and the coefficient of diffusivity, in the form of $\sqrt{Dt}$, and since D itself is a function of temperature
(equation 7), the maximum diffusion distance will be determined by the amount
of time the rock spent at the high temperatures where relatively high D's are
attained. Grant and Weiblen (1971) presented a resorption/diffusion model
very similar to that in Figure 15, in which they postulated interdiffusion of
Fe and Mg over considerable distances in garnet. Cygan and Lasaga (1982)
criticized the Grant and Weiblen model for allowing diffusion to occur over
distances that were too great, but Cygan and Lasaga based their criticism upon
calculations which assume a relatively rapid cooling rate of 100°C per million
years. With such rapid cooling, D would indeed decrease abruptly and diffusion
should be limited to short distances. However, a more reasonable cooling rate
of 5-10° per million years relieves the difficulty cited by Cygan and Lasaga;
with slower cooling, the more gradual decrease in D should allow diffusion
over much greater distances.

The reader may have noted that Ca has been ignored in the above discussion, even though its behavior should parallel that of Mn since neither Ca nor
Mn is accommodated in the product phases biotite or cordierite. It is, however, a troublesome fact that in most of the garnets which have been described,
Ca either remains constant or even decreases near the garnet rim; one excep-
tion is some Ca increase in the edges of high-grade garnets which have been
described as being retrograded and partly resorbed (Arkai et al., 1975).
Hollister (1977) has ascribed the slight decrease of Ca in garnet rims to the

production of some anorthite component in the resorption reaction. It may be that in any but quite calcic rocks, the residual grossular component reacts with sillimanite and quartz in this way to generate a small amount of anorthite which dissolves in plagioclase solid solution. In rapidly cooled rocks, this process might even be detected by locating Ca-enriched rims on plagioclase crystals near garnet.

The concentric zoning that is very apparent in the garnet in Figure 14 is the typical zoning pattern observed in garnets of this grade. However, in high-grade garnets in general, there is a continuum of observed zoning distributions which ranges from perfectly concentric to highly localized. Some garnets which show localized zoning have obviously suffered cation exchange with adjacent ferromagnesian phases (e.g., Lasaga et al., 1977) with no volume loss of garnet, but other garnets apparently have undergone some reaction and been partly resorbed near adjacent ferromagnesian minerals. The degree of concentric zoning in resorbed garnets is apparently a function of the amount of long range intergranular diffusion that can occur in the rock matrix, which is itself a function of the amount and/or composition of intergranular fluid (assuming similar P, T, and mineral assemblage). As an example, pelitic rocks in two adjacent terrains in Massachusetts that reached similar maximum temperatures, but by different P-T paths (Tracy and Robinson, 1980), have quite different zoning distributions in garnets. The rocks from the lower-pressure path seem to have undergone significantly more dehydration than those of the higher-pressure path. Garnets from the lower-pressure terrain show only localized zoning (see Fig. 18), while the higher-pressure garnets typically show more concentric zoning (e.g., Fig. 14). The rocks which have undergone less dehydration appear to have intergranular diffusion characteristics which allow more extensive reaction of garnet edges, while the more dehydrated rocks appear to have allowed more restricted diffusion.

The reaction proposed by Tracy et al. (1976) to explain the concentric zoning is a continuous (Fe-Mg) hydration reaction that progressed upon *cooling*:

$$\text{Garnet} + \text{K-feldspar} + H_2O = \text{Biotite} + Al_2SiO_5 + \text{Quartz} \qquad (8)$$

garnet and biotite compositions both become more Fe-rich with reaction progress. Long-range intergranular diffusion through an intergranular medium (possibly fluid) during slow cooling would allow development of concentric zoning even if the garnet were not completely surrounded by biotite. Less efficient intergranular diffusion (or less long-range), due to such factors as different properties of the intergranular medium or more rapidly decreasing temperature, would necessarily localize the garnet edge zoning to the proximity of adjacent biotite.

Reaction (8) appears to operate not only in the initial stages of cooling after a metamorphic peak, but also during a later remetamorphism of the originally high-grade rock at lower grade. An example of this is discussed by Tracy and Robinson (1980) in which a schist that was metamorphosed to sillimanite + K-feldspar grade in the late Precambrian was remetamorphosed to kyanite grade in the Devonian. Evidence of the earlier metamorphism is preserved as relict orthoclase crystals with sillimanite inclusions and large garnets with very Mg-rich cores and Fe-rich rims, while the rock matrix contains a kyanite-muscovite-biotite assemblage. The garnet zoning is continuous and encompasses a change of more than 20 mole percent pyrope, while also showing a slight increase of Mn at the rim of the garnet. The zoning has been interpreted as a result of reaction (8) in which garnet was resorbed and biotite (plus kyanite) produced. The rock appears to have achieved heterogeneous equilibrium throughout most of the matrix. The limiting kinetic factor on complete reequilibration seems to have been the slow volume diffusion in the large garnets which allowed preservation of the Mg-rich cores.

The zoning trends of high-grade garnets can be viewed in a ternary Fe-Mg-Mn plot exactly as was earlier done with low-grade garnets. Figure 16 shows that garnet zoning trends from a number of different terrains are remarkably consistent. Most of these garnets are from rocks in which reaction (8) can be assumed to have operated, and the zoning trends toward higher Fe/Mg and and higher Mn at the rims are compatible with this reaction, as discussed above. Although the trends in Figure 16 are similar, the actual composition-distance relations, as discerned from profiles or maps, are variable from one terrain to another, reflecting the different volume diffusion characteristics of garnet along different P-T-time metamorphic trajectories. The consistency of trends, however, supports the proposition that the same reaction mechanism occurred in each case.

The shapes of the trends in Figure 16 can perhaps be better understood through a graphical treatment similar to that of ternary liquidus diagrams in igneous petrology. Figure 17 shows a schematic Fe-Mg-Mn pseudoternary diagram on which the composition is shown of the original prograde garnet, G1. We shall assume that the reaction (8) begins immediately upon cooling, and that the biotite is an Fe-Mg solid solution without appreciable Mn. The first biotite produced by reaction of the garnet edge has the composition B1,

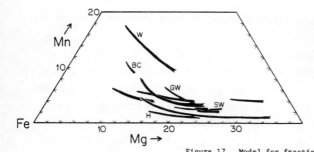

Figure 16. Fe-Mg-Mn ternary atomic percent plot of garnets from sillimanite + K-feldspar (or granulite) grade rocks. Arrows point from core to rim compositions. Unlabelled garnets are from central Massachusetts; W - O'Connor (1973); BC - Woodsworth (1977); H - Hollister (1977); GW - Grant and Weiblen (1971); SW - Schmid and Wood (1976).

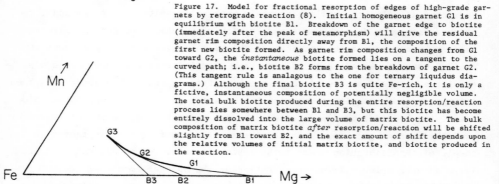

Figure 17. Model for fractional resorption of edges of high-grade garnets by retrograde reaction (8). Initial homogeneous garnet G1 is in equilibrium with biotite B1. Breakdown of the garnet edge to biotite (immediately after the peak of metamorphism) will drive the residual garnet rim composition directly away from B1, the composition of the first new biotite formed. As garnet rim composition changes from G1 toward G2, the *instantaneous* biotite formed lies on a tangent to the curved path; i.e., biotite B2 forms from the breakdown of garnet G2. (This tangent rule is analagous to the one for ternary liquidus diagrams.) Although the final biotite B3 is quite Fe-rich, it is only a fictive, instantaneous composition of potentially negligible volume. The total bulk biotite produced during the entire resorption/reaction process lies somewhere between B1 and B3, but this biotite has become entirely dissolved into the large volume of matrix biotite. The bulk composition of matrix biotite *after* resorption/reaction will be shifted slightly from B1 toward B2, and the exact amount of shift depends upon the relative volumes of initial matrix biotite, and biotite produced in the reaction.

the same as the bulk composition of the matrix biotite. Growth of this biotite will drive the composition of the residual garnet edge along the path G1-G2, that is directly away from the composition of the new biotite. As cooling proceeds and the reaction progresses, the equilibrium compositions of garnet edge and biotite move toward higher Fe/Mg, but at any instant, the garnet edge composition must be moving directly away from the biotite that is forming *at that instant*. The process is analogous to the fractional crystallization of a solid-solution phase from a magma in a ternary liquidus diagram; in Figure 17 the garnet composition path mimics the liquid path on a liquidus diagram. (For a discussion of the geometry of ternary fractional crystallization, see Morse, 1980, pp. 104-112.) The analogy between garnet resorption and igneous fractional crystallization is judged to be appropriate since the biotite components diffused away into the rock, and we are not constrained to a constant composition at the garnet edge.

A series of tie-lines between garnet edge composition and *instantaneous* biotite composition would therefore be tangents to the garnet path, as shown in Figure 17. The final instantaneous biotite composition, B3, in equilibrium with garnet edge G3, is very Fe-rich compared to the initial biotite B1. But since the biotite in the whole rock appears to have been homogenized during and after the reaction (e.g., Grant and Weiblen, 1971; Tracy et al., 1976;

Hollister, 1977), the final bulk rock biotite composition will depend upon the initial volume of biotite in the rock relative to the volume of reacted garnet edge. A very high modal ratio of biotite to garnet in a rock will mean that a significant change in the composition of garnet edges would result in a rather minor change in the Fe/Mg of post-reaction biotite. This of course may have serious implications for the use of biotite and garnet compositions for geothermometry in high-grade rocks.

It is apparent then that the concave-upward curved paths in Figure 16 are a necessary product of the postulated retrograde reaction. Another inter- esting point to be made from Figures 16 and 17 is that the mechanism just propounded for the curved garnet paths requires that the higher in Mn the starting point (G1) is, the more quickly the zoning should move to higher Mn in the garnet edge. This relation is required by the geometry of Figure 17. An examination of the zoning paths of Figure 16 indicates that the natural garnets do show this relationship, at least in a general way.

Diffusion zoning in garnet-cordierite and garnet-biotite pairs

A special case of the diffusion zoning of garnet occurs when the only observed zoning is in the immediate vicinity of grain boundaries between garnet and an adjacent ferromagnesian phase, typically either biotite or cordierite. Such restricted zoning has only been reported in granulite-grade rocks, apparently reflecting the absence in these rocks of any intergranular medium through which longer-range diffusion can occur. Some examples of this type of zoning have been reported by Berg (1977), Blackburn (1969), Hess (1971), Schmid and Wood (1976), Lasaga et al. (1977), and Tracy et al. (1976). Commonly, though not always, this type of zoning occurs in the absence of any reaction in which mineral proportions change, and involves only cation ex- change, typically of Fe and Mg, between adjacent phases. Neglecting for the moment the question of how the cations cross the grain boundary between the two phases, the process seems to be a relatively simple one of volume diffusion in the two crystal lattices in response to decreasing temperature. The driving force for the diffusion is the tendency for the two phases to maintain partition (or distribution) equilibrium of Fe and Mg at their immedi- ate interfaces as the rock cools.

In general, only Fe and Mg are zoned in these garnets -- presumed origi- nal homogeneity seems to be preserved for Ca and Mn. This behavior is to be expected if the process involves essentially no volume reduction of garnet (in contrast to the previously discussed case of concentric zoning) and if

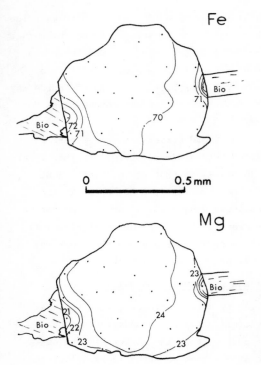

Fe

0   0.5 mm

Mg

Figure 18. Composition "maps" of 100
Fe/(Fe+Mg+Mn+Ca) and 100 Mg/(Fe+Mg+Mn+Ca) in
garnet FW-407 (Tracy et al., 1976) from a near-
granulite grade rock in central Massachusetts.
Dots indicate analysis points. Ca and Mn both
show less than one atom percent variation. Note
the pronounced zoning of Fe and Mg in proximity
to adjacent biotite grains; unzoned garnet edges
are adjacent to sillimanite and quartz.

the ferromagnesian phase adjacent to garnet contains virtually no Ca or Mn,
which is true for biotite and cordierite. A composition map for a cation-
exchanged garnet from central Massachusetts is shown in Figure 18. Only Fe
and Mg are shown; Ca and Mn have total variation of less than one atom percent
and cannot be contoured. Note the localization of Fe-Mg zoning around the
garnet-biotite grain boundaries; except for the two biotite grains, the garnet
is surrounded only by quartz and sillimanite. Microprobe traverses were done
in the biotite to check for zoning, and it was found that biotite at the garnet edge was systematically slightly more magnesian than biotite farther away;
total variation in Fe/(Fe + Mg) was only about 0.02, however. Hess (1971) and
Richardson (1975) studied rocks from the same terrain and found essentially no
zoning in biotite adjacent to garnet. Hess also reported garnet zoning
around large biotite inclusions, but found that the inclusions themselves were
homogeneous. All of these findings suggest that at the temperatures of cation
exchange (probably from 700°C down to about 500°C; see below), D_{Fe-Mg}^{Bio} was
probably significantly larger than D_{Fe-Mg}^{Gar}, so that while the scale of volume
diffusion in garnet was no more than one hundred microns (Lasaga et al., 1977),

388

volume diffusion in biotite was so extensive that entire biotite grains were homogenized, or nearly so.

The relative simplicity of the process described above (for example, (a) the diffusion involves only two expecies, or even only one, $FeMg_{-1}$, if one accepts the premise that Fe and Mg are not independently variable in this case, and (b) it is a fixed rather than moving boundary problem) makes it a tempting target for quantitative treatment. The goal of a quantitative kinetic analysis would be to calculate the time it took for the concentration profile to form, assuming that one has independent information on diffusion rates and that the relevant physical parameter (in this case compositional gradient) can be measured in the rock.

Lasaga et al. (1977) developed a rather elegant mathematical treatment for the above diffusion problem; their paper provides a nice summary of the process of cation exchange during a time interval of decreasing temperature. Using their model, they calculated a series of predicted compositional gradients for the outer part of a garnet adjacent to cordierite, at 25°C steps from their estimated initial temperature of 675°C down to 475°C (see Fig. 19). They note that the profiles converge below about 450°C, presumably because of the exponential dependence on temperature of the diffusion coefficient. One interesting point about Figure 19 is that the solution of Lasaga et al. predicts concentration gradients limited to about 50 microns from the garnet edge. Broader gradients, however, have been observed in garnets in

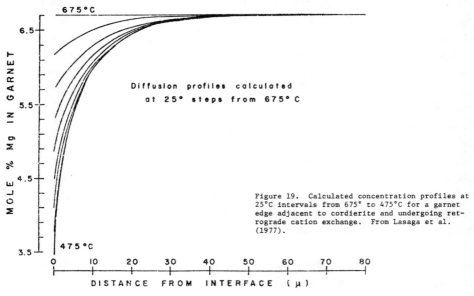

Figure 19. Calculated concentration profiles at 25°C intervals from 675° to 475°C for a garnet edge adjacent to cordierite and undergoing retrograde cation exchange. From Lasaga et al. (1977).

389

this terrain (e.g., Fig, 18). A minor change in the value of the interdiffusion coefficient chosen by Lasaga et al. would solve this problem. In fact, they noted that uncertainty in D greatly affected their calculations. As a result, they effectively inverted the problem and attempted to use geological arguments on the duration of the metamorphic episode to provide constraints on the cooling rate that they calculated. They concluded that the minimum cooling rate allowed by their model is 100°C per million years, a rate that requires extremely rapid uplift and erosion.

A different treatment of garnet-cordierite zoning in the central Massachusetts terrain is given by Tracy and Dietsch (1982). They report the occurrence of large ($\sim$1 cm) cordierite crystals which contain very small euhedral garnets plus intergrowths of sillimanite and quartz in their interiors. They interpret this inclusion texture to have developed through incipient breakdown of cordierite according to the reaction

$$\text{Cordierite} = \text{Garnet} + \text{Sillimanite} + \text{Quartz}. \qquad (9)$$

A sketch of one of these reaction aggregates is shown in Figure 20. Growth of the garnets has been interpreted by Tracy and Dietsch as a retrograde process, and they estimate the temperature interval to be about 150°C (700° to 550°), not very different from the interval over which Lasaga estimated cation exchange to occur.

Figure 21 shows a chemical profile for both garnet and cordierite along a traverse indicated on Figure 20. Both minerals are strongly zoned in Fe and Mg, and it is especially notable that the cordierite is zoned up to 100 microns from the interface with garnet. The authors interpreted the garnet zoning to be growth zoning developed during the growth of garnet while temperature fell. This is apparently the only way to develop garnet growth zoning characterized by strong Fe-enrichment at the edge (Cygan and Lasaga, 1982). The cordierite zoning was interpreted as representing long-range volume diffusion in which Fe diffused through the cordierite toward the growing garnet and Mg diffused away.

Tracy and Dietsch presented a mathematical treatment for the diffusion and growth process with the goal of extracting a cooling rate as Lasaga et al. had. Similar problems regarding the uncertainty in Fe-Mg interdiffusion coefficient existed, and a similar resort to geological arguments was made by Tracy and Dietsch. Their best estimate of a cooling rate was about 5°C per million years, a figure which is in better agreement with the data from other terrains (Jäger et al., 1967) than the 100°C per million years of Lasaga et al. In fact, the reaction of cordierite to garnet + sillimanite + quartz is

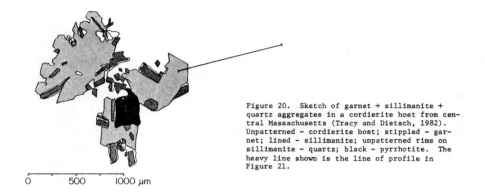

Figure 20. Sketch of garnet + sillimanite + quartz aggregates in a cordierite host from central Massachusetts (Tracy and Dietsch, 1982). Unpatterned - cordierite host; stippled - garnet; lined - sillimanite; unpatterned rims on sillimanite - quartz; black - pyrrhotite. The heavy line shown is the line of profile in Figure 21.

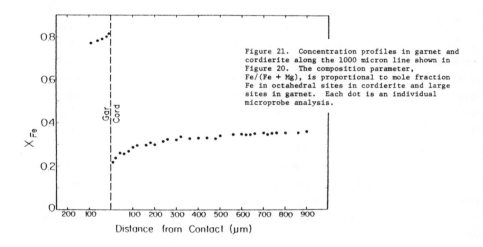

Figure 21. Concentration profiles in garnet and cordierite along the 1000 micron line shown in Figure 20. The composition parameter, Fe/(Fe + Mg), is proportional to mole fraction Fe in octahedral sites in cordierite and large sites in garnet. Each dot is an individual microprobe analysis.

is a likely indication that pressure stayed the same or even increased as cooling commenced (Thompson, 1976) rather than supporting a hypothesis of rapid uplift, erosion, and cooling of the terrain.

The principal purpose of the foregoing discussion is to indicate to the reader the petrologic and even tectonic information that may potentially be extracted from kinetic analysis of zoning. It is unfortunate that at present such kinetic analysis is hampered by our very incomplete knowledge of such important data as diffusion coefficients, both intergranular and intracrystalline, in metamorphic rocks and minerals. Such data are presently being gathered, however, and it is incumbent upon metamorphic petrologists to apply it to rocks as it becomes available.

CONCLUDING REMARKS

This review has covered most of the aspects of zoning in metamorphic minerals: development and preservation of growth zoning, modification of growth zoning, homogenization of zoned grains at high grades, and development of diffusion zoning through reactions. It is not a coincidence that the discussion has centered on garnets from pelitic rocks, since the great bulk of the literature on zoned metamorphic minerals deals with garnets. The reasons for this are several, including the lively interest in metamorphosed aluminous rocks and, more important, the relative commonness of zoning in garnet relative to other minerals. The special combination of growth and volume diffusion characteristics of garnet, along with its widespread occurrence, have made it virtually unique as a tool for studying kinetic processes in metamorphic rocks.

Other metamorphic minerals display zoning more rarely, and generally as a result of special conditions such as rapid heating or cooling. Although some studies have dealt with rocks that have undergone unusual metamorphic conditions, there is much further information to be extracted from these rocks. Better characterization of the diffusion properties of a variety of metamorphic minerals will allow quantitative limits to be placed upon the P-T-time paths taken in different metamorphic terrains, and will provide for a considerably better integration of metamorphism with other plate tectonic processes.

Finally, it is worthwhile stressing the potential value of zoning studies in deducing the *details* of metamorphic evolution of a particular rock. Although considerable consistency of behavior is displayed by garnet zoning, for example, one is still struck by the variety of patterns that emerge from seemingly similar garnets taken from seemingly similar rocks. The fact is that every rock has undergone a unique history and it behooves us not to overlook this in an attempt to generalize. The issue was eloquently stated by F.J. Turner (1970):

> Conformity of rock associations to pattern not only facilitates teaching and learning in the field of petrology but it gives a sound basis for prediction in both pure and applied geology. To many geologists, too, conformity affords a mental satisfaction that comes with the illusion of safety, order and predictability in that part of the universe in which our principal interests lie. I do not wish to deny the existence of discernible order in petrological phenomena -- indeed, much of my own work has been directed toward expounding such order. But today I would simply raise a question frequently discussed in recent years with my colleague J. Verhoogen: Every geological event, like every event in human history, is unique. If

we concentrate too heavily on discernment of order and pattern, may we perhaps overlook or underrate the unique quality of each igneous or metamorphic episode? Possibly uniqueness may have as great significance as conformity to pattern when we attempt to fit the phenomena of petrology into the broad framework of geology.

ACKNOWLEDGMENTS

I would like to acknowledge the assistance of Craig Dietsch and Paul Merewether in the preparation of this review. I would also like to thank Tony Lasaga, Pete Robinson and Ben Harte for many stimulating discussions on the topic, and John Ferry and Frank Spear for helpful reviews of the manuscript. Financial assistance was provided by NSF grants EAR-7819901 and EAR-8120670.

Adams, H.G., L.H. Cohen, and J.L. Rosenfeld (1975a) Solid inclusion piezothermometry: I. Comparison dilatometry. Am. Mineral. 60, 574-583.

_____, _____, and _____ (1975b) Solid inclusion piezothermometry: II. Geometric basis, calibration for the association quartz-garnet, and application to some pelitic schists. Am. Mineral. 60, 584-598.

Akizuki, M. and I. Sunagawa. (1978) Study of the sector structure in adularia by means of optical microscopy, infrared absorption, and electron microscopy. Mineral. Mag. 42, 453-462.

Albee, A.L. (1968) Metamorphic zones in northern Vermont. *In* Zen, E-an et al., Eds., *Studies of Appalachian Geology, Northern and Maritime*. 329-341, John Wiley and Sons, New York.

Anderson, D.E. and G.R. Buckley (1973) Zoning in garnets -- diffusion models. Contrib. Mineral. Petrol. 40, 87-104.

_____ and J.C. Olimpio (1977) Progressive homogenization of metamorphic garnets, South Morar, Scotland: Evidence for volume diffusion. Canadian Mineral. 15, 205-216.

Arkai, P., G. Nagy, and G. Panto (1975) Types of composition zoning in the garnets of polymeta- morphic rocks and their genetic significance. Acad. Sci. Hungarica Acta Geol. 19, 17-42.

Atherton, M.P. (1968) The variation in garnet, biotite and chlorite composition in medium grade pelitic rocks from the Dalradian, Scotland, with particular reference to zonation in garnet. Contrib. Mineral. Petrol. 18, 347-371.

_____ and W.M. Edmunds (1966) An electron-microprobe study of some zoned garnets from metamor- phic rocks. Earth Planet. Sci. Lett. 1, 185-193.

Banno, S. (1965) Notes on rock-forming minerals (34), zonal structure of pyralspite in Sanbagawa Schists in the Bessi area, Shikoku. J. Geol. Soc. Japan 71, 185-188.

_____ and H. Kurata (1972) Distribution of Ca in zoned garnets of low-grade pelitic schists. J. Geol. Soc. Japan 78, 507-512.

Berg, J.H. (1977) Regional geobarometry in the contact aureoles of the anorthisitic Nain Complex, Labrador . J. Petrol. 18, 399-430.

Blackburn, W.H. (1969) Zoned and unzoned garnets from the Grenville Gneisses around Gananoque, Ontario. Canadian Mineral. 9, 691-698.

_____ and E. Navarro (1977) Garnet zoning and polymetamorphism in the eclogitic rocks of Isla de Margarita, Venezuela. Canadian Mineral. 15, 257-266.

Bollingberg, H.J. and I. Bryhni (1977) Minor element zonation in an eclogite garnet. Contrib. Mineral. Petrol. 36, 113.

Boulter, C.A. and A. Raheim (1974) Variation in Si^{4+} content of phengites through a three stage deformation sequence. Contrib. Mineral. Petrol. 48, 57-71.

Brady, J.B. (1975a) Reference frames and diffusion coefficients. Am. J. Sci. 275, 954-983.

_____ (1975b) Chemical components and diffusion. Am. J. Sci. 275, 1073-1088.

Bryhni, I. and W.L. Griffin (1971) Zoning in eclogite garnets from Nordfjord, west Norway. Contrib. Mineral. Petrol. 32, 112.

Cannon, R.T. (1966) Plagioclase zoning and twinning in relation to the metamorphic history of some amphibolites and granulites. Am. J. Sci. 264, 526-542.

Carpenter, M.A. (1980) Composition and cation order variations in a sector-zoned blueschist pyroxene. Am. Mineral. 65, 313-320.

Chinner, G.A. (1962) Almandine in thermal aureoles. J. Petrol. 3, 316-340.

_____ (1965) The kyanite isograd in Glen Clova, Angus, Scotland. Mineral. Mag. 34, 132-143.

_____, J.V. Smith, and C.R. Knowles (1969) Transition-metal contents of Al_2SiO_5 polymorphs. Am. J. Sci. 267-A, 96-113.

Cooper, A.F. (1972) Progressive metamorphism of metabasic rocks from the Haast schist group of southern New Zealand. J. Petrol. 13, 457-492.

Crawford, M.L. (1977) Calcium zoning in almandine garnet, Wissahickon Formation, Philadelphia, Pennsylvania. Canadian Mineral. 15, 243-249.

Cygan, R. and A.C. Lasaga (1982) Crystal growth and the formation of chemical zoning in garnets. Contrib. Mineral. Petrol. 79, 187-200.

DeBethune, P., P. Goossens, and P. Berger (1965) Emploi des grenats zonaires comme indicateurs du degre de metamorphisme. C. R. Acad. Sci. Paris 260, 6946-6949.

_____, D. Laduron, and J. Bocquet (1975) Diffusion processes in resorbed garnets. Contrib. Mineral. Petrol. 50, 197-204.

_____
[*]In addition to references cited in the text, this list contains a number of other references relating to the study of compositional zoning and inclusions in metamorphic minerals.

Donnay, G. (1969) Crystalline heterogeneity: Evidence from electron probe study of Brazilian tourmaline. Carnegie Inst. Wash. Year Book 67, 219-220.

Dudley, P.P. (1969) Electron microprobe analysis of garnet in glaucophane schists and associated eclogites. Am. Mineral. 54, 1139-1150.

Evans, B.W. (1966) Microprobe study of zoning in eclogite garnets. Geol. Soc. Am. Spec. Paper 87, 54.

_____ and C.V. Guidotti (1966) The sillimanite - potash feldspar isograd in Western Maine, USA. Contrib. Mineral. Petrol. 12, 25-62.

Freer, R. (1981) Diffusion in silicate minerals and glasses: A data digest and guide to the literature. Contrib. Mineral. Petrol. 76, 440-454.

Frost, B.R. (1976) Limits to the assemblage forsterite-anorthite as inferred from peridotite hornfelses, Icicle Creek, Washington. Am. Mineral. 61, 732-751.

Grant, J.A. and P.W. Weiblen (1971) Retrograde zoning in garnet near the 2nd sillimanite isograd. Am. J. Sci. 270, 281-296.

Guidotti, C.V. (1974) Transition from staurolite to sillimanite zone, Rangeley Quadrangle, Maine. Geol. Soc. Am. Bull. 85, 475-490.

Harte, B. and K. J. Henley (1966) Occurrence of compositionally zoned almanditic garnets in regionally metamorphosed rocks. Nature 210, 689.

Hess, P.C. (1971) Prograde and retrograde equilibrium in garnet - cordierite gneisses in south-central Massachusetts. Contrib. Mineral. Petrol. 30, 177-195.

Holland, T.J.B. and S.W. Richardson (1979) Amphibole zonation in metabasites as a guide to the evolution of metamorphic conditions. Contrib. Mineral. Petrol. 70, 143.

Hollister, L.S. (1966) Garnet zoning: An interpretation based on the Rayleigh fractionation model. Science 154, 1647-1651.

_____ (1969) Contact metamorphism in the Kwoiek area of British Columbia: An end-member of the metamorphic process. Geol. Soc. Am. Bull. 80, 2464-2494.

_____ (1970) Origin, mechanism, and consequences of compositional sector zoning in staurolite. Am. Mineral. 55, 742-766.

_____ (1977) The reaction forming cordierite from garnet, the Khtada Lake metamorphic complex, British Columbia. Canadian Mineral. 15, 217-229.

Jäger, E., E. Niggli, and E. Wenk (1967) Rb-Sr alterbestimmungen an glimmern der zentralalpen. Beitr. Geol. Karte Schweiz. N.F. 134, 67-80.

Jones, J.W. and E.D. Ghent (1971) Zoned siderite porphyroblasts from the Esplanade Range and Northern Dogtooth Mountains, British Columbia. Am. Mineral. 56, 1910-1916.

Kirkpatrick, R.J. (1981) Kinetics of crystallization of igneous rocks. In Lasaga, A.C. and R.J. Kirkpatrick, Eds., Kinetics of Geochemical Processes, Reviews in Mineralogy 8, 321-398.

Kretz, R. (1973) Kinetics of the crystallization of garnet at two localities near Yellowknife. Canadian Mineral. 12, 1-20.

Kwak, T.A.P. (1970) An attempt to correlate non-predicted variations of distribution coefficients with mineral grain internal inhomogeneity using a field example studied near Sudbury, Ontario. Contrib. Mineral. Petrol. 26, 199-224.

_____ (1981) Sector-zoned annite 85 - phlogopite 15 micas from the Mt. Lindsay Sn-W-F (-Be) deposit, Tasmania, Australia. Canadian Mineral. 19, 643-651.

Laird, J. and A.L. Albee (1981a) High-pressure metamorphism in mafic schist from northern Vermont. Am. J. Sci. 281, 97-126.

_____ and _____ (1981b) Pressure, temperature, and time indicators in mafic schist: Their application to reconstructing the polymetamorphic history of Vermont. Am. J. Sci. 281, 127-175.

Lasaga, A.C., S.M. Richardson, and H.C. Holland (1977) Mathematics of cation diffusion and exchange between silicate minerals during retrograde metamorphism. In Saxena, S.K. and S. Bhattacharji, Eds., Energetics of Geological Processes, 353-388, Springer-Verlag, Berlin.

Linthout, K. and L. Westra (1968) Compositional zoning in almandine-rich garnets and its relation to the metamorphic history of their host rocks. Proc. Koninkl. Ned. Akad. Wetenschop 71 Ser. B, 297-312.

Loomis, T.P. (1972) Contact metamorphism of pelitic rocks by the Rhonda ultramafic intrusion, southern Spain. Geol. Soc. Am. Bull. 83, 2449-2474.

_____ (1975) Reaction zoning of garnet. Contrib. Mineral. Petrol. 52, 285-305.

_____ (1976) Irreversible reactions in high-grade metapelitic rocks. J. Petrol. 17, 559-588.

_____ (1978a) Multicomponent diffusion in garnet: I. Formulation of isothermal models. Am. J. Sci. 278, 1099-1118.

Loomis, T.P. (1978b) Multicomponent diffusion in garnet: II. Comparison of models with natural data. Am. J. Sci. 278, 1119-1137.

Manning, J.R. (1974) Diffusion kinetics and mechanisms in simple crystals. *In* Hoffman, A.W., B.J. Giletti, H.S. Yoder, and R.A. Yund, Eds., *Geochemical Transport and Kinetics*, 3-13, Carnegie Inst. Wash. Publ. 634.

Malcuit, R.J. and R.A. Heimlich (1972) Zircons from Precambrian gneiss, southern Bighorn Mountains, Wyoming. Am. Mineral. 57, 1190-1209.

Morse, S.A. (1980) *Basalts and Phase Diagrams*. Springer-Verlag, Berlin.

Nord, G.L., Jr., J. Hammarstrom, and E-an Zen (1978) Zoned plagioclase and peristerite formation in phyllites from southwestern Massachusetts. Am. Mineral. 63, 947.

Obata, M. (1980) The Ronda peridotite: garnet-, spinel- and plagioclase-lherzolite facies and the P-T trajectories of a high-temperature mantle intrusion. J. Petrol. 21, 523-572.

O'Connor, B.J. (1973) *A Petrologic and Electron Microprobe Study of Pelitic Mica Schists in the Vicinity of the Staurolite Disappearance Isograd in Philadelphia, Pennsylvania and Waterbury, Connecticut*. Ph.D. Thesis, The Johns Hopkins University, Baltimore.

Phillips, R. and I.J. Stone (1974) Reverse zoning between myrmekite and albite in a quartzo-feldspathic gneiss from Broken Hill, New South Wales. Mineral. Mag. 39, 654-658.

Putnis, A. and J.D.C. McConnell (1980) *Principles of Mineral Behavior*. Elsevier Publications, New York.

Ramberg, H. (1952) *The Origin of Metamorphic and Metasomatic Rocks*. University of Chicago Press, Chicago.

Ribbe, P.H. (1980) Chloritoid. *In* Ribbe, P.H., Ed., *Orthosilicates*, Reviews in Mineralogy 5, 155-169.

Richardson, S.M. (1975) *Fe-Mg Exchange among Garnet, Cordierite and Biotite during Retrograde Metamorphism*. Ph.D. Thesis, Harvard University, Cambridge.

Rimsaite, J. (1970) Anionic and cationic variations in zoned phlogopite. Contrib. Mineral. Petrol. 29, 186-194.

Robinson, P., F.S. Spear, J.C. Schumacher, J. Laird, C. Klein, B.W. Evans, and B.L. Doolan (1982) Phase relations of metamorphic amphiboles: Natural occurrence and theory. *In* Veblen, D.R. and P.H. Ribbe, Eds., *Amphiboles: Petrology and Experimental Phase Relations*, Reviews in Mineralogy 9B, 1-226.

Rosenfeld, J.L. (1970) Rotated garnets in metamorphic rocks. Geol. Soc. Am. Spec. Paper 129, 105 pp.

Rumble, D., III and T.A. Finnerty (1974) Devonian grossularite-spessartine overgrowths on Ordovician almandine from eastern Vermont. Am. Mineral. 59, 558-562.

Schmid, R. and B.J. Wood (1976) Phase relationships in granulitic metapelites from the Ivrea-Verbano zone (Northern Italy). Contrib. Mineral. Petrol. 54, 255-279.

Smellie, J.A.T. (1974) Compositional variation within staurolite crystals from the Ardara aureole, County Donegal, Ireland. Mineral. Mag. 39, 672-684.

Springer, R.K. (1974) Contact metamorphosed ultramafic rocks in the western Sierra Nevada foothills, California. J. Petrol. 15, 160-195.

Spry, A. (1969) *Metamorphic Textures*. Pergamon, Oxford.

Tewhey, J.D. and P.C. Hess (1976) Reverse manganese-zoning in garnet as a result of high f_{O_2} conditions during metamorphism. Geol. Soc. Am. Abstr. Prog. 8, 1135.

Thompson, A.B. (1976) Mineral reactions in pelitic rocks. Am. J. Sci. 276, 401-454.

_____, R.J. Tracy, P. Lyttle, and J.B. Thompson, Jr. (1977) Prograde reaction histories deduced from compositional zonation and mineral inclusions in garnet from the Gassetts schist, Vermont. Am. J. Sci. 277, 1152-1167.

Tracy, R.J., P. Robinson, and A.B. Thompson (1976) Garnet composition and zoning in the determination of temperature and pressure metamorphism, central Massachusetts. Am. Mineral. 61, 762-775.

_____, H.W. Jaffe, and P. Robinson (1978) Monticellite marble at Cascade Mountain, Adirondack Mountains, New York. Am. Mineral. 63, 991-999.

_____ and P. Robinson (1980) Evolution of metamorphic belts: Information from detailed petrologic studies. *In* Wones, D.R., Ed., *The Caledonides in the U.S.A.*, Virginia Polytechnic Inst. & State Univ. Memoir 2, 189-195.

_____ and C.W. Dietsch (1982) High temperature retrograde reactions in pelitic gneiss, central Massachusetts. Canadian Mineral. 20, in press.

Trzcienski, W.E., Jr. (1977) Garnet zoning -- product of a continuous reaction. Canadian Mineral. 15, 250-256.

Turner, F.J. (1970) Uniqueness versus conformity to pattern in petrogenesis. Am. Mineral. 55, 339-348.

Vogel, D.E. and C. Bahezre (1965) The composition of partially zoned garnet and zoisite from Cabo Ortegal, Northwest Spain. N. Jahrb. Mineral. Mh. 5, 140-149.

Wolff, R.A. (1978) *Ultramafic lenses in the middle Ordovician Partridge Formation, Bronson Hill Anticlinorium, central Massachusetts.* Contribution 34, Dept. of Geology, Univ. of Massachusetts, Amherst, Mass., 162 pp.

Woodsworth, G.J. (1977) Homogenization of zoned garnet from pelitic schists. Canadian Mineral. 15, 230-242.

Yardley, B.W.D. (1977) An empirical study of diffusion in garnet. Am. Mineral. 62, 793.

Date Due